In vitro Plant Breeding towards Novel Agronomic Traits

Manoj Kumar • Annamalai Muthusamy
Vivek Kumar • Neera Bhalla-Sarin
Editors

In vitro Plant Breeding towards Novel Agronomic Traits

Biotic and Abiotic Stress Tolerance

Springer

Editors
Manoj Kumar
Department of Life Sciences
Central University of Jharkhand
Ranchi, Jharkhand, India

Vivek Kumar
Himalayan School of Biosciences
Swami Rama Himalayan University
Jolly Grant
Dehradun, Uttarakhand, India

Annamalai Muthusamy
Department of Plant Sciences, Manipal
School of Life Sciences
Manipal Academy of Higher Education
Manipal, Karnataka, India

Neera Bhalla-Sarin
School of Life Sciences
Jawaharlal Nehru University
New Delhi, Delhi, India

ISBN 978-981-32-9823-1 ISBN 978-981-32-9824-8 (eBook)
https://doi.org/10.1007/978-981-32-9824-8

This Springer imprint is published by the registered company Springer Nature Singapore Pte Ltd.
The registered company address is: 152 Beach Road, #21-01/04 Gateway East, Singapore 189721, Singapore

Preface

Swelling demands for crop products prolong to rise strongly; agricultural productivity is threatened by various stress factors, often associated with global food for suitability and sustainability. To sustain and advance yield capacity, it is necessary to realize how plants respond to various stresses, and to use the hover knowledge in modern breeding programs. A good number of publications regarding molecular mechanisms associated with stress responses have been obtained from contemporary investigations using the model plant *Arabidopsis thaliana*. Molecular mechanism for stress amelioration is still in fancy where plant hormones, such as abscisic acid, jasmonic acid, and salicylic acid, have been shown to play key roles in defense responses against abiotic and biotic stresses. Epigenetic regulation at the DNA and histone level, and gene regulation by small non-coding RNAs also appear to be significant. Various approaches have been used for mutant screens, as well as next-generation sequencing approaches, to identify key factors and mechanisms of plant responses to the environment. However, it is often unclear to which extent the elucidated mechanisms also operate in crops.

This individual edition consequently seeks contributions to exposure of how crop plant species respond to various abiotic stresses, such as drought, heat, cold, flooding, and salinity, as well as biotic stimuli during microbial infections. The current edition welcomes reviews, perspectives, and original articles, and its focus is on our biochemical and molecular understanding of biotic and abiotic stress responses in crops, highlighting, among other aspects, the role of stress hormonal metabolism, fundamental mutagenesis, and changes in gene expression patterns and their regulation. Approaches and concepts to attain stress tolerance and to uphold yield permanence of agricultural crops during stress periods are of precise interest. These comprise perspectives on how knowledge from model plants can be utilized to facilitate crop-plant breeding and biotechnology.

The purpose of compendium is to conceptualize the innovative technology augmenting traditional-modern plant tissue culture, breeding, and markers approaches for novel agronomic traits with competency of biotic and abiotic stress tolerance. The book illustrates different evolutionary trends with modern concepts from different academic and research conditions to achieve sustainable tissue culture of ignored plants. The contributors highlight the contemporary themes, i.e., Breeding, Biotechnology and Molecular Tools and Volume, Agronomic, Abiotic and Biotic Stress Traits, and so on. The book contains 14 in-depth insight-based chapters from learned researchers, academicians, and scholars who are working strategically for specific plant traits including improved nutritional and pharmaceutical properties as well as enhanced tolerance to insects, diseases, drought, salinity, and temperature extremes expected under predicted global climate change.

Ranchi, India — Manoj Kumar
Udupi, India — Annamalai Muthusamy
Dehradun, India — Vivek Kumar
New Delhi, India — Neera Bhalla-Sarin

Contents

About the Editors

Manoj Kumar is an Associate Professor at the Department of Life Sciences, Central University of Jharkhand, Ranchi, India. Committed to research and development, Dr. Kumar holds a PhD in Plant Biotechnology and is involved in multidisciplinary research in the area of plant developmental biology, plant-microbe interaction and forest molecular genetics, including as a leader of a multidisciplinary research group. He is a referee for various journals, including *Phytoremediation* and the *Journal of Soil Sediments*.

Annamalai Muthusamy is currently a Professor and Head of the Department of Plant Science, School of Life Sciences, Manipal University. Prof. Muthusamy has 22 years of academic and research experience in the field of plant tissue culture and in vitro breeding of plants. He teaches undergraduate and postgraduate courses on Plant Biology and Biotechnology, Cell Biology, and Recombinant DNA Technology and is also involved in research on in vitro breeding of plants for biotic and abiotic stress, induction of genetic variability for improved plant secondary metabolites and DNA methylation, metabolomics and bioactive molecules of plants, and also is coordinator of the Alumni Centre, SLS, MAHE.

Vivek Kumar is an Agricultural Microbiologist with 20 years of experiences in teaching, research and guidance. Dr. Kumar is currently an Associate Professor at the Department of Microbiology, Himalayan School of Biosciences, Swami Rama Himalayan University, India.

He serves on the editorial board of various recognized international journals and is also a peer reviewer. He has published 61 research papers, 19 book chapters, six review articles and two books. For eight years, Dr. Kumar was Microbiologist at the Department of Soil and Water Research, Public Authority of Agricultural Affairs & Fish Resources, Kuwait.

Neera Bhalla-Sarin is an internationally recognized scientist and a consultant for various institutions in India, Europe, and Australia. She was also a Professor, and Dean of the School of Life Sciences, JNU, New Delhi. She was the chairperson of several academic councils at university level, has led an international plant developmental biology group, and completed numerous research projects sponsored by the Government of India and international research funding institutions (Indo-Swiss, Indo-Korea, Indo-Australia, Indo-US).

Withania somnifera (L.) Dunal: An Overview of Bioactive Molecules, Medicinal Properties and Enhancement of Bioactive Molecules Through Breeding Strategies

1

Poornima Poojari, Kodsara Ramachandra Kiran, Puthanvila Surendrababu Swathy, and Annamalai Muthusamy

Abstract

Withania somnifera is a very unique medicinal plant explored by number of orthodox medicinal systems such as Ayurveda, Siddha and Unani. It is commonly known as Ashwagandha, Indian ginseng and winter cherry, and finds its potential medicinal properties in the Ayurvedic Pharmacopoeia of India and Siddha Pharmacopoeia of India. The *Withania* genus is classified under the Solanaceae family and includes around 60 species, among which *W. somnifera* and *W. coagulans* are often mentioned in Ayurveda. Ashwagandha is a treasure house of widespread array of metabolites such as steroids, flavones, alkaloids, carbohydrates, glycosides, saponins, tannins, terpenoids and coumarin. Eight different polyphenols (five phenolic acids, vanillic, benzoic, p-coumaric, gallic and syringic acid, and three types of flavonoids, naringenin, catechin and kaempferol) were reported from Ashwagandha. Each part of this plant holds an assortment of different metabolites, and the metabolite concentrations vary among the different chemotypes. Ashwagandha has been used for a variety of ailment since a long time by traditional medicinal systems. It has been used for a range of diseases such as diabetes, emaciation, arthritis and rheumatoid arthritis-related inflammations, some kinds of seizures, diarrhoea, dermatitis and insect bite and specially used in the treatment of nervous disorder. The importance of Ashwagandha in medicinal treatments has attracted the attention of a large number of scientists; as a result, numerous experiments have been carried out, which have verified the therapeutic properties of Ashwagandha. Further, the

P. Poojari · K. R. Kiran · P. S. Swathy · A. Muthusamy (✉)
Department of Plant Sciences, Manipal School of Life Sciences,
Manipal Academy of Higher Education, Manipal, Karnataka, India
e-mail: a.msamy@manipal.edu

M. Kumar et al. (eds.), *In vitro Plant Breeding towards Novel Agronomic Traits*,
https://doi.org/10.1007/978-981-32-9824-8_1

efforts are being taken for the breeding strategies via conventional breeding as well as plant tissue culture techniques and gene transformation for the improvement of withanolide contents. This chapter gives an extensive insight on the various aspects such as basic introduction, classification, botanical description, bioactive molecules, medicinal properties and commercially available products. Furthermore, this chapter narrates the several breeding approaches for the improvement of withanolide contents of Ashwagandha.

Keywords

Ashwagandha · *Withania somnifera* · Medicinal properties · Bioactive compounds · Pharmacological effects · Improvement of withanolide content

1.1 Introduction

Withania somnifera (*W. somnifera*) is a very significant plant used in a number of orthodox medicinal systems like Ayurveda, Siddha and Unani. Ashwagandha, common name of *W. somnifera*, finds its mention both in the Ayurvedic and Siddha Pharmacopoeia of India (Umadevi et al. 2012). The use of Ashwagandha for its medicinal properties was preached by Punarvasu Atreya around 4000 years ago. Ayurveda described Ashwagandha as a rasayana herb, that is, an herb which has anti-ageing activity and can restore youthfulness by increasing physical strength and immunity. Rasayana also helps in improving memory and intelligence and relieves stress (Kulkarni and Dhir 2008). The renowned Indian literatures like *Charaka Samhita* and *Sushruta Samhita* have appreciated the importance of Ashwagandha for treating emaciation, rheumatoid arthritis-related inflammations and many other ailments. Ashwagandha is commonly used in India as a home therapy for alleviating hand and limb shivering of elderly people. It is also a very potent aphrodisiac. It is used to prepare a number of medicinal preparations, either alone or in combination with other herbs, for various types of nervous disorders as well. The leaves of Ashwagandha are useful in treating all types of skin lesions, boils, swelling, ulcers, pus formation and inflammation (Atal and Schwarting 1961).

1.1.1 Taxonomical Classification

The *Withania* genus is classified under the Solanaceae family and includes around 60 species of plants, among which *W. somnifera* and *W. coagulans* are often mentioned in Ayurveda. The taxonomical classification of Ashwagandha is as given below:

Kingdom: Plantae
Subkingdom: Tracheobionta
Super division: Spermatophyta
Division: Angiosperma

Class: Dicotyledons
Order: Tubiflorae
Family: Solanaceae
Subfamily: Solanoideae
Tribe: Physaleae
Subtribe: Withaninae
Genus: *Withania*
Species: *W. somnifera* Dunal

1.1.2 Botanical Description

The plant is a shrub growing to around 3–4 feet. It is usually erect and tomentose. The leaves of Ashwagandha are simple, ovate, exstipulate, glabrous and petiolate. The margin of the leaves is complete with acute to thick apex and cuneate or oblique base. The leaves are large and arranged alternate on the vegetative shoot but are opposite on floral branches. The flowers of Ashwagandha are pale green, inconspicuous, gamosepalous and arranged in cymose inflorescence. The flowers are characterized by the innate, oval anthers, epipetalous stamens arising from the petal base and slender filaments. Syncarpous gynoecium compelling of small swollen ovary with a long slender style. During the development of fruits, the calyx becomes enlarged, inflated and encloses the fruit. Fruits of this plant are berries having diameter of around 5 mm. The unripe fruits are green and turn red to orange-red on maturity. A single fruit encloses several small reniform seeds (Atal and Schwarting 1961; Mir et al. 2012).

1.1.3 Distribution and Cultivation

Ashwagandha wildly grows in dry regions and is wildly distributed in sub-tropical regions. It is found in Africa, Asia and Europe. It can be seen in African regions like Morocco, Congo, Egypt and South Africa, Middle Eastern region like Jordan as well as Asian regions like Afghanistan, India and Pakistan. In India, Ashwagandha is grown in Gujarat, Punjab, Uttar Pradesh, Maharashtra Haryana and Madhya Pradesh Rajasthan (Kokate 1996; Umadevi 2012,). Ashwagandha is an annual drought-tolerant plant. Semi-arid tropical areas with an average precipitation of 500–750 mm are appropriate for Ashwagandha as far as cultivation is concerned. Ashwagandha requires neutral to slightly basic soil with a pH of 7.5–8.0 and with a good drainage capacity as in sandy and sandy loam soil or red/black soil with light texture are advantageous for the growth of the plant. During the growing stage, dry season is necessary, and one or two late winter rains really favour the growth of the roots. The cultivation is easier and profitable since very less investment is required and the sale of roots fetch good price. The profit can be further boosted by selling leaves and the seeds. Hence it is cultivated on a large scale by small and marginal farmers in drier areas especially in the areas in Karnataka, Rajasthan, Andhra Pradesh, Madhya Pradesh and other states in India (Rao et al. 2012).

1.2 Bioactive Compounds of Ashwagandha

Ashwagandha is a treasure house of wide-ranging metabolites such as steroids, flavones, alkaloids, carbohydrates, glycosides, saponins, tannins, terpenoids and coumarin (Singh et al. 2010a, b; Nasreen and Radha 2011) (Fig. 1.1). Alam et al. (2011) described the occurrence of eight different polyphenols (five phenolic acids: vanillic, benzoic, p-coumaric, gallic and syringic acid and three types of flavonoids: naringenin, catechin and kaempferol). Each part of this plant holds an assortment of different metabolites, and the metabolite concentrations vary among the different chemotypes of Ashwagandha (Kumar et al. 2007; Srivastava et al. 2018). Ashwagandha has physiological effects similar to *Panax ginseng*, with respect to the antistress activity, hence also named Indian ginseng (Grandhi et al. 1994). Currently, leaves are reported to harbour 62 and roots are known to harbour 48 major and minor primary as well as secondary metabolites, of which 29 are common major and minor primary as well as secondary metabolites in both leaves and roots. Among the secondary metabolites, withaferin A are common major bioactive molecules, whereas withanolide D are common minor bioactive molecules in both roots and leaves (Singh et al. 2015) (Fig. 1.2).

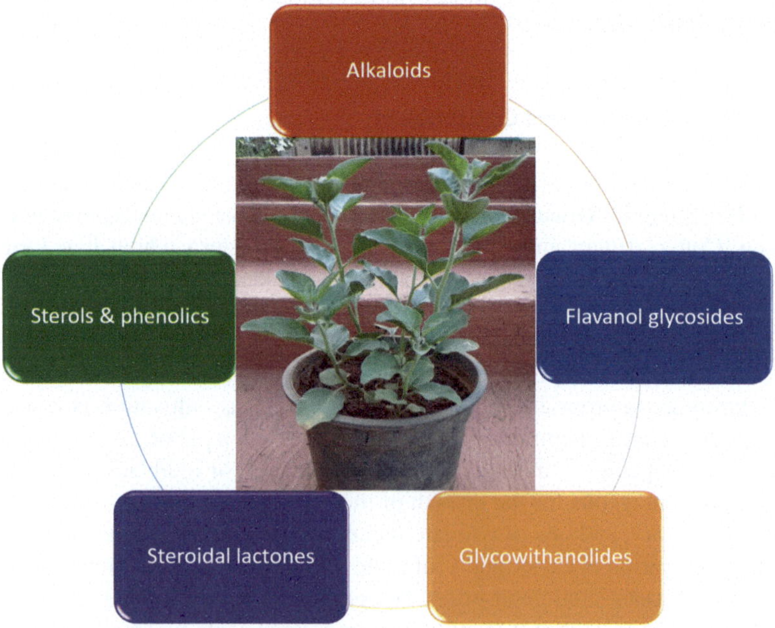

Fig. 1.1 Major groups of secondary metabolites from Ashwagandha

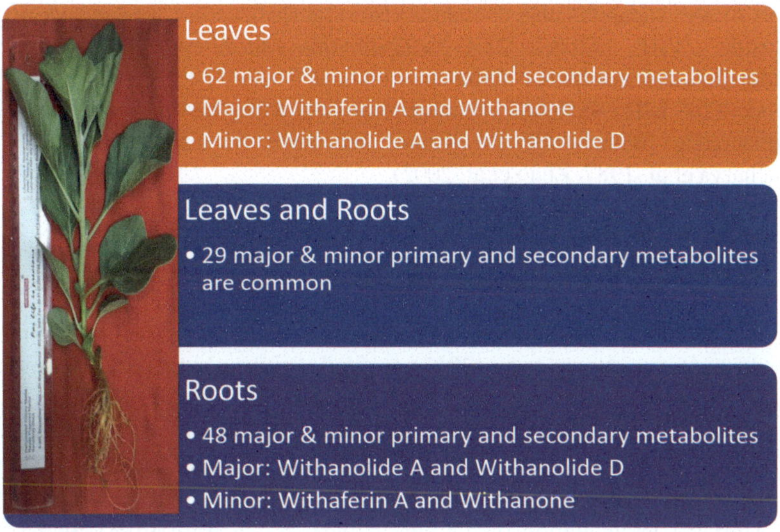

Fig. 1.2 Number of major and minor primary and secondary metabolites from Ashwagandha (Adopted from Singh et al. 2015 and modified)

1.2.1 Roots

Roots are therapeutically and traditionally most used part of the plant. Alam et al. (2011) determined the presence of catechin and benzoic acid in Ashwagandha by HPLC analysis of roots. Misra et al. (2008) were first to report the presence of stigmasterol glucoside, viscosa lactone B, α + β glucose, stigmasterol, β-sitosterol and β-sitosterol glucoside in the roots of Ashwagandha. Sharma et al. (2013a, b) analysed the alkaloid content of the roots of Ashwagandha with the help of GC-MS. They confirmed the presence of 17 alkaloids in the roots of Ashwagandha, which are withasomnine; somniferine; isopelletierine propanone; anaferine; anahygrine; pseudotropine; Iron-pseudotropine; withanine; 1-[(5-Nitro-2-furfurylidene)amino]; oxacyclohexadecane-2,13-dione,13-oxime; scopoletin; Ashwagandhine; 2,4-imidazolidiendione1-{(5-nitro-2-furanyl)methylene]amino}. The alkaloid withasomine and anaferine are reportedly the chief alkaloids the roots of Ashwagandha (Sharma et al. 2013a, b). Misra et al. (2012) isolated stearic fatty acids, sitostanone, oleanolic acid, octacosane, stigmasterone, oleic acids, 1,4-dioxane derivative, stigmasterol and ergosterol from Ashwagandha root extracted using n-hexane. According to Chaurasiya et al. (2008), major withanolides of the roots are withaferin A, 17-hydroxy-27-deoxy withaferin A and 17-hydroxy withaferin A. They reported the presence of 27-deoxy withaferin A and 27 hydroxy withanolide B in the roots as well albeit in lower amount. Two new compounds 6α-hydroxy-5,7α-epoxy as functional groups and 16β-acetoxy-17 (20)-ene from the Ashwagandha roots were isolated by Misra et al. (2008).

Roots of Ashwagandha also contain acylsterylglucosides like sitoindoside VII and sitoindoside VIII; (20S, 22R)-5a,27-dihydroxy-6a,7a-epoxy-1-oxowitha-2,24-dienolide; (20S,22R)-3 alpha, 6 alpha-epoxy-4 beta, 5 beta, 27-trihydroxy-1-oxowitha-24-enolide; (20S,22R)-4b,5b,6a,27-tetrahydroxy-1-oxowitha-2,24-dienolide; physagulin D; withacoagin; withanolide D; lycium substance B; and withanoside II–XI (Bhattacharya et al. 1987, Zhao et al. 2002).

1.2.2 Leaves

Singh et al. (2010a, b) identified amino acids like alanine, aspartate, asparagines and choline from leaves. Nema et al. (2013) profiled various phytoconstituents such as starch, amino acids, carbohydrate, protein, flavonoids, tannins, alkaloids, oxalic acid, steroids, phenolic compounds and inorganic acids from hydroalcoholic extract of Ashwagandha leaves. Bashir et al. (2013) detected the flavonoids, steroids, alkaloids, saponins and tannins in the methanolic crude extract of Ashwagandha leaves. They also isolated three flavonoids, namely, 5, 7, 4′-triahydroxy-methyl-3-O-galactosyl flavonol, 7, 3′, 4′-trihydroxy flavone-3-O-rhmnosyl and quercetin -3-O-galactosyl from the leaf extract and characterized their structures based on NMR, IR and MS spectroscopy.

Jayaprakasam and Nair (2003) discovered four novel withanolide glycosides, namely, 27-O-β-D-glucopyranosyl viscosalactone B; physagulin D (1 → 6)-β-d-glucopyranosyl-(1 → 4)-β-d-glucopyranoside; 4,16-dihydroxy-5β,6β-epoxyphysagulin D and 27-O-β-D-glucopyranosyl physagulin D; and a novel withanolide, viz. 4-(1-hydroxy-2,2-dimethylcyclo-propanone)-2,3-dihydrowithaferin A from Ashwagandha leaves. They also reported seven known withanolides withaferin A, viscosalactone B, sitoindoside IX, physagulin D and withanoside IV, 2,3-dihydrowithaferin A and 27-desoxy-24,25-dihydrowithaferin A from Ashwagandha leaves.

The leaves of Ashwagandha also contain chlorinated withanolides, namely, 6α-chloro-5β,17α-dihydroxywithaferin A and withanolide Z. Furthermore, leaves also contain withaferin A, withanone, 6α-chloro-5β-hydroxywithaferin A, withanolide B, (22R)-5β-formyl-6β,27-dihydroxy-1-oxo-4-norwith-24-enolide, withanolide A, withanoside IV, 2,3-dihydrowithaferin A, 3-methoxy-2,3-dihydrowithaferin A, 2,3-diehydrosomnifericin, withanoside X and 27-hydroxywithanolide B (Pramanick et al. 2008; Tong et al. 2011).

1.2.3 Fruits

Bhatia et al. (2013) determined different aromatic and aliphatic amino acids, fatty acids, phenolic acids, organic acids, sterols, tocopherols, polyols, sugars and withanamides from the fruits of Ashwagandha using ^{1}H NMR spectroscopy and GC-MS. These compounds are mainly involved in the different metabolic pathways like shikimic acid, mevalonate (MVA), non-mevalonate (DOXP) and phenylpropanoid pathway. The metabolite concentrations varied among the fruits of different chemotypes of Ashwagandha. Lal et al. (2006) reported two withanolides, viz. iso-withanone which has an unusual 17α-oriented side chain and 6α,7α-epoxy-1α,3β,5α-trihydroxy-witha-24-enolide possessing an 1α,3β -dihydroxy group for the first time from the fresh berries of a plant in India. These berries also contain coumarins like scopoletin and aesculetin and triterpene like β-amyrin (Abou-Douh 2002). Reportedly the fruits also contain tocopherols like α-tocopherol and β-tocopherol and phytosterols like cholesterol, β-sitosterol, campesterol and stigmasterol (Bhatia et al. 2013; Abou-Douh 2002). Fruits of Ashwagandha contain withanolide A and withanone in substantial amount. Other withanolides present in the fruit are 27-deoxy withaferin A, 27 hydroxy withanolide B and 27-hydroxy withanolide A (Chaurasiya et al. 2008; Lal et al. 2006). Abou-Douh (2002) isolated two novel withanolides 5β, 6α, 14α, 17β, 20β-pentahydroxy-1-oxo-20 S, 22R-witha-2, 24-dienolide and 6α, 7α-epoxy-5α, 14α, 17α, 23β-tetrahydroxy-1-oxo-22R-witha-2, 24-dienolide from the Ashwagandha fruit grown in Southern Egypt. They also contain nine different withanamides, named withanamide A to withanamide I; this class of compound has a unique chemical structure, consisting of long-chain hydroxyl fatty acid moieties, glucose and serotonin (Jayaprakasam et al. 2004).

1.2.4 Bark

Five new withanolides, viz. somnifera withanolide, withanolide, somniwithanolide, withasomniferanolide and somniferanolide, were isolated from the bark of Ashwagandha grown at South Delhi region and characterized through spectroscopic and phytochemical techniques (Ali et al. 1997).

1.2.5 Commercial Products

Currently more than 24 products are available in the market as Ashwagandha as a sole or one of the ingredients. Ashwagandha products are used in wide range from overall health to specific health for organs and glands (Table 1.1).

1.3 Pharmacological Effects of Ashwagandha

Ashwagandha has been used for a variety of ailment since a long time by traditional medicinal systems. It has been used for a range of diseases such as diabetes, emaciation, arthritis and rheumatoid arthritis-related inflammations, seizures, diarrhoea, dermatitis and insect bite and specially used in the treatment of nervous disorder (Verma and Kumar 2011). The importance of Ashwagandha in medicinal treatments has attracted the attention of a large number of scientists; as a result,

Table 1.1 Commercially available products of *Withania somnifera*

S. No.	Product name	Manufacturer	Benefits	References
01	Adrenal Health	Gaia Herbs	This mix of herbs nourishes the adrenals, enabling the body to adapt to stress	https://www.gaiaherbs.com/products/detail/777/Adrenal-Health-Daily-Support
02	Ashwagandha Root	Gaia Herbs	This includes single herb, Ashwagandha, and helps nourish and restore the optimal nervous and immune system, energy levels and overall immune function	https://www.gaiaherbs.com/products/detail/662/Ashwagandha-Root
03	SleepThru	Gaia Herbs	This blend of herbs supports restful sleep and healthy adrenal function	https://www.gaiaherbs.com/products/detail/745/SleepThru
04	Stress Response	Gaia Herbs	This formula of herbs supports a healthy response to stress	https://www.gaiaherbs.com/products/detail/77/Stress-Response
05	Thyroid Support	Gaia Herbs	A combination of Ashwagandha and other herbs, this formulation supports normal thyroid hormone production, which in turn helps maintain proper weight, neuromuscular and cardiovascular health	https://www.gaiaherbs.com/products/detail/81/Thyroid-Support
06	Ashwagandha	Himalaya Herbals	Relieves stress by acting on adrenal glands by normalizing cortisol levels and eases sleeplessness	http://himalayausa.com/products/best-sellers/organic-ashwagandha/
07	Anti-Stress Massage Oil	Himalaya herbals	Useful in overcome stress and fatigue	http://www.himalayaherbals.com/products/healthcare/anti-stress-massage-oil.htm
08	Ashwagandharishta	Baidyanath Ayurved	This liquid medicine helps in alleviating all kinds of vata-related diseases. It is beneficial in controlling diabetes, digestive problems, fatigue and fever. It helps strengthen immunity and memory. It is useful in treatment of men problems like impotency and erectile dysfunction	https://www.baidyanath.com/product/ashwagandharishta/

09	Ashwagandha Capsules	Planet Ayurveda	This medicine, prepared from Ashwagandha root extract, is useful in treating debilitating conditions, fatigue, stress, anxiety, palpitation, ageing, nervous breakdown, neuropathy due to diabetes, chronic fatigue syndrome and loss of weight due to cancer, diabetes or any other reason	https://www.planetayurveda.com/ashwagandha_uses.htm
			Strengthens and relieves muscles. It is a good nerve tonic. It can be given to growing children, as well as to old age people	
10	Male Support	Planet Ayurveda	Improves general health. Beneficial for erectile dysfunction, premature ejaculation, low sperm count, general physical and sexual weakness, and infertility in men	https://www.planetayurveda.com/male-support-formula.htm
11	Aller-G Care Capsule	Planet Ayurveda	Best anti-allergic and anti-inflammatory herbs	https://www.planetayurveda.com/allergformula.htm
12	Aamvatantak Churna	Planet Ayurveda	This is a blend of unique herbs, which also has Ashwagandha as one of its ingredients. Used for controlling rheumatoid arthritis and its symptoms like stiff joints, inflammation, swelling and joint pain	https://www.planetayurveda.com/aamvatantak.htm
13	Total Heart Support Capsules	Planet Ayurveda	A combination of herbs with Ashwagandha as one of its ingredients; this medication is effective in controlling high cholesterol, blocked coronary arteries, congestive heart failures. This strengthens failing heart muscles, eases breathlessness and cleanses blocked arteries. Useful in post-myocardial infarction support and congestive heart failure	https://www.planetayurveda.com/total-heart-support.htm
14	Stress Support Capsules	Planet Ayurveda	Corrects vata imbalances due to stress	https://www.planetayurveda.com/stress-support.htm
15	Dabur Ashwagandha Churna	Dabur India Limited	Improves stamina and energy. Relief from stress, weakness	
16	Dabur Ashwagandharishta	Dabur India Limited	This contains other medicinal plants like Ashwagandha, Mushali, Manjishtha, Haritaki and Nisha. These ingredients together help treating depression and anxiety, improving memory, calming nerves and strengthening digestive system	https://www.dabur.com/in/en-us/ayurvedic-herbal-products/dabur-ashwagandharishta

(continued)

Table 1.1 (continued)

S. No.	Product name	Manufacturer	Benefits	References
17	Sona Chandi Chyawanprash Plus	Zandu	This contains other ingredients like Bala, Guduchi, Haritaki, Jyotishmati, Pippali, Punarnava, Yastimadhu and Amla Pishti, along with Ashwagandha, which are known for their immunity-boosting properties. This preparation improves memory and learning ability and body immunity and controls frequency of cough, cold and allergies	http://www.zanduayurveda.com/products/71/zandu-sona-chandi-chyawanplus.php
18	Vigorex	Zandu	Ashwagandha is one of the ingredients of this product. This product is used to enhance stamina and energy	http://www.zanduayurveda.com/products/70/vigorex.php
19	Zandu Pancharishta	Zandu	Ashwagandha is one of the ingredients of this tonic. This helps in building the entire digestive system and building digestive immunity and reduces problems like acidity, gas, indigestion, flatulence and constipation	http://www.zanduayurveda.com/products/29/zandu-pancharishta.php
20	Zandu Kesari Jivan	Zandu	Also contains Kesar, fresh Amla, exotic herbs, spices and trace minerals	http://www.zanduayurveda.com/products/27/zandu-kesari-jivan.php
			Makes one physically strong and keeps youthful vigour intact	
21	Brento	Zandu	It is a brain tonic which restores brain energy. Useful in treating mental debility, cognitive dysfunction and other mental complications	http://www.zanduayurveda.com/products/31/brento.php
22	Ashwagandha Capsules	Dehlvi Naturals	Beneficial for people with mental problems, arthritis, asthma, sexual problems, bradycardia, bronchitis, cancer, convalescence, dyspepsia, general debility, hydrospermia, hypertension, infertility, insomnia, lack of appetite, leucorrhoea, lumbago, polyuria, premature ageing, stress, suppressed post-partum lactation	http://www.dehlvi.com/medicine-Ashwagandha_Capsules-view-598.html
23	Ashwagandha Ras	Dehlvi Naturals	Helps in the treatment of anxiety, arthritis, convalescence, depression, erectile dysfunction, exhaustion, fatigue, general debility, infertility, lack of vigour, lumbago, rheumatism, rheumatoid arthritis, seminal debility, senile debility, sexual debility and stress	http://www.dehlvi.com/medicine-Ashwagandha_Ras-view-636.html
24	Ashwagandha Tea	Buddha Tea	Ashwagandha tea helps in de-stressing and improving memory	https://www.buddhateas.com/ashwagandha-tea.html

numerous experiments have been carried out to verify the therapeutic properties of Ashwagandha.

The in vitro studies have demonstrated that Ashwagandha possesses anxiolytic (Andrade et al. 2000), cytoprotective (Thiagarajan et al. 2003), antifungal (Singh et al. 2010a, b), antioxidant, pesticidal (Gupta and Srivastava 2008) and antibacterial properties (Mehrotra et al. 2011). Mahdi et al. (2009) stated that this plant could improve fertility by improving sperm quality and stress-related infertility. Various studies on animal models demonstrated that the extracts of Ashwagandha improve sexual health (Borde et al. 2008) and exhibit cardioprotective (Gupta et al. 2004), immunomodulatory (Gupta et al. 2006), hypocholesteremic (Visavadiya and Narasimhacharya 2007), antidiabetic and anti-ageing (Babu et al. 2007), hypoglycaemic (Udayakumar et al. 2009) and antilipidemic effects (Khursheed et al. 2010). The main pharmacological activities of the Ashwagandha are described separately.

1.3.1 Antistress and Antioxidant Property

Various studies have been showed to authenticate the antioxidative property of Ashwagandha, and it has been observed to influence both the non-enzymatic and enzymatic antioxidants. Ashwagandha restores the activity of antioxidant enzymes, viz. glutathione reductase (GR), glutathione- S-transferase (GST) superoxide dismutase (SOD), glutathione peroxidase (GPX) and catalase (CAT) to the near normal levels and decreases lipid peroxidation (LPO) (Anwer et al. 2012; Kumar and Kumar, 2008). Bhattacharya et al. (2001) demonstrated that the glycowithanolides present in the Ashwagandha could normalize the activity of enzymatic antioxidants like CAT, SOD, GPX and LPO in the frontal cortex and striatum of the rat brain with induced chronic footshock stress. Anwer et al. (2012) noted the augmentation in the activity of antioxidant enzymes in the type 2 diabetic rats when administered extract of Ashwagandha, which results in the reinstatement of the form and structure of pancreatic β-cells close to normal form. The intake of Ashwagandha extracts also helped in bringing the blood glucose, tissue LPO levels and glutathione (GSH) contents to a near normal state.

1.3.2 Aphrodisiac Properties

Mahdi et al. (2009) suggested that Ashwagandha could help the male with infertility due to stress. The normozoospermic males suffering from infertility, due to heavy smoking, psychological stress or/and with other unknown reason, witnessed a general improvement in semen quality with a decrease in the stress. The partners of 14% of the individuals could even conceived after the treatment and decrease in the lipid peroxidation and increase in enzymes of antioxidant system, antioxidant vitamins and fructose in the seminal plasma was noted. Ashwagandha controlled protein carbonyl content and enhanced the sperm quality. The treatment stabilized the levels of sex hormones like testosterone, luteinizing hormone (LH), follicle stimulating hormone (FSH) and prolactin in infertile subjects (Ahmad et al. 2010). In a study, Shukla et al. (2011) observed the reduced apoptosis and intracellular level of ROS in spermatozoa and stabilized level of the important metal ions such as Cu^{2+}, Zn^{2+}, Fe^{2+} and Au^{2+} in seminal plasma of infertile individuals after treatment with Ashwagandha. Administration with Ashwagandha improved the levels of citrate, lactate, GPC and aliphatic amino acids like alanine and aromatic amino acids like histidine and phenylalanine in seminal fluid, and the semen quality was found to be recovered in Ashwagandha-treated men (Gupta et al. 2013). The intake of the plant was noticeably increased sex hormones and gonadotropins especially testosterone, oestrogen and luteinizing hormone even in the male rats addicted to morphine. Morphine consumption results in a number of sexual and fertility problems. Intake of Ashwagandha also increased the level of progesterone which stimulates most sex hormones in addicted male rats (Rahmati et al. 2016).

1.3.3 Antidepression Properties

The roots of Ashwagandha have been termed as rasayana herb in Ayurveda, which has a positive effect on the physical as well as mental health. Bhattacharya et al. (2000b) observed that the anxiolytic effect of the glycowithanolides was comparable to that of lorazepam as studied by the tests like elevated plus maze, feeding potential in unacquainted environment and social interaction. Glycowithanolides could also reduce the anxiety by reducing the concentration of tribulin which is a well-known marker of stress and anxiety, like lorazepam. The bioactive component also displayed antidepressant effect in the forced swim-induced 'behavioural despair' and 'learned helplessness' tests. Hence Bhattacharya et al. (2000b) put forward the idea that the glycowithanolides from Ashwagandha root could help in the state of anxiety and depression. Nonetheless, Ashwagandha root extract improved the open field behaviour and emotional stability in normal rats as well. It also enhanced the functional sensitivity of 5 HT2 receptors in the brain and a reciprocal subsensitivity of the 5HT1A receptors. Chronic Ashwagandha treatment also resulted into adaptive supersensitivity of the postsynaptic 5HT2 receptors in the brain. The observations were similar to the ones caused by the chronic electroconvulsive therapy treatment and several antidepressant drugs (Tripathi et al. 1998). Jayanthi et al. (2012) noted the dose-dependent reduction in immobility time in forced swim test (FST) and tail suspension test (TST) in rat administered with fat extract of Ashwagandha, Ayurvedic formulation Ashwagandha ghrita (AGG). In the Reserpine test, the scores of the ptosis, catatonia and sedation were significantly improvised in the groups treated with AGG and combination of AGG with Imipramine. These observations supported the use of Ashwagandha as potential adjuvant in depression disorders.

1.3.4 Neurological Effects

Ashwagandha extracts could suppress the release of corticosterone in the mice, the chemical which can induce neurotoxicity (Bhatnagar et al. 2009; Huang et al. 2009). This in turn activates choline acetyltransferase causing an increase in the level of serotonin in the hippocampus and inhibition of NADPH-d activity of neuronal nitric oxide synthase (nNOS). The key fundamental mechanism of the neuroprotective effects of Ashwagandha can be accredited to its involvement in the selective inhibition of nNOS and variations of definite neurotransmitter systems. Attari et al. (2016) presented the fact that probably the NO (nitric oxide) does not mediate this beneficial effect and Ashwagandha may affect other neurochemical systems and pathways. They found that Ashwagandha extract dose dependently decreased the immobility time comparable to fluoxetine (20 mg/kg). Ashwagandha also lowered the immobility measure in TST. These effects were not related to change in locomotor activity. L-NAME (N omega-nitro-L-arginine), the NOS inhibitor, did not influence the effect of the extract on the behavioural tests. Ashwagandha root exhibited a cytoprotective effect on separated PC12 cells against cytotoxic effects of both H_2O_2- and Ab(1–42). Ashwagandha extract could prevent the maturation of amyloid-β fibrils in vitro, which is the first significant step towards formation of amyloid plaque in vivo (Kumar et al. 2012). Kurapati et al. (2013) further perceived that the Ashwagandha extract could neutralize the toxicity mediated by β-amyloid and HIV-1 Ba-L(clade B) infection in the human neuron-derived cell line -SK-N-MC. Furthermore, Sehgal et al. (2012) put forth that Ashwagandha extract could fix the behavioural abnormality, accumulation of oligomers of beta-amyloid peptides (Aβ) and plaque formation in the brains of APP/PS1 Alzheimer's disease transgenic mice. They also put forth exceptional observation that the treatment with Ashwagandha caused decrease in brain Aβ monomer accompanied with an increase in plasma Aβ. They also noticed that the level of low-density lipoprotein receptor-related protein (LRP) and Aβ-degrading protease neprilysin in brain microvessels and plasma sLRP increased before the transfer of the Aβ peptides into plasma. Ashwagandha could be used to enhance the immediate and general memory in people with mild cognitive impairment was confirmed in a pilot study in India. It was also noticed that it could improve executive function, attention and information processing speed (Choudhary et al. 2017). The neuroprotective effect of the plant extract was studied in human cultured neuron model, which was injured to represent mildly traumatized brain injury. It was observed that there was an increase in the neurites length and neuronal survival (Saykally et al. 2017). Among the different constituents of the Ashwagandha components, the study by Pandey et al. (2018) specifically worked on withanone and found that in Wistar rats, it could boost the cognitive skills by inhibiting amyloid β-42 and controlled the levels of pro-inflammatory cytokines and eased the pathophysiology of the disease. In a randomized clinical trial, it was confirmed that Ashwagandha extract when used as adjunctive treatment benefits the patients with recent exacerbation of schizophrenia, along with decrease in the level of stress in the patients (Chengappa 2018).

1.3.5 Cytoprotective Activity

Bhattacharya et al. (2000a) observed that the treatment of equimolar concentrations of sitoindosides VII–X and withaferin A have stabilized the hepatic lipid peroxidation and levels of specific aminotransferase and lactate dehydrogenase in serum of rats with induced hepatotoxicity. Hence signifying the potential use of Ashwagandha in the treatment of heavy metal as well as other toxins induced hepatic dysfunction as the Ashwagandha extract is rich in antioxidant glycowithanolides. Sumantran et al. (2007) verified the inhibition of gelatinase activity of type 2 collagenase and theory by exhibiting chondroprotective action on diseased osteoarthritic cartilage by aqueous extract of Ashwagandha. Nakajima et al. (2011) observed that Ashwagandha extract alleviates endothelin-1(EDN1)-mediated pigmentation by restraining EDN1-activated protein kinase C (PKC) activity. Withaferin A causes the downregulation of the phosphorylation of the kinases like Raf-1, MEK, ERK, MITF and CREB 15 min after EDN1 treatment. Since the phosphorylation of Raf-1 by PKC activity can be downregulated by withaferin A, these findings indicate that hyperpigmentary disorders can be treated with the help of an antioxidant-rich Ashwagandha extract.

1.3.6 Antitumour Properties

Many researchers have put light on the antitumour and proliferation inhibitory effect of Ashwagandha on the cancer cells both in cell lines as well as animal model system. Osman et al. (2012) reported the cytotoxic and proliferation inhibitory effect of Ashwagandha extract on MCF 7, a human breast cancer cell line, with IC50 0.86% as measured by Trypan blue and MTT assay. Oza et al. (2010) investigated the antitumour effect of purified L-asparaginase from Ashwagandha fruits which showed antitumour activity at very low doses, and subsequently Ashwagandha is explored as a potent source of L-asparaginase. L-Asparaginase from Ashwagandha is known to have structural and functional similarity with bacterial L-asparaginases EC-2. Kataria et al. (2011) showed that the Ashwagandha could be used as safer complimentary therapy for glioma as the aqueous extract of Ashwagandha leaves exhibited proliferation inhibiting, differentiation inducing and anti-metastasis effect. Khazal and Hill (2015) reported the antitumour effect of Ashwagandha root extract on the xenografts model of breast cancer cell line MDA-MB-231 in nude mouse. The substantial increase in the sub-G1 phase cells and cytotoxic effect on MDA-MB-231 cells in dose-dependent manner indicate the potential complimentary therapy for breast cancer using Ashwagandha extract. In totality Ashwagandha root extract could inhibit the proliferation of breast cancer cells in vitro as well as in vivo, accompanied by significant downregulation of CCL2, a chemokine. The study by Henley et al. (2017) revealed that Ashwagandha root extract could 'prime' the cancer cells, specifically HT-29 colon cells, to the chemotherapy. This increased the effectiveness of the chemotheraupetic agent used (cisplatin) through enhancement of the mitochondrial dysfunction.

1.3.7 Antimicrobial Activity

The extracts of Ashwagandha have been demonstrated to have antifungal, antibacterial and pesticidal properties. Khan and Nasreen (2010) found that the protein fraction of Ashwagandha exhibited higher inhibition of mycelia growth of *Bipolaris oryzae* and *Colletotrichum lindemuthianum* than nonprotein fractions. Similarly, Singh et al. (2010a, b) used the solvent extract of Ashwagandha for agar diffusion test against a number of bacteria. The Ashwagandha extract using ethyl acetate and hexane exhibited higher inhibition zone against *Aspergillus niger*, whereas methanolic extract inhibited the growth of *Fusarium oxysporum* and *A. flavus*, while the growth of *F. moniliformis* was inhibited by aqueous extract. Singariya et al. (2012) also reported that Ashwagandha extracts had antimicrobial activity against important human pathogens. Furthermore, the polar and non-polar solvent extracts of immature and mature fruit and calyx of Ashwagandha were screened for their antimicrobial activity against *Bacillus subtilis* (Gram-positive bacteria), *Pseudomonas aeruginosa* and *Enterobacter aerogenes* (Gram-negative bacteria) and *A. flavus* (fungus) by agar diffusion test. Chloroform extract of Ashwagandha calyx revealed maximum activity against *B. subtilis*. Similarly Gupta and Srivastava (2008) found that different extracts of (aqueous suspension and ether) various parts of Ashwagandha showed varying effect on the adult mortality of pest *Callosobruchus chinensis* L. The ether extract resulted in higher adult mortality followed by aqueous extract. Recently there have been increased cases of resistance of pathogenic bacteria to antibiotics, pushing the researchers to find herbal alternatives. Bisht and Rawat (2014) noticed that the methanolic leaf extract of Ashwagandha could actively hinder the pathogens, including a total of 20 isolates including methicillin-resistant *Staphylococcus aureus* and methicillin-sensitive *S. aureus*, *Enterococcus* and *Streptococcus* spp. Datta et al. (2011) also reported growth inhibitory effect of Ashwagandha root extracts against all the multidrug-resistant strains of *S. aureus* isolated from local and patient sources. Bokaeian et al. (2015) remarked that Ashwagandha extracts could inhibit the drug-resistant *Escherichia coli* strains collected from patients with infections in urinary tract.

1.4 Breeding Strategies for Improvement of Withanolide Content in *W. somnifera*

1.4.1 Plant Cell, Tissue and Organ Culture

Many researchers have established protocols for propagation of Ashwagandha and analysed the amount of withanolides in the cultured plants and different parts of the plant. Roja et al. (1991) produced Ashwagandha plantlets and callus from axillary meristems. They observed that the amount of withanolides varies at the different morphogenetic stages. At callus stage they did not detect withanolides, whereas the multiple shoot cultures could synthesize a substantial amount of withanolides. They also noticed that the amount of withanolides varied with different hormonal combinations, illustrating the importance of hormones in the biosynthesis of withanolide. Further Sangwan et al. (2007) described that the genotype of the plant along with the hormonal combination, influenced the withanolide A accumulation in shoot cultures. They established the shoot cultures from nodal segments of Ashwagandha plants from two experimental lines, i.e. *RS-Selection-1* and *RS-Selection-2*, using different combinations of plant growth regulators. Using [2-14C] acetate as a precursor, they also demonstrated that the in vitro shoots synthesized withanolide A de novo. Awasthi et al. (2008) reported elevated production of withanolide A, withanolides and withaferin A, in the plantlets regenerated in vitro from axillary bud explants by using additional carbohydrates in the rooting media. Multiple shoots were developed from axillary bud explants on Murashige and Skoog (MS) media fortified with 0.5 mg/1 of 6-benzylaminopurine (BAP) and 1.0 mg/1 of KN. The shoots proliferated on 3/4 MS medium complemented with BAP and KN (0.1 mg/l and 0.05 mg/l, respectively) with mean of 8 shoots per explants. The shoots were excised and subcultured onto ½ MS with 2.0 mg/L IBA with additional carbohydrate (4% sucrose +2% mannitol) for rooting. The rooted shoots thus obtained were then transferred to the soil. It was reported that the roots of these micropropagated plant contained more withaferin A (0.047% w/w) and withanolide A (0.0401% w/w), respectively, as compared to the seed raised plants (0.0379% w/w and 0.0386% w/w, respectively). A group of scientists demonstrated that the concentration of macroelements present in the media, too, influenced the accumulation of withanolides A and biomass in the cell suspension and adventitious root suspension culture of Ashwagandha. They varied the concentration of calcium chloride, potassium sulphate, ammonium nitrate, magnesium sulphate and potassium nitrate and strength of nitrogen source [NH_4^+/NO_3^-]. In the cell suspension culture, maximum of withanolide A production (4.36 mg g^{-1} DW) was achieved with 2.09 KNO_3 in media, and both the parameters were increased when nitrate concentration was higher than ammonia. In case of adventitious root suspension culture, maximum biomass (127.52 g/L FW and 12.45 g/L) was recorded in 0.5x concentration of NH_4NO_3, and maximum accumulation (14.00 mg/g DW) of withanolide A was observed with 2.0x KNO_3. Higher biomass of Ashwaganda was noted when NO_3^- was greater than the NH_4^+, whereas absence of NH_4^+ was favor withanolide A production. Similarly, the treatment ratio of NH_4^+/NO_3^- at 0.00:18.80 mM showed maximum production of withanolide A (Murthy and Praveen 2011; Murthy and Praveen 2012).

1.4.2 Biotic and Abiotic Elicitors

Despite the optimization of various media constituents, the amount of withanolides produced was not as high. Many researchers successfully used the elicitors to further increase the content of withanolides in Ashwagandha through in vitro culture. Elicitors are microbial, physical or chemical factors which invoke physiological and morphological response in plants. Elicitation is a technique to fortify survival endurance and competitiveness of plants by enhancing the synthesis of secondary metabolites. The secondary metabolites production was successfully achieved in many medicinal plants by using the elicitors (Patel and Krishnamurthy 2013).

The elicitation of withanolides production in Ashwagandha was first demonstrated by Ciddi (2006). They used biotic elicitor salacin, methyl jasmonate (MJ) and arachidonic acid for the elicitation experiments in the cell suspension culture. Salacin (750 mM) enhanced the synthesis of withaferin A by 50-fold (25 ± 2.9 mg/l) as compared to control (0.47 ± 0.03 mg/l), whereas elicitation of cell suspension culture by MJ and arachidonic acid did not result in biosynthesis of withaferin A. Baldi et al. (2008) later put forth a dual elicitation strategy wherein two different elicitors together could be used together to enhance the biosynthesis of withaferin A. They successfully used *Agrobacterium tumefaciens*-mediated transformation technique for establishing high-yielding callus of Ashwagandha. Different biotic (cell extracts and culture filtrates of *Fusarium solani*, *Alternaria alternata* and *Verticillium dahliae*) and abiotic elicitors (arachidonic acid, calcium chloride, MJ and copper sulphate) were used to elevate the synthesis of withaferin A in the cell suspension culture. Copper sulphate and *V. dahliae* were able to enhance the production to an extent of 5.4 and 9.7 times, respectively, when added solely to suspension cultures as compared to the control (2.65 mg/l). The combination of the two elicitors, that is, dual elicitation method, increased the yield by 13.8-fold.

Chitturi et al. (2010) used fungal culture filtrate and cell extracts of *F. solani* MTCC 350, *A. alternata* MTCC 1779 and *V. dahliae* MTCC 2063 and the abiotic elicitors, viz. calcium chloride, copper sulphate and cinnamic acid, to the suspension cultures of Ashwagandha. They also fed the cell suspension culture with different concentrations of precursors of withanolide biosynthesis like sodium acetate, mevalonolactone, squalene in alcohol, squalene in colloidal form, cholesterol in alcohol and cholesterol in colloidal form for the improvement in the yield of withanolides. Precursors like sodium acetate (tenfold), mevalonolactone (14-fold), squalene in colloidal form (23-fold) and cholesterol in colloidal form (30.5-fold) and elicitors like cells extract of *V. Dahlia* (tenfold) and copper sulphate (2.5-fold) increased the bioproduction of withaferin A. The bioproduction of withaferin A was maximum when precursors and elicitors were added to 3-day-old suspension cultures. Sivanandhan et al. (2012) observed superior production of some major and minor withanolides, like withaferin A, withanolide A, withanone, withanolide B, 12-deoxywithastramonolide, withanolide V and withanoside IV in the adventitious root cultures, when exposed to the biotic elicitors like MJ and salicyclic acid (SA). They optimized the concentration of elicitor, growth index, time of contact with the elicitor as well as the age of adventitious roots for elicitation treatment to enhance withanolides biosynthesis in the Ashwagandha.

The withanolide content increased significantly in the 40 days old culture when the adventitious root cultures were subcultured to media containing150 μM SA with 4 h exposure period on the 30th day, as compared to untreated cultures. The major withanolides were increased considerably: 17.47 mg g/l DW of withaferin A (20-fold), 42.88 mg g/l DW of withanone (37-fold), 33.74 mg g/l DW of withanolide B and 64.65 mg g/l DW of withanolide A (48-fold increase). Recently Sivanandhan et al. (2014a) successfully enhanced the withanolides content and biomass in shoot suspension culture of Ashwagandha by using the extract of *Gracilaria edulis* and *Sargassum wightii*. They discovered that the addition of the extract of *G. edulis* enhanced the synthesis of the withanolides and biomass by nearly 1.45–1.58-fold as compared to the control culture. Further optimization showed that by exposing the culture with 40% *G. edulis* extract for 24 h in the culture leads to the maximum accumulation of biomass (62.4 g FW and 17.82 g DW) and different withanolides after 5 weeks. This increase in the withanolides was accompanied with increased expression of the enzymes like SE, SS, HMGR and FPS. Sivanandhan et al. (2014b) also studied the influence of addition of precursor molecules and elicitors to cell suspension culture maintained in bioreactor and shake flask, in order to achieve an increased production of Ashwagandha. They used elicitors, namely, cadmium chloride, precursors mevalonic acid and squalene and aluminium chloride and chitosan. Maximum amount of withanolide A, withanolide B, withanone, withaferin A, 12 deoxy withanstramonolide, withanoside IV and V was obtained with combination of chitosan and squalene with KN (0.5 mg/l), picloram (1 mg/L), L-glutamine (200 mg/L) and sucrose (5%) upon culturing for 28 days. The concentration of total withanolides showed 2.13- and 1.66-fold increase in shake-flask culture and bioreactor, respectively. This protocol can be employed on suspension culture for the production of higher yield of withanolides in a short culture period using industrial bioreactors (Sivanandhan et al. 2014b). An increased withanolides production was observed in the hairy root culture of Ashwagandha found that addition of biotic elicitors like *G. edulis* and *S. wightii* extracts and abiotic elicitors such as SA and MJ (Sivanandhan et al. 2013). Kannan and Kulandaivelu (2011) in their experiment on 1-month-old Ashwagandha seedlings in pots found that drought or water stress resulted in an increase of withaferin A by 5%. Various growth characteristic such as root and shoot lengths, leaf area, photosynthetic activity and photosynthetic pigments suffered significant reduction. They proposed that these changes could be due to the variation in enzymatic activity occurring under stress conditions which in turn lead to the production of different compounds through different pathways. These compounds could be different proteins and bioactive compounds which help Ashwagandha tolerate mild drought stress.

1.4.3 Genetic Engineering

Ray and Jha (1999) transformed wild strains of *A. tumefaciens* to the culture of Ashwagandha. Some of the galls infected with N2/73 nopaline strain induced spontaneous teratomas in shoots with abnormal morphology, and these teratomas synthetized higher amount of withanolide D and withaferin A than the non-transformed shoot cultures. Joshi and Dahake (2009) initiated cell suspension cultures of Ashwagandha by transforming the hypocotyls explant with MTCC-2250 strain of *A. tumefaciens*. The cell suspension culture produced withanolides, out of which they isolated withaferin A in the form of white crystals by column chromatography using chloroform-methanol (97:3). Pandey et al. (2010) attempted to transform Ashwagandha with LBA4404 strain of *A. tumefaciens* having binary vector pIG121Hm containing reporter gene (gusA) and intron under transcriptional regulation of the 35S promoter of cauliflower mosaic virus (CaMV). Their observation led to the conclusion that the leaf fragments obtained from 10-week-old seedling grown in green house were highly competent for the transformational events as compared to in vitro shoots. They also stated that highest transient transformation could be obtained in the second extended leaf from the shoot tip. Grover et al. (2013) overexpressed an enzyme named squalene synthase, involved in the production of withanolides, in Ashwagandha. The transformation was mediated by *A. tumefaciens*. Cell suspension cultures showed substantial increase in squalene synthase activity and withanolide A content by 4- and 2.5-fold, respectively, when compared to the non-transformed cell cultures. Furthermore, withaferin A was produced in cell suspension culture, but non-transformed cultures were unable to produce. Udayakumar et al. (2014) were first in carrying out the genetic transformation of the nodal explants of Ashwagandha. They used *A. tumefaciens* strain EHA105 possessing a binary vector pGA492. The transformation was confirmed using npt II gene PCR and histochemical β-glucuronidase (Gus) assay. The explants were transformed with a success frequency of 3.16%. Their protocol opened up the arena for improving therapeutic value of Ashwagandha using transformation.

Numerous scientists have attempted to enhance the synthesis of secondary metabolites at large scale using hairy root cultures in variety of plants. The hairy roots are the transformed roots resulting from the transfer of a segment of Ri plasmid of *Agrobacterium rhizogenes* to plant genome. The hairy root culture of a plant has an array of advantages like elevated growth rate, profuse branching and biochemical and genetic stability. They also have relatively low-cost cultural requirements. Due to these favourable characteristics, hairy roots have been developed in vitro for an array of plants, especially for the synthesis of secondary metabolites produced by the roots (Banerjee et al. 2012; Sharma et al. 2013a, b).

Many researchers have come forward to synthesize withanolides using hairy root cultures of Ashwagandha. Banerjee et al. (1994) induced hairy root culture in Ashwagandha using three different strains of *A. rhizogenes*, i.e. LBA 9360, A4 and LBA 9402. *A. rhizogenes* A4 showed the best transformation capability and induced faster-growing hairy roots, which was succeeded by *A. rhizogenes* LBA 9402. *A. rhizogenes* LBA 9360 did not induce hairy roots and analysed the synthesis of withanolides, especially withaferin A, in the A4-induced hairy root cultures and the culture media, which was detected both in liquid media and in the hairy roots of two and a half month cultures. In another study, Doma et al. (2012) increased the amount of withanolides such as withaferin A and withanolide A synthesized by the hairy root cultures of Ashwagandha using carbohydrates and elicitors. They used different concentrations of sucrose (2, 3, 4 and 6% w/v). The sucrose (3%) elevated the accumulation of withanolide A and withaferin A (886 and 890 mg/g dry tissue, respectively). The amount of withaferin A increased with the increasing concentrations of glucose and was raised by 34% as compared to untransformed roots and leaves. It was observed that the acetyl salicylic acid stimulated accumulation of both withanolide A and withaferin A at 2.0 mg/land 1.0 mg/l concentrations and other biotic elicitors like nitric oxide and chitosan enhanced withaferin A, while JA decreased it. 50 mg/l and 100 mg/l of fungicide triadimefon when added to the culture media increased the accumulation of withaferin A by 1626 to 3061%, respectively; however it was not able to induce an overproduction of withanolide A.

Ara and Choudhary (2014) checked the transformation frequency, number of root induction and lateral branching in roots of Ashwagandha by two *A. rhizogenes* strains ATCC 15834 and MTCC 532. The two strains exhibited variation despite of the same type of explants, i.e. shoot tip. ATCC 15834 gave better results than MTCC 532 when all the parameters were considered. Transformation frequency, number of root/explant and number of lateral branching in roots/explants induced by strain ATCC 15834 were 91.29 ± 1.75%, 11.15 ± 0.30 and 46.09 ± 0.16, respectively, in 3 h of co-culture period, while that of MTCC 532 were 85.41 ± 2.08%, 08.27 ± 0.35 and 35.12 ± 0.23 in 4 h of co-culture period. After successful achievement of genetic transformation, it was shown that hairy roots induce directly from the subcultured part of the plant.

Increased withanolides production was observed in the hairy root culture of Ashwagandha on addition of abiotic elicitors like SA and MJ. The highest biomass and withanolides (withanolide A, withanone and withaferin A) production was obtained when hairy roots were treated with 150 μM SA with exposure time of 4 h and harvested on 40 days (Sivanandhan et al. 2013; Sivanandhan et al. 2014a, b).

Acknowledgements The authors are grateful to Department of Science and Technology (DST), Govt. of India, New Delhi, for the financial assistance as Women Scientist (Ref. No. SR/WOS-A/LS-564/2011 (G) dated 26.07.2012) to Mrs. P. Poornima. We would like to thank Prof. K. Satyamoorthy, Director, Manipal School of Life Sciences for his constant encouragement, support and critical comments. We are grateful to Dr. K.K. Mahato, Professor and Head, Dept. of Biophysics for his constant support and Manipal Academy of Higher Education (MAHE), Manipal for Dr. T.M.A. Pai PhD scholarship to P.S. Swathy and K.R. Kiran. We thank Ms. P. Bhagyashree and Ms. K. Arya for their help in preparation and format of the article.

References

Abou-Douh AM (2002) New withanolides and other constituents from the fruit of *Withania somnifera*. Arch Pharm 335(6):267–276

Ahmad MK, Mahdi AA, Shukla KK, Islam N, Rajender S, Madhukar D, Shankhwar SN, Ahmad S (2010) *Withania somnifera* improves semen quality by regulating reproductive hormone levels and oxidative stress in seminal plasma of infertile males. Fertil Steril 94(3):989–996

Alam N, Hossain M, Khalil MI, Moniruzzaman M, Sulaiman SA, Gan SH (2011) High catechin concentrations detected in *Withania somnifera* (Ashwagandha) by high performance liquid chromatography analysis. BMC Complement Altern Med 11(1):65

Ali M, Shuaib M, Ansari SH (1997) Withanolides from the stem bark of *Withania somnifera*. Phytochemistry 44(6):1163–1168

Andrade C, Aswath A, Chaturvedi SK, Srinivasa M, Raguram R (2000) A double-blind, placebo-controlled evaluation of the anxiolytic efficacy of an ethanolic extract of *Withania somnifera*. Indian J Psychiatry 42(3):295–301

Anwer T, Sharma M, Pillai KK, Khan G (2012) Protective effect of *Withania somnifera* against oxidative stress and pancreatic beta-cell damage in type 2 diabetic rats. Acta Pol Pharm 69(6):1095–1101

Ara T, Choudhary AK (2014) Study on efficacy of two strains (ATCC 15834 and MTCC 532) of *Agrobacterium rhizogenes* on hairy root induction of *Withania somnifera*. Int J Adv Biotechnol Res 4(3):1–8

Atal C, Schwarting AE (1961) Ashwagandha- an ancient Indian drug. Econ Bot 15(3):256–263

Attari M, Jamaloo F, Shadvar S, Fakhraei N, Dehpour AR (2016) Effect of *Withania somnifera* Dunal root extract on behavioural despair model in mice: a possible role for nitric oxide. Acta Med Iran 54(3):165–172

Awasthi D, Nigam RK, Mishra R, Ali SA (2008) Secondary metabolite enhancement through elicitation in micropropagated plants of Ashwagandha (*Withania somnifera* Dunal). Biosci Biotech Res Commun 1(2):173–180

Babu PV, Gokulakrishnan A, Dhandayuthabani R, Ameethkhan D, Kumar CV, Ahamed MI (2007) Protective effect of *Withania somnifera* (Solanaceae) on collagen glycation and cross-linking. Comp Biochem Physiol B Biochem Mol Biol 147(2):308–313

Baldi A, Singh D, Dixit VK (2008) Dual elicitation for improved production of withaferin A by cell suspension cultures of *Withania somnifera*. Appl Biochem Biotechnol 151(2–3):556–564

Banerjee S, Naqvi A, Mandal SP, Ahuja PS (1994) Transformation of *Withania somnifera* (L) Dunal by *Agrobacterium rhizogenes*: infectivity and phytochemical studies. Phytother Res 8(8):452–455

Banerjee S, Singh S, Ur Rehman L (2012) Biotransformation studies using hairy root cultures- a review. Biotechnol Adv 30(3):461–468

Bashir HS, Mohammed AM, Magsoud AS, Shaoub AM (2013) Isolation of three flavonoids from *Withania somnifera* Leaves (Solanaceae) and their Antimicrobial Activities. JFPI 2(5):39–45

Bhatia A, Bharti SK, Tewari SK, Sidhu OP, Roy R (2013) Metabolic profiling for studying chemotype variations in *Withania somnifera* (L.) Dunal fruits using GC-MS and NMR spectroscopy. Phytochemistry 93:105–115

Bhatnagar M, Sharma D, Salvi M (2009) Neuroprotective effects of *Withania somnifera* Dunal.: A possible mechanism. Neurochem Res 34(11):1975–1983

Bhattacharya SK, Goel RK, Kaur R, Ghosal S (1987) Anti – stress activity of sitoindosides VII and VIII, new acylsterylglucosides from *Withania somnifera*. Phytother Res 1(1):32–37

Bhattacharya A, Ramanathan M, Ghosal S, Bhattacharya SK (2000a) Effect of *Withania somnifera* glycowithanolides on iron-induced hepatotoxicity in rats. Phytother Res 14(7):568–570

Bhattacharya SK, Bhattacharya A, Sairam K, Ghosal S (2000b) Anxiolytic-antidepressant activity of *Withania somnifera* glycowithanolides: an experimental study. Phytomedicine 7(6):463–469

Bhattacharya A, Ghosal S, Bhattacharya SK (2001) Anti-oxidant effect of *Withania somnifera* glycowithanolides in chronic footshock stress-induced perturbations of oxidative free radical scavenging enzymes and lipid peroxidation in rat frontal cortex and striatum. J Ethnopharmacol 74(1):1–6

Bisht P, Rawat V (2014) Antibacterial activity of *Withania somnifera* against Gram-positive isolates from pus samples. Ayu 35(3):330–332

Bokaeian M, Fakheri B, Najed NM, Zarei H, Saeidi S, Mohasseli T (2015) The effect of *Withania somnifera* extract on drug resistant strains of *Esherichia coli*. Int J Infect 2(2):e23036

Borde AU, Athawaley AM, Mendhe MS, Patil MK, Lokhande PR, Jaiswal SA (2008) Ameliorating potential of Ashwagandha on cadmium chloride induced changes in weights of visceral organs. Vet World 1(11):343–345

Chaurasiya ND, Uniyal GC, Lal P, Misra L, Sangwan NS, Tuli R, Sangwan RS (2008) Analysis of withanolides in root and leaf of *Withania somnifera* by HPLC with photodiode array and evaporative light scattering detection. Phytochem Anal 19(2):148–154

Chengappa KNR, Brar JS, Gannon JM, Schlicht PJ (2018) Adjunctive use of a standardised extract of *Withania somnifera* (Ashwagandha) to treat symptom exacerbation in schizophrenia: A randomized, double-blind, placebo-controlled study. J Clin Psychiatry 79(5):17m11826

Chitturi D, Venisetty RK, Molmoori RK, Kokate CK, Apte SS (2010) Enhanced bioproduction of Withaferin A from suspension cultures of *Withania somnifera*. Ann Biol Res 1(2):77–86

Choudhary D, Bhattacharyya S, Bose S (2017) Efficacy and safety of Ashwagandha (*Withania somnifera* (L.) Dunal) root extract in improving memory and cognitive functions. J Diet Suppl 14(6):599–612

Ciddi V (2006) Withaferin A from cell cultures of *Withania somnifera*. Indian J Pharm Sci 68(4):490–492

Datta S, Kumar Pal NK, Nandy AK (2011) Inhibition of the emergence of multi drug resistant *Staphylococcus aureus* by *Withania somnifera* root extracts. Asian Pac J Trop Med 4(11):917–920

Doma M, Abhayankar G, Reddy VD, Kavi Kishor PB (2012) Carbohydrate and elicitor enhanced withanolide (withaferin A and withanolide A) accumulation in hairy root cultures of *Withania somnifera* (L.). Indian J Exp Biol 50(7):484–490

Grandhi A, Mujumdar AM, Patwardhan B (1994) A comparative pharmacological investigation of Ashwagandha and Ginseng. J Ethnopharmacol 44(3):131–135

Grover A, Samuel G, Bisaria VS, Sundar D (2013) Enhanced withanolide production by overexpression of squalene synthase in *Withania somnifera*. J Biosci Bioeng 115(6):680–685

Gupta L, Srivastava M (2008) Effect of *Withania somnifera* extracts on the mortality of *Callosobruchus chinensis* L. J Biopestic 1(2):190–192

Gupta SK, Mohanty I, Talwar KK, Dinda A, Joshi S, Bansal P, Saxena A, Arya DS (2004) Cardioprotection from ischemia and reperfusion injury by *Withania somnifera*: a hemodynamic, biochemical and histopathological assessment. Mol Cell Biochem 260(1–2):39–47

Gupta MS, Shivaprasad HN, Kharya MD, Rana AC (2006) Immunomodulatory activity of the Ayurvedic formulation 'Ashwagandha Churna'. Pharm Biol Pharm Biol 44(4):263–265

Gupta A, Mahdi AA, Shukla KK, Ahmad MK, Bansal N, Sankhwar P, Sankhwar SN (2013) Efficacy of *Withania somnifera* on seminal plasma metabolites of infertile males: a proton NMR study at 800 MHz. J Ethnopharmacol 149(1):208–214

Henley AB, Yang L, Chuang KL, Sahuri-Arisoylu M, Wu LH, Bligh SW, Bell JD (2017) *Withania somnifera* root extract enhances chemotherapy through priming. PLoS One 12(1):e0170917

Huang Z, Mao Q, Zhong X, Feng C, Pan A, Li Z (2009) Herbal formula SYJN protect PC12 cells from neurotoxicity induced by corticosterone. J Ethnopharmacol 125(3):456–460

Jayanthi MK, Huralikuppi JC, Suresha RN, Dhar M (2012) Anti-depressant effects of *Withania somnifera* fat (Ashwagandha Ghrita) extract in experimental mice. Int J Pharma Bio Sci 3(1):33–42

Jayaprakasam B, Nair MG (2003) Cyclogenase-2 inhibitory withanolides from *Withania somnifera* leaves. Tetrahedron 59(6):841–849

Jayaprakasam B, Strasburg GA, Nair MG (2004) Potent lipid peroxidation inhibitors from *Withania somnifera* fruits. Tetrahedron 60(13):3109–3121

Joshi D, Dahake AP (2009) Withaferin-A isolated from the cell cultures of *Withania somnifera*. Der Pharma Chemica 1(2):109–112

Kannan ND, Kulandaivelu G (2011) Drought induced changes in physiological, biochemical and phytochemical properties of *Withania somnifera* Dun. J Med Plants Res 5(16):3929–3935

Kataria H, Shah N, Kaul SC, Wadhwa R, Kaur G (2011) Water extract of ashwagandha leaves limits proliferation and migration and induces differentiation in glioma cells. Evid Based Complement Alternat Med 2011:267614. 12 pgs

Khan ZS, Nasreen S (2010) Phytochemical analysis, antifungal activity and mode of action of methanol extracts from plants against pathogens. J Agric Technol 6(4):793–805

Khazal KF, Hill DL (2015) *Withania somnifera* extract reduces the invasiveness of MDA-MB-231 breast cancer and inhibits cytokines associated with metastasis. J Cancer Metastasis Treat 1(2):94–100

Khursheed F, Rehman K, Akhtar MS, Dogar MZ, Khalil B (2010) Comparative antilipidemic effects of native and gemmotreated *Withania somnifera* (Asghand) extracts. J Appl Pharm 1(2):47–59

Kokate CK, Purohit AP, Gokhale SB (1996) Pharmacognosy, 4th edn. Nirali Prakashan, Pune, pp 624–629

Kulkarni SK, Dhir A (2008) *Withania somnifera*: an Indian ginseng. Prog Neuropsychopharmacol Biol Psychiatry 32(5):1093–1105

Kumar P, Kumar A (2008) Effects of root extract of *Withania somnifera* in 3-nitropropionic acid-induced cognitive dysfunction and oxidative damage in rats. Int J Health Res 1(3):139–149

Kumar A, Kaul MK, Bhan MK, Khanna PK, Suri KA (2007) Morphological and chemical variation in 25 collections of the Indian medicinal plant, *Withania somnifera* (L.) Dunal (Solanaceae). Genet Resour Crop Evol 54(3):655–660

Kumar S, Harris RJ, Seal CJ, Okello EJ (2012) An aqueous extract of *Withania somnifera* root inhibits amyloid β fibril formation *in vitro*. Phytother Res 26(1):113–117

Kurapati KRV, Atluri VS, Samikkannu T, Nair MP (2013) Ashwagandha (*Withania somnifera*) reverses β-amyloid1-42 induced toxicity in human neuronal cells: Implications in HIV-associated neurocognitive disorders (HAND). PLoS ONE 8(10):e77624

Lal P, Misra L, Sangwan RS, Tuli R (2006) New withanolides from fresh berries of *Withania somnifera*. Z Natur für sch 61(9):1143–1147

Mahdi AA, Shukla KK, Ahmad MK, Singh R, Shankhwar SN, Singh V, Dalela D (2009) *Withania somnifera* improves semen quality in stress-related male fertility. Evid Based Complement Alternat Med 2011:576962

Mehrotra V, Mehrotra S, Kirar V, Shyam R, Misra K, Srivastava AK, Nandi SP (2011) Antioxidant and antimicrobial activities of aqueous extract of *Withania somnifera* against methicillin-resistant Staphylococcus aureus. J Microbiol Biotech Res 1(1):40–45

Mir BA, Koul S, Kumar A, Sushant S, Kaul MK, Soodan AS (2012) Reproductive behaviour and breeding system of wild and cultivated types of *Withania somnifera* (L.) Dunal. J Med Plants Res 6(5):754–762

Misra L, Mishra P, Pandey A, Sangwan RS, Sangwan NS, Tuli R (2008) Withanolides from *Withania somnifera* roots. Phytochemistry 69(4):1000–1004

Misra L, Mishra P, Pandey A, Sangwan RS, Sangwan NS (2012) 1,4-Dioxane and ergosterol derivatives from *Withania somnifera* roots. J Asian Nat Prod Res 14(1):39–45

Murthy HN, Praveen N (2011) Effects of macroelements and nitrogen source on biomass accumulation and withanolide-A production from cell suspension cultures of *Withania somnifera* (L.) Dunal. Plant Cell Tiss Organ Cult 104(1):119–124

Murthy HN, Praveen N (2012) Influence of macro elements and nitrogen source on adventitious root growth and withanolide-A production in *Withania somnifera* (L.) Dunal. Nat Prod Res 26(5):466–473

Nakajima H, Wakabayashi Y, Wakamatsu K, Imokawa G (2011) An extract of *Withania somnifera* attenuates endothelin-1-stimulated pigmentation in human epidermal equivalents through the interruption of PKC activity within melanocytes. Phytother Res 25(9):1398–1411

Nasreen S, Radha R (2011) Assessment of quality of *Withania somnifera* Dunal (Solanaceae) pharmacognostical and phyto physicochemical profile. Int J Pharm Pharm Sci 3(2):152–155

Nema R, Khare S, Jain P, Pradhan A (2013) Anticancer Activity of *Withania somnifera* (Leaves) Flavonoids Compound. Int J Pharm Sci Rev Res 9(1):103–106

Osman M, Wafaa A, Nadia E, Amany A, Doa E (2012) Potential role of *Withania somnifera* on human breast cancer. Nat Sci 10(11):53–59

Oza VP, Parmar PP, Kumar S, Subramanian RB (2010) Anticancer properties of highly purified L-asparaginase from *Withania somnifera* L. against acute lymphoblastic leukemia. Appl Biochem Biotechnol 160(6):1833–1840

Pandey V, Misra P, Chaturvedi P, Mishra MK, Trivedi P, Tuli R (2010) Agrobacterium tumefaciens-mediated transformation of *Withania somnifera* (L.) Dunal: an important medicinal plant. Plant Cell Rep 29(2):133–141

Pandey A, Bani S, Dutt P, Kumar Satti N, Avtar Suri K, Nabi Qazi G (2018) Multifunctional neuroprotective effect of Withanone, a compound from *Withania somnifera* roots in alleviating cognitive dysfunction. Cytokine 102:211–221

Patel H, Krishnamurthy R (2013) Elicitors in plant tissue culture. J Pharmacogn Phytochem 2(2):60–65

Pramanick S, Roy A, Ghosh S, Majumder HK, Mukhopadhyay S (2008) Withanolide Z, a new chlorinated withanolide from *Withania somnifera*. Planta Med 74(14):1745–1748

Rahmati B, Moghaddam MHG, Khalili M, Enayati E, Maleki M, Rezaeei S (2016) Effect of *Withania somnifera* (L.) dunal on sex hormone and gonadotropin levels in addicted male rats. Int J Fertil Steril 10(2):239–244

Rao R, Rajput DK, Nagaraju G, Adinarayana G (2012) Opportunities and challenges in the cultivation of Ashwagandha (*Withania somnifera* (L.) Dunal). J Pharmacogn 3(2):88–89

Ray S, Jha S (1999) Withanolide synthesis in cultures of *Withania somnifera* transformed with *Agrobacterium tumefaciens*. Plant Sci 146(1):1–7

Roja G, Heble R, Sipahimalani AT (1991) Tissue cultures of *Withania somnifera*: morphogenesis and withanolide synthesis. Phytother Res 5(4):185–187

Sangwan RS, Chaurasiya ND, Lal P, Misra L, Uniyal GC, Tuli R, Sangwan NS (2007) Withanolide A biogeneration in *in vitro* shoot cultures of Ashwagandha (*Withania somnifera* Dunal), a main medicinal plant in Ayurveda. Chem Pharm Bull 55(9):1371–1375

Saykally JN, Hatic H, Keeley KL, Jain SC, Ravindranath V, Citron BA (2017) *Withania somnifera* extract protects model neurons from in vitro traumatic injury. Cell Transplant 26(7):1193–1201

Sehgal N, Gupta A, Valli RK, Joshi SD, Mills JT, Hamel E, Khanna P, Jain SC, Thakur SS, Ravindranath V (2012) *Withania somnifera* reverses Alzheimer's disease pathology by enhancing low density lipoprotein receptorrelated protein in liver. Proc Natl Acad Sci 109(9):3510–3515

Sharma P, Padh H, Shrivastava N (2013a) Hairy root cultures: a suitable biological system for studying secondary metabolic pathways in plants. Eng Life Sci 13(1):62–75

Sharma RA, Goswami M, Yadav A (2013b) GC-MS screening of alkaloids of -*Withania somnifera*. L. in vivo and in vitro. Ind J Appl Res 3(8):43

Shukla KK, Mahdi AA, Mishra V, Rajender S, Sankhwar SN, Patel D, Das M (2011) *Withania somnifera* improves semen quality by combating oxidative stress and cell death and improving essential metal concentrations. Reprod Biomed Online 22(5):421–427

Singariya P, Kumar P, Mourya KK (2012) Evolution of antimicrobial activity of leaf extracts of winter cheery (*Withania somnifera*). Int J Pharm Tech Res 4(3):1247–1253

Singh BK, Gahoi R, Sonkar A (2010a) A quality assessment and phytochemical screening of selected region of *Withania somnifera* Dunal. Int J Pharm Sci Res 1(7):73–77

Singh SP, Tanwer BS, Khan M (2010b) Antifungal potential of Ashwagandha against some pathogenic fungi. Int J Biopharm 1(2):72–74

Singh P, Guleri R, Singh V, Kaur G, Kataria H, Singh B, Kaur G, Kaul SC, Wadhwa R, Pati PK (2015) Biotechnological interventions in *Withania somnifera* (L.) Dunal. Biotechnol Genet Eng Rev 1(1–2):1–20

Sivanandhan G, Arun M, Mayavan S, Rajesh M, Jeyaraj M, Dev GK, Manickavasagam M, Selvaraj N, Ganapathi A (2012) Optimization of elicitation conditions with methyl jasmonate and salicylic acid to improve the productivity of withanolides in the adventitious root culture of *Withania somnifera* (L.) Dunal. Appl Biochem Biotechnol 168(3):681–696

Sivanandhan G, Dev GK, Jeyaraj M, Rajesh M, Arjunan A, Muthuselvam M, Manickavasagam M, Selvaraj N, Ganapathi A (2013) Increased production of withanolide A, withanone and withaferin A in hairy root cultures of *Withania somnifera* (L.) Dunal elicited with methyl jasmonate and salicyclic acid. Plant Cell Tissue Organ Cult 114:121–129

Sivanandhan G, Selvaraj N, Ganapathi A, Manickavasagam M (2014a) Improved production of withanolides in shoot suspension culture of *Withania somnifera* (L.) Dunal by seaweed extracts. Plant Cell Tissue Organ Cult 119(1):221–225

Sivanandhan G, Selvaraj N, Ganapathi A, Manickavasagam M (2014b) Enhanced biosynthesis of withanolides by elicitation and precursor feeding in cell suspension culture of *Withania somnifera* (L.) Dunal in shake-flask culture and bioreactor. PLoS ONE 9(8):e104005

Srivastava A, Gupta AK, Shanker K, Gupta MM, Mishra R, Lal RK (2018) Genetic variability, associations, and path analysis of chemical and morphological traits in Indian ginseng [*Withania somnifera* (L.) Dunal] for selection of higher yielding genotypes. J Ginseng Res 42(2):158–164

Sumantran VN, Kulkarni A, Boddul S, Chinchwade T, Koppikar SJ, Harsulkar A, Patwardhan B, Chopra A, Wagh UV (2007) Chondroprotective potential of root extracts of *Withania somnifera* in osteoarthritis. J Biosci 32(2):299–307

Thiagarajan G, Venu T, Balasubramanian D (2003) Approaches to relieve the burden of cataract blindness through natural antioxidants: use of Ashwagandha (*Withania somnifera*). Curr Sci 85(7):1065–1071

Tong X, Zhang H, Timmermann BN (2011) Chlorinated withanolides from *Withania somnifera*. Phytochem Lett 4(4):411–414

Tripathi AK, Dey S, Singh RH, Dey PK (1998) Alterations in the sensitivity of 5th receptor subtypes following chronic Asvagandha treatment in rats. Anc Sci Life 17(3):169–181

Udayakumar R, Kasthurirengan S, Mariashibu TS, Rajesh M, Anbazhagan VR, Kim SC, Ganapathi A, Choi CW (2009) Hypoglycaemic and hypolipidaemic effects of *Withania somnifera* root and leaf extracts on alloxan-induced diabetic rats. Int J Mol Sci 10(5):2367–2382

Udayakumar R, Kasthurirengan S, Mariashibu TS, Sahayarayan J, Ganapathi A, Kim SC, Kim JJ, Choi CW (2014) *Agrobacterium*-mediated genetic transformation of *Withania somnifera* using nodal explants. Acta Physiol Plant 36(8):1969–1980

Umadevi M, Rajeswari R, Rahale CS, Selvavenkadesh S, Pushpa R, Kumar KPS, Bhowmik D (2012) Traditional and Medicinal Uses of *Withania somnifera*. Pharm Innov 1(9):102–110

Verma SK, Kumar A (2011) Theraupetic Uses of *Withania somnifera* (Ashwagandha) with a note on withanolides and its pharmacological actions. Asian J Pharm Clin Res 4(1):1–4

Visavadiya NP, Narasimhacharya AVRL (2007) Hypocholesteremic and antioxidant effects of *Withania somnifera* (Dunal) in hypercholesteremic rats. Phytomedicine 14(2–3):136–142

Zhao J, Nakamura N, Hattori M, Kuboyama T, Tohda C, Komatsu K (2002) Withanolide derivatives from the roots of *Withania somnifera* and their neurite outgrowth activities. Chem Pharm Bull 50(6):760–765

Somatic Embryogenesis and Plant Regeneration in *Gloriosa superba* L.: An Endangered Medicinal Plant

2

Venkatachalam Balamurugan, Thomas Cheeran Amal, Palanisamy Karthika, Subramanian Selvakumar, and Krishnan Vasanth

Abstract

Gloriosa superba L. is an endangered ornamental medicinal plant which is in great demand for pharmaceutical use. Thus, there is overexploitation of this important species in cancer treatments due to its important alkaloid colchicine content. The development of efficient protocol for indirect organogenesis and somatic embryogenesis is necessary for its conservation. In vitro propagation was standardized using MS medium supplemented with various hormone concentrations and combinations of BAP (3 mg/l). The maximum shoot regeneration was 28 days, whereas prolonged duration of shoot in the medium leads to the development of micro tubers. The regenerated shoots were transferred to root induction medium supplemented with BAP + NAA (5 ± 0.81); development occurred in 34 days. The somatic embryogenesis showed high frequency of cell suspension culture in different cytokines such as BAP and KIN. The maximum numbers of somatic embryos were significantly highest in liquid MS medium containing BAP 2 mg/l (23.25 ± 0.95). The plantlets were successfully acclimatized into the greenhouse conditions. This protocol provides a system for very large-scale propagation and leads to the development and conservation of plants.

Keywords

G. superba · Callus induction · Shoot regeneration · Somatic embryogenesis

V. Balamurugan · T. C. Amal · P. Karthika · K. Vasanth (✉)
Department of Botany, Bharathiar University, Coimbatore, Tamil Nadu, India

S. Selvakumar
Department of Biochemistry, Bharathiar University, Coimbatore, Tamil Nadu, India

M. Kumar et al. (eds.), *In vitro Plant Breeding towards Novel Agronomic Traits*,
https://doi.org/10.1007/978-981-32-9824-8_2

2.1 Introduction

Gloriosa superba L. belonging to the family Colchicaceae is one of the important endangered ornamental medicinal plants. The plant is being exploited extensively by the tribal and local people for the preparation of traditional medicine. It is native to tropical and sub-tropical regions of Africa and Southeast Asia including India, Sri Lanka, Burma and Malaysia. In India, it is mainly found in southern parts of Tamil Nadu, Kerala, Andhra Pradesh and Karnataka (Jayaweera 1982; Chandel et al. 1996; Ghosh et al. 2002; Sivakumar and Krishnamurthy 2002). It is designated as the national flower of Zimbabwe and the state flower of Tamil Nadu. In India, *G. superba* and *G. rothschildiana* are its commonly found species (Tarar and Vishwakarma 1995), whereas in Africa, several species like *G. superba*, *G. simplex*, *G. grandiflora*, *G. lutea*, *G. planti*, *G. lipolidi*, *G. longifolia*, *G. rothschildiana*, *G. virenses* and *G. sudanica* are distributed. The plant is famous particularly in South India for its high colchicine content. Due to its overexploitation, a major problem during the field cultivation was distinguished. Thus, there is a very urgent need to conserve and standardize in vitro propagation protocol for mass production of high-yield traits. This may help to promote the cultivation of this crop on commercial scale.

2.1.1 Phenotypic Traits

This is an herbaceous, branched, summer-growing climber reaching up to 1.5 m tall with tuberous roots and stems that can rise up to 4 m long. It has mature tubers having a V-shaped structure. The leaves are greenish in colour and tipped with tendrils. It has strong leaf blade with parallel nerves. Leaves are sessile, lanceolate, acuminate, entire, glabrous, alternate and lance-shaped. The margins of the flowers are generally wavy of fiery colours like dork in yellow and orange scarlet in the period of November to March. The flowers are usually long, pedicellate with 20 cm length having 3 to 6 free petals generally fiery colors like dusky yellow with scarlet orange during the period of November to March. The prominent stamens are 3–4 cm long consisting large anther at the tip that drops large amount of yellow pollens. The flowers are made up of large, shiny, greenish gynoecium sharply outer point of ovary tip. The fruits are large, oblong capsules and 6–12 cm long containing 20 globose red-coloured seeds on each valve (Fig. 2.1).

2.1.2 Pollination

The attracting look like flowers and its hercogamous nature help for cross-pollination. The peculiar structure of the large flowers with 6 perianth lobes bent backwards, 6 radiating anther and style bent 90 at the point of attachment to the ovary does not make them suitable for pollination by tiny insects (Anandhi and

Fig. 2.1 Florescence in *G. superba*

Rajamani 2012). It they have been both self- and cross-pollinated, and seed set is related to the time of pollination and pollinator activity. Even though there were no barriers for self- or cross-incompatibility. The fruit set was 90% under artificial cross-pollination, 100% under artificial self-pollination and only 40% under open pollination which indicates the need for artificial pollination (Mamtha et al. 1993). The highest pollination per pod (~70.93%) followed by air blowing polllination (~65.52 %) was recorded for each event. Maximum pod set was observed in artificial cross-pollination within the species followed by self-pollination. There was minimum pod set recorded in natural self- or cross-pollination (Nagajothi et al. 2008).

2.1.3 Cultivation and Harvest

It grows in tropical warm and humid regions with an annual rainfall of 350–400 cm which is sufficient enough for cultivation. The soil and water testing is a very essential criteria before selecting the area for the cultivation. It is basically propagated from its underground V-shaped rhizomes. Healthy tubers alone should be selected for tubers before sprouting. The plant is grown from small rhizomes and there is no flowering during the first year. On an average, 600 kg of rhizome can be acquired to maintain the plant population of 12,000–13,000 plants per acre. The mature fruits have been harvested after a period of 160–180 days. The fruit skin becomes shrunken and light green to dark green, and mature pods that are weightless and give a crinkling sound are the best choice for harvesting. On an average, the seed yield is 300–500 kg/hectare (Padmapriya et al. 2015). The yield loss of this plant is mainly affected by the insects like *Polytela gloriosae*, *Plusia signata* and *Spodoptera litura* which causes a 25.8–100% economic loss (Dhir et al. 1992; Suganthy and Sakthivel 2013) (Fig. 2.2).

Fig. 2.2 Infestation of *Polytela gloriosae*

2.1.4 Conventional Breeding

Traditional or conventional breeding has not been attempted so far as there is only one ecotype introduction of new variability. And there is only one option for breeding techniques to create a new variability for high-yielding cultivars (Rajadurai 2001; Selvarasu and Kandhasamy 2013). Vegetative propagation through tubers resulted in reduced biotic tolerance, abiotic stress and crop yields. This may be due to the genetic similarities: parents (2n = 22) that enable the hybridization in both directions. Similarly, 90% of pod set is under controlled cross-pollination (Gupta and Raina 2001; Sudhendra and Rudre Gowda 1997; Omolaja 2009). The seeds after germination produce the microtuber during the first growing season F1 seedling and the initial mean value of seed germination rate was 18.28%, seedling height was 11.01 cm, root length was 6.06 cm, microtuber length was 1.74 cm, microtuber girth was 2.12 cm, then microtuber weight was 1.04 g. (Selvarasu and Kandhasamy 2013). The colchicine's used for plant breeding and biological studies aspect to produced polyploidy plants in G. superba contains about 0.1-0.8% of colchicine in bulb. Which is used for plant breeding to induce high yield polyploidy or breeding plants (Ravindra and Mahendra 2009).

2.1.5 Natural Bioactive Compounds

The seeds are best source of colchicines as their content is 2–5 times higher than in tubers, and other compounds like lumicolchicine, 3-demethyl-N-deformyl-Ndeacetylcolchicine, 3-demethylcolchicine and N-formyldeacetylcolchcine have been isolated from *G. superba* (Sugandhi 2000; Suri et al. 2001). The colchicine's contents varied among different parts like seeds (0.6%), tuber (0.3%), stem (0.33–0.41%), flower (1.18%) and ovary (0.08%) (Clewer et al. 1915; Finnie and Staden 1994; Sarin et al. 1974). Colchicine has the potential to alter the cell membranes' structure indirectly by inhibiting the synthesis of membrane constituents. It attaches to the tubulin to prevent polymerization into microtubules. Gloriosine also has an antimitotic effect that disrupts the spindle fibre apparatus and separates chromosomes during the metaphase involved by the arrest of mitosis. It also causes inhibition on various phosphatases (Gooneratne 1966; Craker and Simson 1987; Gautam et al. 2017).

2.1.6 Potential Medicinal Values

As a tradition since 2000 B.C., it has been widely used as a medicinal plant by tribes (Ade and Rai 2009). All the plant parts were used as medicine by Siddha, Ayurveda and Yunani. Ayurvedic medicinal system highlights the use of entire plant parts for treating itches and skin diseases including wounds and ailments caused by vitiated Kapha and Vata. Sushruta Samhita and Chakradatta highlighted its application. The literatures such as Dhanvantari Nighantu, Madanadi Nighantu and Bhavaprakasha exhibited the use of different parts to cure dropsy, piles, wounds and placenta and treat skin diseases (Jain and Suryavanshi 2010). The ointments made from the roots were used to treat labour pains, expulsion of the placenta, paralysis, rheumatism, snake bites and insect bites (Patil and Gavade 2012). Generally, tubers are used for abortion purposes, fevers, wounds, anti-fertility, gonorrhoea, leprosy and piles. Even the leaves have been more used for curing asthma and skin disorders. And the flowers have analgesic, anti-inflammatory, antimicrobial, larvicidal, antipoxviral and anti-tumour properties (Rao et al. 2014). Seeds were used for rheumatic pains and as muscle relaxants (Nadkarni 2002).

2.1.7 Plant Propagation Through In Vitro Method

G. superba is an important medicinal plant that has various therapeutic applications, and over-exploitation of this plant leads to local extinction. Using conventional method to propagate this plant has many disadvantages (Mrudul et al. 2001); thus, biotechnological approach would be a better alternative (Sivakumar and Krishnamurthy 2002). Hence, in vitro techniques played a valuable tool to overcome conventional breeding problems. Minimal seed-set with poor germination and propagation by corm is the major limiting factor. An efficient and simple micropropagation protocol is required to meet the demand for pharmaceutical applications (Custers and Bergervoet 1994; Sayeed and Shyamal 2005).

2.1.8 In Vitro Breeding Experiments Towards the Improvement of Secondary Metabolites

In vitro culture techniques embrace a vital key for the production of beneficial secondary metabolites. The synthesis by chemical method becomes a difficult task and economically not feasible because of their diverse and complex structure (Pickens et al. 2011). Hence, isolation and characterization of powerful biologically active compounds from natural sources are more commonly adopted. Conversely, in vitro breeding experiment becomes a more predominant method for the large-scale production of specific metabolites on pharmaceutical demand. The appropriate selection of suitable plant development through organ or callus culture method could be useful for the enormous production of typical secondary metabolites (Atanasov et al. 2015; Ochoa-Villarreal et al. 2016). The appropriate medium composition and PGRs improve the callus growth and metabolite production. Susceptible actively dividing callus can significantly improve the production of important metabolites through cell suspension cultures (Verpoorte et al. 2002). It alters the biosynthetic cycle of particular metabolite that could be useful for synthesis of highly valuable medicinal bioactive compounds in good quantities (Singh and Chaturvedi 2012). The callus initiated from the rhizome explant used for cell suspension cultures for the biosynthesis of improved production of colchicine and thiocolchicoside and also phycomolecule-coated silver nanoparticle was effective biostimulants to enhance the rate of somatic embryo production, maturation and plant regeneration in *G. superba* (Mahendran et al. 2018). There are several precursors and nutritional factors used for the improvement of colchicine content in callus cultures (Sivakumar et al. 2004). The medium supplemented with sucrose and ammonium nitrate showed greatest potential of biomass and colchicine content from the callus. In addition to this, sulphate ions markedly increased the formation of colchicine (Pandurangan and Philomina 2010). Elicitation is one of the methodologies adopted for enhanced commercial production of secondary metabolites from plant cell culture system. They are chemical compounds from abiotic and biotic sources that can trigger the stress responses in plants, leading to the enhanced synthesis and accumulation of secondary metabolites or the induction of novel secondary metabolites. Root cultures of *G. superba* were treated with methyl jasmonate and aluminium chloride to significantly improve the intracellular colchicine content (Ghosh et al. 2006). A comparative analysis was drawn for colchicine content in Colchicum and Gloriosa in vitro and in vivo conditions both, and, measured around 10–25 times high concentration of colchicine in the case of in vitro grown plants (Hayashi et al. 1988, Finnie and van Staden 1991). The medium supplemented with combination of p-coumaric acid and tyramine increased colchicine content from the root cultures of *G. superba* (Ghosh et al. 2002). These findings justify that in vitro breeding experiments could be a valuable tool for the synthesis of pharmacologically important secondary metabolites in *G. superba*.

2.2 Materials and Methods

2.2.1 Explant Preparation and Surface Sterilization

The plant in a field grown on the campus of Bharathiar University, Coimbatore, Tamil Nadu, India, was collected in the month of December. The plant specimen was referred and authenticated by the Department of Botany, Bharathiar University,

Coimbatore – 641,046. The leaf explants were initially washed in running tap water for 30 min followed by 1% sodium hypochlorite for 5 min. It was then sterilized with 70% ethanol for 10 min. After that, explants were treated with 0.2% mercuric chloride ($HgCl_2$) for 10 min. Finally, explants were washed three or four times in sterile double distilled water. The surface-sterilized explants were transferred to a medium supplemented with various hormonal combinations under aseptic conditions.

2.2.2 Media and Culture Condition

The basic culture medium consisting of MS medium (Murashige and Skoog 1962) was supplemented with 30 g/l sucrose and solidified with 0.8% agar. The pH of the medium was regulated to 5.8 ± 0.2 with 0.1 N KOH or 0.1 N HCl after adding plant growth regulators. The media was autoclaved at 121 °C for 15 min. The thermo-liable growth regulators (filter sterilized) were added to the media succeeding autoclaving. After sterilization the medium was allowed to solidify in test tubes under aseptic conditions. The inoculated culture tubes were incubated at 25 ± 2 °C under 16 h light and 8 h dark conditions irradiance provided by cool white fluorescence tubes (Philips, India) and 55–60% relative humidity. The culture was observed at regular intervals.

2.2.3 Callus Induction

The leaf segments which are about 3–5 mm in length long were excised to be employed as explants for callus formation. The concentration of cytokines and auxins varied between BAP (0.5–5 mg/l), BAP + NAA, (1–5 mg/l), 2,4-D (1–5 mg/l) and picloram (2–10 mg/l). These combinations were tested for the induction of different types of callus development. Data of frequency (%) of callus induction was measured after 4 weeks of culture.

2.2.4 Shoot Organogenesis from Callus

Well-established hard and compact callus (0.5 g fresh weight) was grown on MS medium supplemented with BAP (0.5–5 mg/l) for shoot organogenesis. Cultures were transferred on fresh media after the 15th day of inoculation. The percentage of shoot organogenesis from callus and average number of shoots and length were recorded on the 45th day after transferring the callus on shoot organogenesis media.

2.2.5 Adventitious Root Formation

The regenerated shoots (3–5 cm long) obtained from micropropagated plantlets were cultured on MS basal medium fortified with NAA (1–5 mg/l), BAP + NAA (1–2.5 mg/l) and IBA (1–4 mg/l) for adventitious rooting. Data were recorded on percentage of rooting, and the length of the roots was calculated after 5 weeks of transfer into the rooting media.

2.2.6 Somatic Embryo Induction and Maturation

MS basal media supplemented with various concentrations and combinations of BAP (1–5 mg/1) and KIN (1–5 mg/l) were used for induction of somatic embryogenesis. Torpedo stage (3–5 mm in length) embryos were separated and transferred to MS medium supplemented with BAP and NAA (1–5 mg/l). Somatic embryos (torpedo stage) were transferred to growth regulator-free MS basal medium to evaluate the effect of the above growth regulators on maturation of somatic embryos from torpedo stage to cotyledonary stage. The embryos were then subcultured after 4 weeks in MS media for further conversion into plantlets.

2.2.7 Hardening and Acclimatization

The plantlets regenerated through both indirect and somatic embryogeneses were gently washed with sterile distilled water to remove the adhering agar and transferred into the plastic nursery tray containing a mixture of sterile soil/vermicompost/sand (1:1:1) and were kept in the greenhouse for acclimatization. The plants were watered twice and maintained at 70–75% relative humidity. After acclimatization in the greenhouse for 2 months, plantlets were successfully transferred to the field.

2.2.8 Observation and Statistical Analysis

Each treatment consists of 30 explants, and the experiment was repeated thrice. Observations on the frequency and the number of shoots and roots were recorded. The data was presented as mean ± SD for all the treatment of replication. The analysis of variance (ANOVA) appropriate for the design was carried out to detect the significance of difference among the treatment means. The treatment means were compared using Duncan's multiple range test (DMRT) at $P < 0.05\%$ level. Data was analysed statistically using STATISTICA 6.0 and SPSS 11.5.

2.3 Results and Discussion

2.3.1 Effect of Plant Growth Regulators on Callus Induction

For callus initiation, leaf explants inoculated in MS basal medium with different plant growth regulators such as picloram (1.0–8.0 mg/l), BAP (1.0–5.0 mg/l), 2,4-D (1.0–5.0 mg/l) and BAP + NAA (0.5–5.0 mg/l) were used. The callus formation was observed within 35 days of inoculation as shown in Table 2.1. The variation in callus response like morphology, colour and callus types depends on the plant growth regulator. The callus showed slightly greenish-yellow compact in BAP + NAA combination. Among various concentrations tested, maximum callus regeneration was observed in the 2,4-D 3 mg/l (84.5 ± 3.31). In this experiment, most of the callus showed compact in texture and colour varied with each hormonal concentration, whereas friable callus procured from picloram suitable for the cell suspension

Table 2.1 Effect of different concentrations on auxins and cytokinin callus induction from leaf explant

Plant growth regulators (mg/l)				
Auxins	Cytokinins	Number of days	Callus type	Callus percentage
2,4-D				
1.0		32.5 ± 2.08	YC	70.75 ± 2.98^{e}
2.0		30.75 ± 0.95	YC	73.25 ± 2.36^{d}
3.0		28.5 ± 0.57	YC	84.5 ± 3.31^{a}
4.0		29.75 ± 1.70	YC	78.5 ± 1.91^{c}
5.0		30 ± 0.95	YC	74.25 ± 1.70^{d}
Picloram				
2.0		35 ± 0.81	YF	41.25 ± 0.95^{k}
4.0		33.25 ± 0.5	YF	70 ± 0.81^{e}
6.0		33 ± 0.81	YF	63.5 ± 1.29^{h}
8.0		36.25 ± 0.5	YF	55.5 ± 2.081^{i}
10.0		39.25 ± 0.95	YF	53 ± 1.41^{i}
	BAP			
	0.5	29 ± 1.15	GYC	52.66 ± 2.51^{i}
	1.0	27.75 ± 0.5	GYC	67 ± 2.64^{f}
	2.0	36 ± 1.41	GYC	53.33 ± 3.51^{i}
	3.0	35.26 ± ±0.5	GYC	43.66 ± 3.21^{j}
	4.0	33.25 ± 0.95	GYC	45 ± 5^{j}
NAA	BAP			
1.0	1.0	34.25 ± 0.95	GYC	63 ± 2.64^{h}
2.0	2.0	28.5 ± 0.57	GYC	81.66 ± 2.88^{b}
3.0	3.0	33.25 ± 0.5	GYC	74 ± 1.73^{d}
4.0	4.0	37.7 ± 1	GYC	65 ± 5^{g}
5.0	5.0	40.75 ± 0.957	GYC	50 ± 5^{ij}

GYC greenish-yellow compact callus, *YC* yellow compact callus, *YF* yellow friable callus

Means (M) ± standard error followed by the common letters is not significantly different at the $p \leq 0.05\%$ level by DMRT

culture has the ability to produce somatic embryos. MS medium supplemented with NAA and BAP induced 96% of callus proliferation from internode explant within 12 days (Nikhila et al. 2014). Similarly, yellowish callus was initiated in 2,4-D and IAA on fourth week of the culture (Arumugam and Gopinath 2012).

2.3.2 Effect of Plant Growth Regulators on Shoot Induction

Various hormone concentrations of BAP were used for shoot induction. Among different hormone concentrations tested, MS medium supplemented with BAP 2 mg/l obtained a maximum number of shoot per explants (5.25 ± 0.5) after 28 days as shown in Table 2.2. There was a significant shoot initiation on BAP (1–3 mg/l) concentration as shown in Fig. 2.3. Hormone concentration increased significantly, resulting in decreased shoot development. After 6 weeks of shoot culture, there was microtuber formation without any additional supplements. Similarly, BAP resulted in the highest number of shoot from callus without any other growth regulators (Kahate 2017). The microtuber formation was reported in BAP within 8 weeks (Kumar et al. 2015). Similarly, BAP and NAA showed a maximum number of primary and secondary microtuber formations (Anandhi and Rajamani 2012), whereas BAP alone and NAA combination obtained very effective shoot initiation (Yadav et al. 2015).

2.3.3 Effect of Plant Growth Regulators on Root Induction

The MS medium without any plant growth regulators did not produce any roots from the in vitro-regenerated shoots. Hence, well-developed shoots were transferred into rooting medium containing various combinations of BAP and NAA. MS

Table 2.2 Effect of cytokinin and auxin on shoot induction from callus

Plant growth regulators (mg/l)				
Cytokinins	Auxins	Number of days	Number of shoots	Shoots length (cm)
BAP				
0.5		30.75 ± 0.9	2.25 ± 0.5^{c}	1.36 ± 0.1^{e}
1.0		27.5 ± 0.5	3.5 ± 0.5^{bc}	2.9 ± 0.2^{bcd}
2.0		28 ± 0.8	5.25 ± 0.5^{a}	3.1 ± 0.8^{bc}
3.0		31.75 ± 0.9	3 ± 0.8^{bc}	4.4 ± 0.4^{a}
4.0		35 ± 0.8	2.5 ± 0.5^{c}	2.2 ± 0.2^{d}
BAP	NAA			
1.0	1.0	28.75 ± 0.5	2.75 ± 0.5^{c}	1.4 ± 0.2^{e}
2.0	2.0	26.5 ± 0.5	4 ± 0.8^{ab}	2.8 ± 0.2^{bc}
3.0	3.0	25 ± 0.8	4.47 ± 0.5^{a}	3.3 ± 0.1^{b}
4.0	4.0	27.25 ± 0.9	3.75 ± 0.9^{b}	2.7 ± 0.2^{bcd}
5.0	5.0	29.5 ± 0.5	2.5 ± 0.5^{c}	2.4 ± 0.1^{cd}

Means (M) ± standard error followed by the common letters is not significantly different at the $p \leq 0.05\%$ level by DMRT

Fig. 2.3 The figure shows the different stages of indirect organogenesis. (**a**) Leaf explant, (**b**) compact callus, (**c** & **d**) shoot initiation at different stages, (**e**) shoot elongation and (**f**) root development

Table 2.3 Effect of root induction from different auxins and cytokinin concentrations

Plant growth regulators (mg/l)				
Auxins	Cytokinins	Number of days	Number of Shoots	Shoots Length (cm)
IBA				
1.0		37 ± 1.8	3.5 ± 0.5[b]	1.6 ± 0.5[c]
2.0		38 ± 2.0	2 ± 0.8[c]	3.3 ± 0.5[ab]
3.0		44.25 ± 2.6	1.25 ± 0.5[c]	3 ± 1.0[ab]
NAA	BAP			
1.0	1.0	33 ± 0.8	3.5 ± 0.5[b]	2.3 ± 0.5[bc]
1.5	1.5	34.25 ± 1.5	5 ± 0.8[a]	2.6 ± 0.5[abc]
2.0	2.0	36.75 ± 2.6	4.75 ± 0.9[a]	3.3 ± 0.5[ab]
2.5	2.5	34.75 ± 3.0	3.25 ± 1.2[bc]	4 ± 1[a]

Means (M) ± standard error followed by the common letters is not significantly different at the $p \leq 0.05\%$ level by DMRT

medium consisting of BAP (1.5 mg/l) in combination with NAA (1.5 mg/l) obtained a maximum number of roots per shoot (5 cm) after 35 days as shown in Table 2.3. Similarly, shoot developed from internode explant shifted into rooting medium containing BAP and NAA observed a maximum root development (Nikhila, et al. 2014). Venkatachalam et al. (2012) reported that MS medium supplemented with

IBA had a possible root induction in *G. superba*. Similarly, IBA was effective in root development on *Swainsona formosa* (Jusaitis 1997). MS medium containing BAP+ NAA + zeatin with activated charcoal had more number of root regeneration from shoots (Arumugam and Gopinath 2012).

2.3.4 Effect of Plant Growth Regulators on Cell Suspension Culture

The cell suspension culture was used effectively for somatic embrogenesis, especially friable calli was found to be responsive at different combinations of BAP, KIN, and 2,4- D respectively. The overall response for somatic embryogenesis was observed in BAP 2 mg/l (23± 0.95) against total numbers of testing sets kept in the range of BAP (1–5 mg /l) and KIN (1–4 mg /l) for 23 days event (Table 2.4). There was minimal somatic embryo formation rather than callus development in 2,4-D (1–5 mg/l) as shown in Fig. 2.4. The maximum number of somatic embryos was developed on MS medium containing 2, 4-D (4 mg/l) + kinetin (5 mg/l) + CH (10 mg/l) + CW (20%) in *G. superba* (Jadhav and Hegde 2001). The maximum somatic embryogenesis was observed on 2, 4-D (0.5 mg/l) + kinetin (0.25 mg/l) (Nikhila et al. 2014). Padilha et al. (2015) reported that somatic embryogenesis in *Acrocomia aculeate* using the thin cell layer technique with low cytokinin concentration resulted in maximum somatic embryos with an average of 38 per nodular callus on medium containing 2-iP and 25 per nodular callus on medium with BAP. Half strength of MS liquid medium fortified with decimal concentration of auxin and cytokinin was found encouraging in the case of cell suspension culture for embryogenesis (Sharma et al. 1986). The cell suspension cultures were used good quantity of callus mass, continuous agitation and activity dividing and newly formed the callus themselves from the inoculation (Parekh et al. 2008). The somatic

Table 2.4 Effect of cytokinins on induced somatic embryogenesis from friable callus

Plant growth regulators (mg/l) cytokinins	Number of days	Somatic embryo induction percentage	Germination percentage
BAP			
1.0	26.75 ± 1.5	62.66 ± 2.5^{d}	$25 \pm 3.0d^{e}$
2.0	23.25 ± 0.95	69.66 ± 2.0^{c}	28.6 ± 3.5^{d}
3.0	27.75 ± 1.25	77.33 ± 1.1^{b}	43.6 ± 1.5^{b}
4.0	29.5 ± 1.29	86.66 ± 3.2^{a}	51.3 ± 3.0^{a}
5.0	33 ± 0.81	70 ± 3.4^{c}	42.6 ± 2.0^{b}
KIN			
1.0	31.75 ± 1.25	42.66 ± 1.5^{f}	15 ± 3.0^{f}
2.0	34 ± 1.15	46.66 ± 2.8^{ef}	18.3 ± 0.5^{f}
3.0	34.75 ± 0.5	50.66 ± 3.7^{e}	36.3 ± 1.5^{c}
4.0	35 ± 0.81	49.33 ± 1.5^{e}	23.3 ± 1.5^{e}

Means (M) ± standard error followed by the common letters is not significantly different at the $p \leq 0.05\%$ level by DMRT

Fig. 2.4 The figure shows the different stages of somatic embryogenesis. (**a**) Friable callus, (**b**) cell suspension culture, (**c** & **d**) different stages of somatic embryo development, (**e** & **f**) germination of somatic embryos and (**g**) regenerated plant

embryos were determined by incorporating the polyethylene glycol (PEG) at different concentrations used for callus induction medium. Medium with PEG (2.5 %) has considerable impact on embryogenesis of *Saccharum officinarum* (cv. L. Co 86032) (Thorat et al. 2017).

2.4 Conclusion

Traditional medicinal plant farming and collection faced some challenges, such as lack of resources, deterioration of environment, germplasm of recession and many other problems. It is thus a matter of utmost concern to human life that urgent action is to be taken to prevent further diminution in availability of medicinal plants. The seeds and flowers were traditionally claimed for a large number of medicinal and pharmacological applications. The exploitation of colchicine content by mankind is accountable for the current status of this plant. The seed has poor germination capacity that limits the reproduction of this plant. The cultivation using corm is not enough for the production for commercial purposes. The present study was to optimize in vitro culture and somatic embryogenesis in leaf explants. These experiments were carried out on MS medium with different concentrations and in accordance with the plant growth regulators. The different concentrations of cytokines and auxins were tested, and BAP 2.0 mg/l showed a maximum somatic embryogenesis. The regenerated plants were successfully acclimatized. In this regard, biotechnological application can play an important role in protecting a large planting area of medicinal plants. Therefore, plant genetic engineering which eventually takes place for the manipulation of plant secondary metabolism is now feasible.

Acknowledgements The authors are thankful to Tamil Nadu State Council for Science and Technology (TNSCST) for providing research financial assistance and also Bharathiar University for providing facility.

References

Ade R, Rai MK (2009) Review: current advances in *Gloriosa superba* L. Biodiversitas 10:210–214

Anandhi S, Rajamani K (2012) Studies on seed germination and growth in *Gloriosa superba* L. Global J Res Med Plants Indigen Med 1:524–528

Arumugam A, Gopinath K (2012) *In vitro* micropropagation using corn bud explant: an endangered medicinal plant of *Gloriosa superba* L. Asian J Biotechnol 4:120–128

Atanasov AG, Waltenberger B, Pferschy-Wenzig AM, Linder T, Wawrosch C, Uhrin P, Temml V, Wang L, Schwaiger S, Heiss EH, Rollinger JM, Schuster D, Breuss JM, Bochkov V, Mihovilovic MD, Kopp B, Bauer R, Dirsch VM, Stuppner H (2015) Discovery and resupply of pharmacologically active plant-derived natural products: a review. Biotechnol Adv 33:1582–1614

Chandel KPS, Shukla G, Sharma N (1996) Biodiversity in medicinal and aromatic plants in India: conservation and utilization. NBPGR, New Delhi

Clewer HWV, Green SS, Tutin F (1915) The constituents of *G. superb*. J Chem Soc 107:835–846

Craker LE, Simson JC (1987) Herbs, spices, and medicinal plants: recent advances in botany, horticulture, and pharmacology, vol 2. Oryx Press, Phoenix, pp 255–280

Custers JBM, Bergervoet JHW (1994) Micropropagation of *Gloriosa*: towards a practical protocol. Sci Hortic 57:323–334

Dhir BC, Mohapatra HK, Senapati B (1992) Assessment of crop loss in groundnut due to tobacco caterpillar, *Spodoptera litura* (F.). Indian J Plant Prot 20:215–217

Finnie JF, van Staden J (1991) Isolation of colchicine from *Sandersonia aurantica* and *Gloriosa superba*. Variation in the alkaloid levels of plants grown *in vivo*. J Plant Physiol 138:691–695

Finnie JF, van Staden J (1994) *Gloriosa superba* L. (flame lily): micropropagation and *In vitro* production of colchicines. In: Bajaj YPS (ed) Biotechnology in agriculture and forestry: medicinal and aromatic plants VI, vol 26. Springer Verlag, Berlin, Heidelberg, pp 146–166

Gautam A, Rao CHV, Paswan SK, Verma P (2017) Poisonous plant (*Gloriosa superb* L.): its pharmacological and therapeutic profile. Elixir Appl Bot 104:69–72

Ghosh B, Mukherjee S, Jha TB, Jha S (2002) Enhanced colchicine production in root cultures of *G. superba* by direct and indirect precursors of the biosynthetic pathway. Biotechnol Lett 24:231–234

Ghosh S, Ghosh B, Jha S (2006) Aluminium chloride enhances colchicine production in root cultures of *Gloriosa superba*. Biotechnol Lett 28:497–503

Gooneratne BWM (1966) Massive generalized alopecia after poisoning by *G. superba*. Br Med J 1:1023–1024
Gupta LM, Raina R (2001) Significance of sequential opening of flowers in *Gloriosa superba* L. Curr Sci 80:1266–1267
Hayashi T, Yoshida K, Sano K (1988) Formation of alkaloids in suspension cultured *Colchicum autumnale*. Phytochemistry 27:371–374
Jadhav SY, Hegde BA (2001) Somatic embryogenesis and plant regeneration in *Gloriosa superba* L. Fitoterapia 82:293–301
Jain AP, Suryavanshi S (2010) *Gloriosa superba* linn. – a pharmacological review. Int J Pharma Res Dev 2:24–26
Jayaweera DMA (1982) Medicinal plants used in Ceylon, vol 3. National Science Council of Sri Lanka, Colombo
Jusaitis M (1997) Micropropagation of adult *Swainsona formosa* (leguminosae: papilionoideae: galegeae). In Vitro Cell Dev Bio Plant 33:213–220
Kahate PM (2017) *In Vitro* shoots multiplication through callus culture of *Gloriosa superba* L., a threatened medicinal plant of Melghat Tiger Reserve, Maharashtra, India. Int J Life Sci 5:33–36
Kumar CN, Jadhav SK, Tiwari KL, Afaque Q (2015) *In vitro* tuberization and colchicine content analysis of *Gloriosa superba* L. Biotechnology 14:142–147
Mahendran D, Kavi Kishor PB, Geetha N, Venkatachalam P (2018) Phycomolecule-coated silver nanoparticles and seaweed extracts induced high-frequency somatic embryogenesis and plant regeneration from *Gloriosa superba* L. J Appl Phycol 30:1425–1436
Mamtha H, Farooqi AA, Joshi SS, Prasad TG (1993) Pollen studies in *Gloriosa superba* Linn. Acta Hortic 33:331–371
Mrudul V, Shirgurkar CK, John RS (2001) Factors affecting *in vitro* microrhizome production in turmeric. Plant Cell Tissue Organ Cult 64:5–11
Murashige T, Skoog F (1962) A revised medium for rapid growth and bio-assay with tobacco tissue culture. Physiol Plant 15:473–497
Nadkarni AK (2002) Indian materia medica. Popular Prakashan Limited, Mumbai, pp 579–581
Nagajothi V (2008) Studies on induction of pollination and improvement of fruit set and yield through growth promoting substances in glory lily (*Gloriosa superba* L.). M.Sc. (Hort) Thesis, Tamil Nadu Agricultural University, Coimbatore
Nikhila GS, Sangeetha G, Archana GN, Pradeesh S, Swapna TS (2014) High frequency embryogenesis and organogenesis in *Gloriosa superba* L. A plant in need of conservation. J Aquat Biol Fish 2:398–402
Ochoa-Villarreal M, Howat S, Hong S, Jang MO, Jin YW, Lee EK, Loake GJ (2016) Plant cell culture strategies for the production of natural products. BMB Rep 49:149–158
Omolaja SS (2009) Floral and hybridization studies in selected Dioscorea species. Biosci Res Commun 21:99–105
Padilha JHD, Ribas LLF, Amano E, Quoirin M (2015) Somatic embryogenesis in *Acrocomia aculeata* Jacq. (Lodd.) ex Mart using the thin cell layer technique. Acta Bot Brasilica 29:516–523
Padmapriya S, Rajamani K, Sathyamurthy VA (2015) Glory lily (*Gloriosa superba* L.) a review. Int J Curr Pharm Res 7:43–49
Pandurangan B, Philomina D (2010) Effect of nutritional factors and precursors on formation of colchicine in *Gloriosa superba in vitro*. Res Biotechnol 1:29–37
Parekh S, Srinivasan V, Horn M (2008) Bioprocessing using novel cell culture systems. Adv Appl Microbiol 63:105–143
Patil PB, Gavade RT (2012) *Gloriosa superba* Linn – a medicinally important plant. Ann Pharm Pharm Sci 3:72–74
Pickens LB, Tang Y, Chooi YH (2011) Metabolic engineering for the production of natural products. Annu Rev Chem Biomol Eng 2:211–236
Rajadurai KR (2001) Enhancing bio productivity of *Gloriosa superba* L. through mutatic genetic manipulation. PhD thesis, Tamil Nadu Agricultural University, Coimbatore, 186 pp

Rao KB, Ramesh NP, Lakshmana Swamy P, Muralinath E (2014) *In vitro* investigation on antimicrobial activity of *Gloriosa superba* Linn tubers against major food borne pathogens. Am J Drug Deliv Ther 1:9–20

Ravindra ADE, Mahendra KR (2009) Review: current advances in *Gloriosa superba* L. Biodiversitas 10:211–215

Sarin YK, Jamwal PS, Gupta RK, Atal CK (1974) Colchicine from seeds of *G. superba*. Curr Sci 43:87–90

Sayeed HAKM, Shyamal KR (2005) Micropropagation of *Gloriosa superba* L. through high frequency shoot proliferation. Plant Tissue Cult 15:67–74

Selvarasu A, Kandhasamy R (2013) Intraspecific hybridization in glory lily. Floriculture Ornamental Biotech 7:99–102

Sharma DR, Deepak S, Chowdhury JB (1986) Regeneration of plantlets from somatic tissues of date palm (*Phoenix dactylifera* L.). Indian J Exp Biol 24:763–766

Singh M, Chaturvedi R (2012) Evaluation of nutrient uptake and physical parameters on cell biomass growth and production of spilanthol in suspension cultures of *Spilanthes acmella* Murr. Bioprocess Biosyst Eng 35:943–951

Sivakumar G, Krishnamurthy KV (2002) *Gloriosa superba* L.- a very useful medicinal plant. Ethnomedicine and pharmacognosy, Part II. Series Sci Tech Pub, Texas, pp 465–482

Sivakumar G, Krishnamurthy KV, Hahn EJ, Paek KY (2004) Enhanced *in vitro* production of colchicine in *Gloriosa superba* L. an emerging industrial medicinal crop in South India. J Hortic Sci Biotechnol 79(4):602–605

Sudhendra, Rudre Gowda H (1997) Final report – the ADHOC scheme collection and evaluation in *Gloriosa superba* L. and germplasm for identifying superior lines for domestication. UAS, GKVK, Bangalore

Sugandhi R (2000) Biodiversity conservation and patenting and property right of tribal medicine of medicinal plants of India. In: 10th Asian symposium on medicinal plants, spices and other natural products (ASOMPS X). Dhaka, Bangladesh, pp 18–23

Suganthy M, Sakthivel P (2013) Field evaluation of biopesticides against tobacco caterpillar, *Spodoptera litura* Fab infesting *Gloriosa superba* (Linn.). J Biopest 6:90–95

Suri OP, Gupta BD, Suri KA (2001) A new glycoside, 3-Odemethylcolchicine-3-O-alpha-d-glucopyranoside from Gloriosa seeds. Nat Prod Lett 15:217–219

Tarar JL, Vishwakarma M (1995) Chromosome of diploid and tetraploid *Gloriosa superba* Linn. In: Padhye MD, Mukherjee PK, Khalatkar AS (eds) Botany towards 2000 AD (Prof. VR Dnyansagar Commemoration Volume). Vedams Book, New Delhi, p 280

Thorat AS, Sonone NA, Choudhari VV, Devarumath RM, Babu KH (2017) Plant regeneration from cell suspension culture in *Saccharum officinarum* L. and ascertaining of genetic fidelity through RAPD and ISSR markers. 3 Biotech 7:1–12

Venkatachalam P, Ezhili N, Thiyagarajan M (2012) *In vitro* shoot multiplication of *Gloriosa superba* L. – an important anticancer medicinal herb. In: The proceeding of international conference on biotechnology, biological and biosystems engineering (ICBBBE'2012), December 18–10, Phuket, Thailand

Verpoorte R, Contin A, Memelink J (2002) Biotechnology for the production of plant secondary metabolites. Phytochem Rev 1:13–25

Yadav K, Aggarwal A, Singh N (2015) Influence of arbuscular mycorrhizal fungi (AMF) on acclimatization and growth enhancement of micropropagated medicinal plants. In: Hemantaranjan A (ed) Advances in plant physiology, volume 15 (2014–15). Scientific Publishers, Jodhpur, pp 495–512

Micropropagation of *Enicostemma littorale* Blume

3

M. Vigneswaran and N. Jayabalan

Abstract

An efficient in vitro plant regeneration protocol was established as a high-frequency in vitro mass propagation system for *Enicostemma littorale* Blume, an important diabetic medicinal plant. Decontaminated shoot tip and nodal segments (2 cm in length) were established on MS medium supplemented with different concentrations (0.5–2.5 mg/L) of 6-benzyladenine (BA), kinetin (Kin), and thidiazuron (TDZ) alone and in combination with different concentrations (0.3–1.8 mg/L) of indole-3-acetic acid (IAA), indole-3-butyric acid (IBA), and α-naphthalene acetic acid (NAA). The highest frequency (90%) of shoot regeneration and number of multiple shoots (179 shoots per explant) were achieved with a combination of Kin (1.5 mg/L) and NAA (0.6 mg/L). The shoot length (14.3 cm in length) was the best on the GA_3 (1.2 mg/L) medium. Elongated shoots were rooted with different concentrations (0.2–1.0 mg/L) of α-naphthalene acetic acid (NAA), indole-3-acetic acid (IAA), and indole-3-butyric acid (IBA) alone. The maximum number of roots (17.4 roots per shoot) with a mean root length (15.2 cm in length) was obtained with MS medium containing IBA (0.4 mg/L). Plantlets were established in vermiculite, sand, and soil (1:1:2 w/v) mixture and acclimatized successfully in plant growth chamber. The acclimatized plants were established successfully in the field during winter season (19–25 °C), and the survival rate was 90%.

Keywords

Enicostemma littorale · Diabetic medicinal plant · Shoot tip and nodal explant · Cytokinin · Auxin

M. Vigneswaran (✉) · N. Jayabalan
Department of Botany, Bharathidasan University, Tiruchirappalli, Tamil Nadu, India

M. Kumar et al. (eds.), *In vitro Plant Breeding towards Novel Agronomic Traits*,
https://doi.org/10.1007/978-981-32-9824-8_3

3.1 Introduction

Enicostemma littorale Blume is an important medicinal plant for curing diabetes and widely distributed throughout India and Japan (Saranya et al. 2013; Sasidharan and Jayachitra 2017). Various ecological imbalances such as global warming, increase in acidity of the soil texture, uneven rainfall, and use of chemical fertilizers are factors that play a major role in the productivity of the plant. Micropropagation is a method of vegetative propagation conducted in the laboratory condition, and it has a significant impact on plant breeding, horticulture, and medicine. It is the ever-ready tool for specialization in hybridization either by sexual or asexual means. It is a suitable method for obtaining a large quantity of genetically homogenous and healthy plant materials which can be used for planting (Pierik 1987). This technique is an alternative method of propagation as there is an increase in the propagation rate of plants, availability of plants throughout the year, production of plants against pests and pathogens under controlled conditions, and the availability of uniform clones and uniform production of secondary metabolites (Bajaj et al. 1988). Through the conventional breeding techniques that have considerably increased the productivity of modern crops, the application of biotechnology could speed up further crop improvement. It overcomes the barriers in conventional vegetative propagation and fulfills the demand for large-scale cultivation in a short period by rapid mass multiplication. The techniques are being used globally for the ex situ conservation of plants. Conservation of endangered medicinal plants has also been achieved through tissue culture (Rao et al. 1996).

Micropropagation is the most widely used method of plant tissue culture, and micropropagation protocols for a large number of medicinal plants have been reported. Micropropagation has many advantages over conventional methods of vegetative propagation, which suffers from several limitations (Nehra and Kartha 1994). With micropropagation, the multiplication rate is greatly increased. It also permits the production of pathogen-free plants. Micropropagation of various plant species, including many medicinal plants, has been reported (Saxena et al. 1998; Faisal and Anis 2003; Baskaran and Jayabalan 2005; Siva et al. 2015; Veraplakorn 2016; Mohan et al. 2017; Purohit et al. 2017).

The purpose of this study is to develop an in vitro micropropagation method from different explants of *E. littorale*. The present study has established a rapid and reproducible method for conservation and high frequency of shoot proliferation from various explants of in vivo-grown plants of *E. littorale*, followed by the successful outdoor establishment of regenerated plant.

3.2 Materials and Methods

3.2.1 Collection of Explants

The mature *Enicostemma littorale* Blume was collected from Puduchatram area of Namakkal, Tamil Nadu, India. The plants were raised in the medicinal plant garden of the Department of Botany, Bharathidasan University, Tiruchirappalli, Tamil Nadu, India. The healthy in vivo-raised plants were used as explant source. Shoot tip and node of 15-day-old plants were used as explants. The explants were further subjected to sterilization process.

3.2.2 Preparation of Glasswares for Media Preparation

The glasswares such as test tubes (150 × 25 mm, Borosil, India), culture bottles (130 × 60 mm, Sri Sai Scientific, Coimbatore, India), Petri plates (Borosil, India), measuring cylinders (100, 250, 500, and 1000 mL, Tarson, India), Erlenmeyer flask (50, 100, 250, 500, 1000, and 2000 mL, Schoot Duran, Germany), and beakers (50, 100, 250, 500, 1000, and 2000 mL, Tarson, India) were washed with liquid detergent (Triton-X-100) and washed in running tap water and then rinsed with distilled water. Highly contaminated glasswares were soaked in chromic acid solution and washed with liquid detergent in running tap water and then finally rinsed with distilled water. The glasswares were dried in hot air oven at 80 °C and were used for the preparation of stock solutions, media, and plant growth hormones for the tissue culture studies.

3.2.3 Preparation of mMS Medium

The modified Murashige and Skoog (mMS) medium was used for the optimization of tissue culture protocols throughout the study. The mMS medium was composed of MS salts (Murashige and Skoog 1962) and B_5 vitamins (Gamborg et al. 1968). An aliquot of frozen stock solutions was thawed at room temperature just before use; to make 1 L of medium, about 500 mL of distilled water was taken in a sterile Erlenmeyer flask (2 L) along with the appropriate aliquots of stock solutions. Carbon sources such as glucose, fructose, sucrose, and maltose (HiMedia, Mumbai, India) (10–50 mg/L) were tested at various concentrations and were added slowly to avoid clumping of the carbon source at the bottom of the flask followed by addition of plant growth hormones. It was then made up to 1 L in volume, and the pH was adjusted to 5.8 with 0.1 N HCl or 0.1 N NaOH. For solidifying agent, 0.8% agar (HiMedia, Mumbai, India) was added and boiled in a conical flask and distributed evenly about 10 mL into test tubes and about 50 mL into culture bottles. The test tubes were plugged with nonabsorbent cotton plugs and wrapped with cheese paper, and culture bottles were screw capped with hand tightly. The media-containing containers were autoclaved at 121 °C for 15 min with 15 lb pressure. The sterilized media were used for the inoculation and culturing of explants.

3.2.4 Preparation of Hormones

Phytohormones or plant growth-promoting substances/chemicals were used to regulate plant growth. Auxins such as indole-3-acetic acid (IAA), indole-3-butyric acid

(IBA), α-naphthalene acetic acid (NAA), 2,4-dichlorophenoxyacetic acid (2,4-D), and 4-amino-3,5,6-trichloro-2-pyridinecarboxylic acid (picloram) of HiMedia company were weighed and dissolved separately (10 mg/100 mL for stock solution) in 0.1 N NaOH solution. The dissolved chemicals were made up to 100 mL with sterile distilled water and stored at refrigerator at 4 °C for further use. Cytokinins, like kinetin (Kin), zeatin, 6-benzyladenine (BA), thidiazuron (TDZ), and 6-(γ,γ-dimethylallylamino)purine (2iP), of HiMedia company were weighed and dissolved separately (10 mg/100 mL for stock solution) in 0.1 N HCl solution. The dissolved cytokinins were made up to 100 mL with sterile distilled water (1.0 mg/mL) and stored. Gibberellic acid (GA_3) was dissolved (10 mg/mL) in sterile distilled water and made up to 100 mL. The plant growth regulators were added to the medium prior to autoclaving.

3.2.5 Preparation of Explants

Shoot tip and node segments were initially disinfected with 5% Teepol solution (Reckitt Benckiser, India) and then washed under running tap water for 10 min. The explants were transferred to aseptic conditions and surface sterilized in 70% ethanol for 30 s and 0.1% mercuric chloride for 2 min and washed with three rinses in sterile distilled water. The surface-sterilized explants were inoculated into an autoclaved 500 mL culture bottle containing mMS basal medium.

3.2.6 Inoculation of Explants and Initial Culture Establishment

Shoot tip and node explants excised from in vitro-grown 15-day-old seedlings were used as explant sources. Shoot tip and node explants were cultured in vertical orientation on MS salts supplemented with B_5 vitamins and various growth hormones. The nutrient medium consisted of major and minor salts, 3% sucrose as a carbon source and 0.8% agar (w/v). The effects of BA (0.5–2.5 mg/L), TDZ (0.5–2.5 mg/L), and zeatin (0.5–2.5 mg/L) were tested for the induction of multiple shoots from shoot tip and node explants that were added to the MS medium in various concentrations. The pH of the medium was adjusted to 5.8 prior to autoclaving at 121 °C for 15 min. The cultured explants were maintained at 25 ± 2 °C under Philips fluorescent lamps (80 μmol photons m^{-2} s^{-1}) and a 16-h light/8-h dark cycle. The explants were subcultured every 15 days. There were ten replicates per treatment, and each experiment was repeated three times (n = 30).

3.2.7 Multiplication of Shoots

To attain high-frequency shoot proliferation, 6-benzylaminopurine (BA), thidiazuron (TDZ), and zeatin (0.5–2.5 mg/L) concentrations were individually tested for multiple shoot induction. To improve the frequency of shoot induction and shoot growth, the auxins were tested in combination with cytokinin. The best cytokinin concentration was used as a standard concentration, and it was tested against various concentrations of NAA (0.3–1.8 mg/L), IAA (0.3–1.8 mg/L), and IBA (0.3–1.8 mg/L) in MS medium. Carbon sources such as glucose, fructose, sucrose, and maltose at various concentrations (10–50 mg/L) were tested for the efficient multiplication of shoots. To maintain growth and vigor, cultures were subcultured on a fresh medium every 15 days. Light and temperature conditions were maintained as described for the initial culture establishment system. The average numbers of shoots per explants were recorded at the third subculture, and the experiment was repeated thrice.

3.2.8 Elongation of Shoots

After proliferation, mini shoots of about 2–4 cm were excised from multiple shoots and cultured on MS medium supplemented with 2iP (0.4–2.0 mg/L), zeatin (0.4–2.0 mg/L), and 1.5–3% (w/v) sucrose and solidified with 0.8% (w/v) agar. The response toward these growth regulators was observed in a 1-week interval. Culture conditions were maintained as described previously. The average highest elongated shoots were recorded after 30 days of subculture. MS medium without any growth regulators is used as a control.

3.2.9 Root Induction from Elongated Shoots

In order to induce high-frequency in vitro rooting, healthy elongated shoots were tested in basal MS medium (without any plant growth regulators) and auxins (NAA, IAA, and IBA) at between 0.2 and 1.0 mg/L. The pH of the medium was adjusted to 5.75 ± 2 before autoclaving for 15 min at 121 °C. The cultures were incubated under a 16-h photoperiod at 25 ± 2 °C with 70% relative humidity and an irradiance of 80 μmol photons m^{-2} s^{-1}. The successfully rooted plantlets were transferred to a plant growth chamber.

3.2.10 Acclimatization of Plantlets

Plantlets were removed from the culture bottles. The agar and other adherents from the in vitro-cultured plantlets were washed in tap water and then hardened in plastic pots containing autoclaved vermiculite, sand, and soil in a 1:1:2 ratio. Then the hardened plantlets were transferred to the plant growth chamber and grown for 7 days at 70% relative humidity at about 25 °C. The plants were watered for 2 days once in the growth chamber; the plantlets were covered with transparent plastic cups to avoid drying of young leaves. After 1 week, the plants were transferred to greenhouse conditions, and 25–30 °C temperature was maintained. In the greenhouse, the plants were watered every 24 h, and humidity was maintained by water sprinkling.

3.2.11 Statistical Analysis

The statistical significance of the data obtained from this study (mean number of shoots per shoot tip and node explants, mean shoot and root length, and mean number of shoots per elongated shoots) was determined by one-way analysis of variance (ANOVA) (SPSS v. 17 for Windows 7). The mean values were compared by using Duncan's multiple range test ($P < 0.05$).

3.3 Results and Discussion

3.3.1 Effect of Basal Media and Explants

Mature plant shoot tip and node and young plant shoot tip and node explants were excised from in vivo-raised plant of *E. littorale* in the medicinal plant garden of the Department of Botany, Bharathidasan University (Fig. 3.1). All disinfected explants were cultured on four types of media (B5, L2, MS, and SH) and supplemented with 5.0 μM BA for multiple shoot induction to determine the multiple shoot potential. Media leaking growth regulators were used as control. Among the different explants and media supplemented with 5.0 μM BA tested, young plant nodes and MS medium were found to be the best response for multiple shoot induction than other explants and media, while no shoot was observed on control media. MS medium was more effective than that of any other media (Gao et al. 1999; Komalavalli and Rao 2000; Baskaran and Jayabalan 2005; Shaik et al. 2010; Ślesak et al. 2014; Cheruvathur et al. 2015; Vijendra et al. 2017; Al-Quraainy et al. 2018). The shoot multiplication significantly varied to a greater extent according to the explant type. A similar phenomenon was observed by Komalavalli and Rao (2000). The mean number of shoot was obtained in MS medium supplemented with 5.0 μM BA from mature plant shoot tip and node and young plant shoot tip and node explants after 8 weeks of culture (Fig. 3.2d,e). The maximum number of shoots was obtained in MS medium supplemented with BA, with a mean of 5.7 shoots per young plant node explants and 5.6 cm shoot length (Table 3.1; Fig. 3.2f). All the media and explants induced little white compact callus at the base of the explants after 30 days of culture. A similar response was observed by Baskaran and Jayabalan (2005), Veraplakorn (2016), Verma et al. (2016), and Ghaffarzadeh-Namazi et al. (2017).

3.3.2 Effect of Plant Growth Regulators on Shoot Initiation

In our study, we tested synthetic hormones for the induction of multiple shoots from shoot tip and node explants. Shoot tip and node explants that were collected from the 15-day-old seedlings showed the best results (Fig. 3.2a). MS media containing BA, TDZ, and Kin were tested individually for shoot induction. Medium supplementation with 1.5 mg/L BA and 1.5 mg/L Kin gave the best response in multiple shoot induction (Fig. 3.2b). Shoot initiation was observed in the second subculture

Fig. 3.1 Micropropagation of *E. littorale* using shoot tip explant. (**a**) Shoot tip cultured on MS medium fortified with plant growth regulators. (**b** and **c**) Multiple shoot induction. (**d** and **e**) Multiple shoots. (**f**) Multiple shoots with roots. (**g**) Hardened plant in growth chamber. (**h**) Hardened plant in greenhouse

after 10 days. Multiple shoot induction frequency was increased in the third subculture. Kin-supplemented medium produced 13.4 shoots per explant at 1.5 mg/L concentration, and the highest shoot length of 2.2 cm was observed (Table 3.1). The highest number of shoots was observed in a medium containing 1.5 mg/L BA which was 13.4 shoots per explants with a maximum shoot length of about 2.9 cm. To support those results, Haddadi et al. (2013) reported that zeatin and TDZ induced high-frequency regeneration in strawberry. The optimum shoot proliferation was achieved

Fig. 3.2 Micropropagation of *E. littorale* using nodal explant. (**a**) Nodal explant cultured on MS medium fortified with plant growth regulators. (**b, c, and d**) Multiple shoot induction. (**e**) Multiple shoots. (**f**) Elongated shoots. (**g**) Hardened plant in growth chamber. (**h**) Hardened plant in greenhouse

in the medium fortified with 4 μM zeatin and μM TDZ. Shoot organogenesis of some crops in tissue culture has been achieved recently using thidiazuron, a substituted phenyl urea compound with cytokinin activity (Baalma et al. 2008). Many reports state that TDZ induces shoot regeneration better than other cytokinins (Thomas 2003; Thomas and Puthur 2004; Husain et al. 2007). The node explant of *E. littorale* which promoted the highest six shoots per explant was obtained on MS medium supplemented with 3.0 mg/L TDZ (Osman et al. 2010). In our study, TDZ produced a maximum of 14 shoots per DCN explant on MS medium supplemented

Table 3.1 Effect of plant growth regulators on multiple shoot initiation from shoot tip and nodal explants of *E. littorale*

	Shoot tip			Node		
PGR's	% of response	Mean number of shoots	Mean shoot length (cm)	% of response	Mean number of shoots	Mean shoot length (cm)
BA (mg L^{-1})						
0.5	70	06.3 ± 0.20^{d}	1.0 ± 0.15^{d}	68	07.2 ± 0.13^{d}	1.7 ± 0.15^{b}
1.0	81	10.2 ± 0.26^{b}	1.5 ± 0.15^{b}	73	08.1 ± 0.23^{c}	$2.0 \pm 0.17b^{ab}$
1.5	86	11.9 ± 0.18^{a}	1.9 ± 0.23^{a}	86	13.4 ± 0.25^{a}	2.2 ± 0.53^{a}
2.0	74	08.6 ± 0.39^{c}	1.3 ± 0.28^{c}	85	11.9 ± 0.28^{b}	1.7 ± 0.15^{b}
2.5	63	03.9 ± 0.15^{e}	0.6 ± 0.16^{e}	83	10.9 ± 0.17^{b}	1.2 ± 0.16^{c}
Kin (mg L^{-1})						
0.5	66	07.4 ± 0.18^{d}	0.9 ± 0.27^{d}	76	10.3 ± 0.30^{c}	1.2 ± 0.32^{c}
1.0	75	11.1 ± 0.39^{c}	1.7 ± 0.22^{c}	81	11.0 ± 0.15^{b}	1.7 ± 0.16^{b}
1.5	89	14.6 ± 0.13^{a}	2.3 ± 0.45^{a}	87	11.4 ± 0.16^{a}	2.9 ± 0.10^{a}
2.0	82	10.4 ± 0.26^{b}	2.0 ± 0.17^{b}	84	11.1 ± 0.10^{b}	1.8 ± 0.13^{b}
2.5	63	05.9 ± 0.10^{e}	0.6 ± 0.25^{e}	70	09.2 ± 0.13^{d}	2.6 ± 0.26^{ab}
TDZ (mg L^{-1})						
0.5	74	09.4 ± 0.16^{c}	1.6 ± 0.16^{c}	75	07.9 ± 0.25^{c}	1.2 ± 0.17^{d}
1.0	83	10.7 ± 0.38^{a}	1.9 ± 0.19^{a}	83	09.7 ± 0.15^{a}	2.0 ± 0.26^{b}
1.5	80	10.2 ± 0.26^{b}	1.8 ± 0.15^{b}	78	08.4 ± 0.17^{b}	2.6 ± 0.12^{a}
2.0	66	08.6 ± 0.13^{c}	1.2 ± 0.10^{d}	71	07.5 ± 0.42^{d}	1.7 ± 0.15^{c}
2.5	71	09.3 ± 0.50^{d}	0.8 ± 0.22^{e}	64	05.8 ± 0.33^{e}	1.1 ± 0.36^{e}

The shoot induction data was evaluated and scored after 60 days. The control was supplemented with MS salts and B_5 vitamins without any plant growth regulators. Shoot tip and node explants were cultured on MS medium fortified with different concentrations of BA, Kin, and TDZ. Values represent the mean SE. Means followed by the same letter within columns are not significantly different, according to Duncan's multiple range test ($P < 0.05$)

with 1.5 mg/L TDZ, and 2 μM TDZ enhanced shoot bud development from cotyledonary nodal explant of mung bean (*Vigna radiata* L.) (Kumar et al. 2003).

3.3.3 Effect of Cytokinin and Auxin on Adventitious Shoot Multiplication

The highest mean shoot induction was observed in the medium supplemented with 1.5 mg/L Kin for shoot tip explant and 1.5 mg/L TDZ for node explant, at which concentration was tested against auxins (NAA, IAA, and IBA) for the highest multiple shoot induction capability. To identify the best responding concentration, Kin was tested in combination with auxins. Of the auxins tested, Kin 1.5 mg/L and NAA 0.6 mg/L showed the best response, and it produced a maximum of 179.8 shoots per node explant, and maximum shoot length of about 4.0 cm was achieved. It was reported that Kin in combination with auxin produced maximum number of multiple shoots in *E. littorale* (Godishala et al. 2012). Individual TDZ-supplemented media produced a mean of 13.4 shoots per explant. Thus to test the capability of

TDZ in combination with auxins on the induction of multiple shoots, three auxins (NAA, IAA, and IBA) at a concentration range of 0.3 to 1.8 mg/L were tested. BA in combination with NAA produced 11.4 shoots per explant, and the highest shoot length of about 3.5 cm was observed with 0.6 mg/L NAA. Subsequently, response of BA in combination with IAA was tested; the highest mean number of shoots (89.7 shoots/node explant) was produced in the medium supplemented with 1.2 mg/L IAA and 1.5 mg/L BA. The highest mean shoot length of about 3.5 cm was observed in the medium supplemented with 0.3 mg/L IAA in combination with BA. The frequency of multiple shoot induction increased to 95.5 shoots per explant in the medium with 1.5 mg/L BA and 1.5 mg/L IBA (Fig. 3.2d). Among the three auxins tested, IBA showed the best response compared to NAA and IAA. BA with IBA at 1.5 mg/L concentration responded well and recorded a maximum of 95.5 shoots per explant, with an average mean shoot length of about 4.1 cm (Table 3.2). Cocu et al. (2004) stated that combination of BA and IBA induced shoot regeneration from cotyledonary nodes in *Calendula officinalis*. It has been reported that the combination of NAA with BA or kinetin negatively affected the multiplication rate

Table 3.2 Effect of combination of Kin with NAA, IAA, and IBA on multiple shoot initiation of *E. littorale*

	% of response	Mean number of shoots	Mean shoot length (cm)
NAA + Kin (mg/L)			
0.3 + 1.5	84	156.4 ± 0.16^{b}	3.2 ± 0.18^{b}
0.6 + 1.5	90	179.8 ± 0.19^{a}	4.0 ± 0.16^{a}
0.9 + 1.5	81	145.5 ± 0.18^{c}	2.7 ± 0.13^{c}
1.2 + 1.5	77	122.7 ± 0.30^{d}	1.6 ± 0.12^{c}
1.5 + 1.5	62	118.3 ± 0.51^{e}	1.5 ± 0.30^{d}
1.8 + 1.5	55	95.1 ± 0.57^{f}	0.9 ± 0.52^{e}
IAA + Kin (mg/L)			
0.3 + 1.5	66	128.1 ± 0.19^{cd}	1.0 ± 0.30^{de}
0.6 + 1.5	78	121.6 ± 0.15^{d}	1.4 ± 0.16^{c}
0.9 + 1.5	86	165.3 ± 0.22^{a}	2.5 ± 0.10^{a}
1.2 + 1.5	80	153.1 ± 0.16^{b}	1.8 ± 0.16^{b}
1.5 + 1.5	72	149.5 ± 0.32^{c}	1.1 ± 0.13^{d}
1.8 + 1.5	52	116.6 ± 0.56^{e}	0.6 ± 0.36^{e}
IBA + Kin (mg L^{-1})			
0.3 + 1.5	58	88.2 ± 0.16^{f}	1.6 ± 0.31^{dc}
0.6 + 1.5	70	122.9 ± 0.10^{d}	1.7 ± 0.29^{d}
0.9 + 1.5	85	155.3 ± 0.13^{b}	2.7 ± 0.15^{b}
1.2 + 1.5	88	167.5 ± 0.30^{a}	3.3 ± 0.19^{a}
1.5 + 1.5	74	144.9 ± 0.33^{c}	2.5 ± 0.10^{c}
1.8 + 1.5	65	111.1 ± 0.36^{de}	1.2 ± 0.13^{e}

The control was supplemented with 1.5 mg/L of Kin. Media were supplemented with 1.5 mg/L of Kin and different concentrations of auxins for the enhancement of multiple shoot induction. Data were collected after 6 weeks of culture. Values represent the mean SE. Means followed by the same letter within columns are not significantly different, according to Duncan's multiple range test $P < 0.05$)

of the *E. littorale* compared with cytokinin used singly (Ishag et al., 2009). It also has been stated that the addition of NAA to a medium containing cytokinin did not improve shoot multiplication rate. The type and concentration suitable for one genotype may not be optimal for others (Bhatia et al. 2004). The mini shoots of about 2–4 cm were excised and used for further experiments.

3.3.4 Effect of Carbon Source on Multiple Shoot Induction

The response of explant to different carbon source has been frequently tested. The type of explant and carbon source greatly influences the mass propagation of cultures. So, optimization of each and every parameter is absolutely necessary in the tissue culture studies. It is well established that carbohydrate requirements depend on the stage of culture and may show differences according to the species (Thomson and Thrope 1987). The node and shoot tip explants cultured on the medium containing various concentrations (10–50 mg/L) of glucose, fructose, sucrose, and maltose were tested. However, ribose, glucose, palatinose, and furanose have also been tried by some researchers. When *E. littorale* callus or cell cultures were placed on media containing ribose as the sole carbon source, the tissues turned dark brown and ceased growth. However, after about 60 days, bright green tissues emerged from about 3% of the brown necrotic callus tissue pieces (Locy 1995; Veraplakorn 2016; Jun-jie et al. 2017). Of the carbon sources examined, sucrose 30 g/L showed the best response, and it produced a maximum of 23.5 shoots per node explant and 29.3 shoots per shoot tip explant, whereas the medium fortified with 50 mg/L maltose produced the least response, and it produced a maximum of 13.2 shoots per node explant and 15 shoots per shoot tip explant. The majority of researchers have used this sucrose concentration in their initiation and multiplication media (Costa et al. 2000; Venkatachalam et al. 2000).

3.3.5 Effect of 2iP and GA_3 in Shoot Elongation

Excised shoots from the multiple shoots were cultured in the MS medium supplemented with N6-2-isopentenyl adenine (2iP) and GA_3 and tested for elongation in different concentrations (Fig. 3.2e). As formation of basal callus during elongation was noticed and as the formation of basal callus may reduce the survival rate of the plant during rooting, the experiment was designed to minimize the basal callus in elongated shoots. Three percent and 1.5% sucrose as carbon source were tested in 2iP- and GA_3-supplemented media. The shoot length recorded at 1.2 mg L^{-1} 2iP was 14.7 cm, and at 1.6 mg/L GA3, it was 12.5 cm (Table 3.3). The medium supplemented with 9 μM of 2iP was found to be best in enhancing the elongation of 2 cm shoots in *Capsicum annuum* (Mohamed and Alsadon 2011; Vijendra et al. 2017). Basal callus formation was observed when the medium was supplemented with 3% sucrose. Formation of basal callus was observed at the base of the elongated shoots, and these basal callus formations reduced the survival rate of the *E. littorale* plant.

Table 3.3 Effect of combination of BA with NAA, IAA, and IBA on multiple shoot initiation of *E. littorale*

	% of response	Mean number of shoots	Mean shoot length (cm)
NAA + BA (mg/L)			
0.3 + 1.5	73	72.8 ± 0.13[c]	2.8 ± 0.13[b]
0.6 + 1.5	84	88.4 ± 0.16[a]	3.5 ± 0.16[a]
0.9 + 1.5	76	84.5 ± 0.17[b]	2.5 ± 0.16[b]
1.2 + 1.5	73	71.7 ± 0.15[c]	2.2 ± 0.32[c]
1.5 + 1.5	60	63.1 ± 0.10[d]	1.8 ± 0.13[c]
1.8 + 1.5	57	57.9 ± 0.10[d]	1.3 ± 0.15[d]
IAA + BA (mg/L)			
0.3 + 1.5	67	48.9 ± 0.10[e]	3.6 ± 0.16[a]
0.6 + 1.5	78	59.5 ± 0.16[d]	2.5 ± 0.16[b]
0.9 + 1.5	83	61.3 ± 0.15[d]	2.1 ± 0.10[c]
1.2 + 1.5	92	89.7 ± 0.15[a]	1.6 ± 0.16[d]
1.5 + 1.5	89	87.1 ± 0.10[ab]	1.2 ± 0.13[e]
1.8 + 1.5	85	76.6 ± 0.16[c]	0.9 ± 0.36[e]
IBA + BA (mg/L)			
0.3 + 1.5	80	41.7 ± 0.15[f]	2.1 ± 0.13[d]
0.6 + 1.5	85	52.1 ± 0.10[e]	2.9 ± 0.10[c]
0.9 + 1.5	90	62.9 ± 0.10[d]	3.4 ± 0.16[b]
1.2 + 1.5	91	86.3 ± 0.15[b]	3.7 ± 0.15[b]
1.5 + 1.5	95	95.5 ± 0.16[a]	4.1 ± 0.10[a]
1.8 + 1.5	93	77.7 ± 0.15[c]	2.6 ± 0.16[c]

The control was supplemented with 1.5 mg/L of BA. Media were supplemented with 1.5 mg/L of BA and different concentrations of auxins for the enhancement of multiple shoot induction. Data were collected after 6 weeks of culture. Values represent the mean SE. Means followed by the same letter within columns are not significantly different, according to Duncan's multiple range test ($P < 0.05$)

Therefore, various parameters were tested as sucrose concentration plays a major role in the formation of basal callus.

3.3.6 Induction of Rooting from Elongated Shoots

The frequency and nature of the roots induced from elongated shoots varied with the type of auxin used. The shoots were blot dried on the sterile filter paper, and then they were transferred into the MS medium supplemented with auxins (NAA, IAA, and IBA) at a concentration range of 0.2–1.0 mg/L. Rooting frequency was optimal, when the medium was supplemented with 0.4 mg/L IBA. Of the three auxins tested, the IBA-supplemented medium produced a maximum of 17.4 roots per shoot and also produced the maximum root length of 15.2 cm (Tables 3.4 and 3.5). Our results are in correlation with the results of Shah et al. (2013). They reported that minimal concentration of IBA (0.2–0.4 mg/L) induced maximum percentage of response in *E. littorale*. The medium supplemented with 1.0 mg/L IAA produced 10.1 roots per

Table 3.4 Effect of 2iP and GA_3 on shoot elongation of *E. littorale*

	% of response	Mean shoot length (cm)	Basal callus formation at 3% sucrose	Basal callus formation at 1.5% sucrose
GA_3 (mg/L)				
0.4	80	8.1 ± 0.10[e]	++	++
0.8	90	8.6 ± 0.16[d]	++	--
1.2	100	14.7 ± 0.15[a]	++	--
1.6	100	12.5 ± 0.16[b]	++	--
2.0	95	9.5 ± 0.22[c]	++	--
2iP (mg/L)				
0.4	85	7.1 ± 0.52[d]	++	--
0.8	90	8.3 ± 0.15[c]	++	--
1.2	100	11.7 ± 0.15[b]	++	--
1.6	100	12.5 ± 0.16[a]	++	--
2.0	100	11.4 ± 0.16[b]	++	--

The control was supplemented with MS salts and B_5 vitamins without any plant growth regulators. MS media were supplemented with GA_3 and 2iP with 1.5% and 3% sucrose for shoot elongation. Shoot length and basal callus formation: ++indicates that basal callus formed at the base of the shoot; -- indicates that no basal callus formation was recorded after two subcultures. Values represent the mean SE. Mean followed by the same letter within columns is not significantly different, according to Duncan's multiple range test ($P < 0.05$)

Table 3.5 Effect of auxins on rooting of elongated shoots

Concentrations	% of root induction	Average no. of roots/shoot	Average root length (cm)
MS basal medium	90	10.6 ± 0.17	9.4 ± 0.15
NAA (mg/L)			
0.2	80	10.8 ± 0.13[e]	11.8 ± 0.13[d]
0.4	90	12.2 ± 0.13[d]	12.5 ± 0.17[c]
0.6	100	13.7 ± 0.15[c]	13.7 ± 0.15[b]
0.8	100	15.3 ± 0.15[a]	14.1 ± 0.10[a]
1.0	100	14.5 ± 0.17[b]	12.2 ± 0.13[c]
IAA (mg/L)			
0.2	100	15.8 ± 0.13[a]	13.8 ± 0.13[a]
0.4	100	13.2 ± 0.13[b]	12.5 ± 0.17[b]
0.6	100	11.5 ± 0.17[c]	10.8 ± 0.13[c]
0.8	90	10.8 ± 0.13[d]	9.2 ± 0.13[d]
1.0	90	10.1 ± 0.10[e]	8.7 ± 0.15[e]
IBA (mg/L)			
0.2	100	15.2 ± 0.13[c]	14.5 ± 0.17[b]
0.4	100	17.4 ± 0.16[a]	15.2 ± 0.13[a]
0.6	100	16.1 ± 0.10[b]	13.6 ± 0.16[c]
0.8	95	14.3 ± 0.15[d]	12.0 ± 0.00[d]
1.0	80	12.8 ± 0.13[e]	10.3 ± 0.15[e]

Data on number and length of roots produced at MS basal medium, NAA, IAA, and IBA were scored after 30 days. Values represent the mean SE. Mean followed by the same letter within columns is not significantly different, according to Duncan's multiple range test ($P < 0.05$)

shoot. Roots were very visible after a week of culture, and primary and secondary roots were produced after 10 days of culture. No basal callus formation in the rooting was observed. IBA was found to be more effective than other auxins (IAA, NAA) tested for root induction in other plant species (Quraishi et al. 2004; Uranbey et al. 2005; Shekhawat et al. 2015; Jun-jie et al. 2017).

3.4 Conclusion

In conclusion in the present study, we have established a high-frequency in vitro mass propagation system for *E. littorale*. Higher frequency of micropropagation depends on type of explants, duration of collection of explants, concentration and combinations of PGRs, culture conditions, and additives added to the medium.

References

Al-Qurainy F, Nadeem M, Khan S, Alansi S, Tarroum M, Al-Ameri AA, Gaafar AR, Alshameri A (2018) Rapid plant regeneration, validation of genetic integrity by ISSR markers and conservation of *Reseda pentagyna* an endemic plant growing in Saudi Arabia. Saudi J Biol Sci 25(1):111–116

Baalma D, Uranbey S, Mirici S, Kolsarici O (2008) TDZ × IBA induced shoot regeneration from cotyledonary leaves and *in vitro* multiplication in safflower (*Carthamus tinctorius* L.). Afr J Biotechnol 7:960–966

Bajaj YP, Furmanowa M, Olszowska O (1988) Biotechnology of the micropropagation of medicinal and aromatic plants. In: Medicinal and aromatic plants I. Springer, Berlin, Heidelberg, pp 60–103

Baskaran P, Jayabalan N (2005) An efficient micropropagation system for *Eclipta alba*–a valuable medicinal herb. In Vitro Cell Dev Biol-Plant 41(4):532–539

Bhatia P, Ashwath N, Senaratna T, Midmore D (2004) Tissue culture studies of tomato (*Lycopersicon esculentum*). Plant Cell Tiss Org 78:1–21

Cheruvathur MK, Abraham J, Thomas TD (2015) *In vitro* micropropagation and flowering in *Ipomoea sepiaria* Roxb. An important ethanomedicinal plant. Asian Pac J Reprod 4(1):49–53

Cocu S, Uranbey S, Pek A, Khawar KM, Sarıhan EO, Kaya MD, Parmaksız OS (2004) Adventitious shoot regeneration and micropropagation in *Calendula officinalis* L. Biol Plant 48:449–451

Costa GM, Nogueira FTS, Otoni WC, Brommonschenkel SH (2000) In vitro regeneration of processing tomato (*Lycopersicon esculentum* Mill.) 'IPA-5' and 'IPA-6'. Ciencia Agrotecnol 24:671–678

Faisal M, Anis M (2003) Rapid mass propagation of *Tylophora indica* Merrill via leaf callus culture. Plant Cell Tissue Organ Cult 75(2):125–129

Gamborg OL, Miller RA, Ojima K (1968) Nutrient requirement of suspension cultures of soybean root cell. Exp Cell Res 50:151–158

Gao SL, Zhu DN, Cai ZH, Jiang Y (1999) Xu DR. Organ culture of a precious Chinese medicinal plant–*Fritillaria unibracteata*. Plant Cell Tissue Organ Cult 59(3):197–201

Ghaffarzadeh-Namazi L, Keller EJ, Senula A, Babaeian N (2017) Investigations on various methods for cryopreservation of callus of the medicinal plant *Satureja spicigera*. J Appl Res Med Arom Plants 5:10–15

Godishala V, Kairamkonda M, Kagithoju S, Mangamoori L, Nanna Rama S (2012) Zeatin induced direct multiple shoots development and plant regeneration from cotyledon explants of cultivated tomato (*Solanum lycopersicum* L.). Aust J Crop Sci 6(1):31

Haddadi F, Abd Aziz M, Kamaladini H, Ali Ravanfar S (2013) Thidiazuron- and zeatin-induced high-frequency shoot regeneration from leaf and shoot-tip explants of strawberry. Hort Technol 23(3):276–281

Husain MK, Anis M, Shahzad A (2007) *In vitro* propagation of Indian Kino (*Pterocarpus marsupium* Roxb.) using thidiazuron. In Vitro Cell Dev-Plant 43:59–64

Ishag S, Osman MG, Khalafalla MM (2009) Effects of growth regulators, explant and genotype on shoot regeneration in tomato (*Lycopersicon esculentum* c.v. Omdurman). Int J Sustain Crop Prod 4:7–13

Jun-jie Z, Yue-sheng Y, Meng-fei L, Shu-qi L, Yi T, Han-bin C, Xiao-yang C (2017) An efficient micropropagation protocol for direct organogenesis from leaf explants of an economically valuable plant, drumstick (*Moringa oleifera* lam.). Ind Crop Prod 103:59–63

Komalavalli N, Rao MV (2000) *In vitro* micropropagation of *Gymnema sylvestre*–a multipurpose medicinal plant. Plant Cell Tissue Organ Cult 61(2):97

Kumar P, Srivastava GC, Panwar JDS, Chandra R, Raghuveer P (2003) Efficiency of thidiazuron on *in vitro* shoot regeneration from cotyledonary node explant in mung bean. Indian J Plant Physiol 8:398–401

Locy RD (1995) Selection of tomato tissue cultures able to grow on ribose as the sole carbon source. Plant Cell Rep 14:777–780

Mohamed MAH, Alsadon AA (2011) Effect of vessel type and growth regulators on micropropagation of *Capsicum annuum*. Biol Plant 55(2):370–374

Mohan C, Naresh B, Kumar BK, Reddy V, Manjula P, Keerthi B, Sreekanth D, Manzelat SF, Cherku PD (2017) Micropropagation studies and phytochemical analysis of the endangered tree *Commiphora wightii*. J Appl Res Med Arom Plants 6:70–79

Murashige T, Skoog F (1962) A revised medium for rapid growth and bioassays with tobacco tissue cultures. Physiol Plantarum 15:473–497

Nehra NS, Kartha KK (1994) Meristem and shoot tip culture: requirements and applications. In: Plant cell and tissue culture. Springer, Dordrecht, pp 37–70

Osman MG, Elhadi EA, Khalafalla MM (2010) Callus formation and organogenesis of tomato (*Lycopersicon esculentum* Mill, C.V. Omdurman) induced by thidiazuron. Afr J Biotechnol 9(28):4407–4413

Pierik RLM (1987) *In vitro* culture of higher plants. Martinus Nijjhoff Publishers, Dordrecht

Purohit S, Nandi SK, Paul S, Tariq M, Palni LM (2017) Micropropagation and genetic fidelity analysis in *Amomum subulatum* Roxb.: a commercially important Himalayan plant. J Appl Res Med Arom Plants 4:21–26

Quraishi A, Koche V, Sharma P, Mishra SK (2004) *In vitro* clonal propagation of neem (*Azadirachta indica*). Plant Cell Tissue Organ Cult 46:1–4

Rao PS, Suprasanna P, Ganapathi T (1996) Plant biotechnology and agriculture: prospects for crop improvement and increasing productivity. Sci Cult 62(7–8):185–191

Saranya R, Thirumalai T, Hemalatha M, Balaji R, David E (2013) Pharmacognosy of *Enicostemma littorale*: a review. Asian Pac J Trop Biomed 3(1):79–84

Sasidharan P, Jayachitra A (2017) Direct shoot bud regeneration from shoot tip explants of *Enicostema axillare*: an important medicinal plant. Agrofor Syst 91:471–477

Saxena C, Rout GR, Das P (1998) Micropropagation of *Psoralea corylifolia*. J Med and Arom Plant Sci 20:15–18

Shah SH, Ali S, Ali GM (2013) A novel approach for rapid in vitro morphogenesis in tomato (*Solanum lycopersicum* Mill.) with the application of cobalt chloride. Eur Acad Res 1(9):2702–2721

Shaik S, Dewir YH, Singh N, Nicholas A (2010) Micropropagation and bioreactor studies of the medicinally important plant *Lessertia (Sutherlandia) frutescens* L. S Afr J Bot 76(2):180–186

Shekhawat MS, Kannan N, Manokari M (2015) *In vitro* propagation of traditional medicinal and dye yielding plant *Morinda coreia* Buch.–Ham. S Afr J Bot 100:43–50

Siva G, Sivakumar S, Prem Kumar G, Vigneswaran M, Vinoth S, Arunachalam S, Elango B, Senthil Kumar T, Jayabalan N (2015) Multiple shoot production from nodal explant and FTIR analysis of *in vitro* regenerated plants of *Psoralea corylifolia* L. Pharm Biol Eval 2:105–109

Ślesak H, Liszniańska M, Popielarska-Konieczna M, Góralski G, Sliwinska E, Joachimiak AJ (2014) Micropropagation protocol for the hybrid sorrel *Rumex tianschanicus* × *Rumex patientia*, an energy plant. Histological, SEM and flow cytometric analyses. Ind Crop Prod 62:156–165
Thomas TD (2003) Thidiazuron induced multiple shoot induction and plant regeneration from cotyledonary explants of mulberry. Biol Plantarum 46:529–533
Thomas TD, Puthur JT (2004) Thidiazuron induced high frequency shoot organogenesis in callus from *Kigelia pinnata* L. Bot Bull Acad Sin 45:307–313
Thompson MR, Thorpe TA (1987) Metabolic and non-metabolic roles of carbohydrates. In: Cell and tissue culture in forestry. Springer, Dordrecht, pp 89–112
Uranbey S, Sevimay CS, Ozcan S (2005) Development of high frequency multiple shoot formation in Persian clover (*Trifolium resupinatum* L.). Plant Cell Tissue Organ Cult 80:229–232
Venkatachalam P, Geetha N, Priya P, Rajaseger G, Jayabalan N (2000) High frequency plantlet regeneration from hypocotyls explants of tomato (*Lycopersicon esculentum* Mill.) via organogenesis. Plant cell Biotechnol. Mol Biol 1:95–100
Veraplakorn V (2016) Micropropagation and callus induction of *Lantana camara* L.–a medicinal plant. Agric Nat Resour 50(5):338–344
Verma SK, Das AK, Cingoz GS, Uslu E, Gurel E (2016) Influence of nutrient media on callus induction, somatic embryogenesis and plant regeneration in selected Turkish crocus species. Biotechnol Rep 10:66–74
Vijendra PD, Jayanna SG, Kumar V, Gajula H, Rajashekar J, Sannabommaji T, Basappa G, Anuradha CM (2017) Rapid *in vitro* propagation of *Lucas aspera* Spreng. A potential multipurpose Indian medicinal herb. Ind Crop Prod 107:281–287

Role of Silver Nitrate and Silver Nanoparticles on Tissue Culture Medium and Enhanced the Plant Growth and Development

4

Durai Mahendran, Natesan Geetha, and Perumal Venkatachalam

Abstract

The objective of this review is to critically study the role of silver nitrate ($AgNO_3$) and silver nanoparticles (Ag NPs) in changing plant growth and development. In recent years, basic studies on ethylene regulation opened novel outlooks for applied research in the field of micropropagation, somatic embryogenesis, and secondary metabolites production. Silver nitrate has evidenced to be a very effective inhibitor of ethylene action and is widely used in plant tissue culture. Nanotechnology opens a large scope of novel application in the field of biotechnology and agricultural, because nanoparticles (NPs) have unique physicochemical properties. Plant tissue cultures are the core of plant biology, which is important for conservation, mass propagation, genetic manipulation, bioactive compounds production, and plant improvement. Nowadays, the application of silver nanoparticles (Ag NPs) has successfully controlled the microbial contaminants from explants and proved the positive role of Ag NPs in callus induction, organogenesis, somatic embryogenesis, somaclonal variation, genetic transformation, and secondary metabolites production. Therefore, the present chapter highlights the significant role of $AgNO_3$ and Ag NPs in plant growth and development. Also, nanoscience contributes new ideas leading us to understand the suitable mode of action of nanoparticles in plants which lead to better plant growth and development.

D. Mahendran · P. Venkatachalam (✉)
Department of Biotechnology, School of Biosciences, Periyar University, Salem, Tamil Nadu, India

N. Geetha
Department of Biotechnology, School of Biosciences, Periyar University, Salem, Tamil Nadu, India

Department of Botany, Bharathiar University, Coimbatore, Tamil Nadu, India

M. Kumar et al. (eds.), *In vitro Plant Breeding towards Novel Agronomic Traits*, https://doi.org/10.1007/978-981-32-9824-8_4

Keywords
Tissue culture · Micropropagation · Silver nitrate · Silver nanoparticles

4.1 Introduction

Silver ions in the form of nitrate, such as $AgNO_3$, play a major role in influencing somatic embryogenesis, shoot formation, and efficient root formation which are the prerequisites for successful genetic transformation (Bais et al. 2001). Silver ions are also employed in the form of silver thiosulfate in several tissue culture studies (Eapen and George 1997). Previous study suggests that the silver nitrate was used as an elicitor on biosynthesis of secondary metabolites in plants, where Ag + ion is present. Silver nitrate can inhibit the action of ethylene hormone; this may increase the production of secondary plant products, plant regeneration, and somatic embryogenesis rate (Mahendran et al. 2018a, b). The positive effect of Ag + may be attributed to enhance the secondary metabolites production in suspension culture (Vildová et al. 2016).

Nanoparticles interact with plants causing many morphological and physiological changes, depending on their properties. Efficacy of NPs is determined by their chemical composition, size, surface covering, reactivity, and most importantly the dose at which they are effective (Khodakovskaya et al. 2012). Researchers from their findings suggested both positive and negative effects on plant growth and development and the impact of engineered nanoparticles (ENPs) on plants depends on the composition, concentration, size, physical and chemical properties, as well as plant species (Ma et al. 2010). Efficacy of NPs depends on their concentration and varies from plants to plants. Silver nanoparticles (Ag NPs) have various applications in consumer product inventories like personal care products, food service, building materials, medical instruments, cosmetics, and textiles due to their known antimicrobial properties (Park et al. 2010). It has also been suggested to have applications in management of plant diseases due to its multiple inhibitory effects toward various plant pathogens in a relatively safer way compared to synthetic fungicides. Also nano silver and TiO_2 are even suggested to have potential for removing the bacterial contaminants in plant tissue culture system (Safavi 2012). Most importantly, treatment of fenugreek seedlings with microbially synthesized silver nanoparticles has shown to have positive effects on its growth parameters such as leaf number, root length, shoot length, and wet weight (Jasim et al. 2017). However, widespread use of Ag NPs might expect to result in toxic effects to living organisms in a concentration-dependent manner (Griffitt et al. 2008). Nanotoxicity studies in plants have generally focused on acute effects (Kumari et al. 2012). Such studies can only evaluate immediate nanotoxicity effects without any insight into chronic effect, which can expect to have different end result. Very few studies have taken place to assess the toxicity of nanoparticles in tissue-cultured plants under in vitro conditions which has the promises to assess the long-term effects. Plant tissue culture system provides an ideal environment to study the effect of Ag NPs on plants. Recent studies have shown that the MS medium containing plant growth hormones in combination with Ag NPs were significantly increased the rate of somatic embryogenesis and shoot multiplication in various plants (Mahendran et al. 2018a; Venkatachalam et al. 2017a). The sterile environment and the chemically defined culture medium avoid most of the extraneous factors which could influence nanoparticle interaction with plants. So the effect of Ag NPs on tissue cultured plantlets by applying in the tissue culture media. In the present chapter, we compile the recent research and developments in nano-mediated plant science that focuses on the positive role of silver and Ag NPs in plant growth, development, and also on plant mechanism.

4.2 Effect of Silver Nitrate ($AgNO_3$) on Tissue Culture Medium

Addition of $AgNO_3$ in to the tissue culture media significantly improved the regeneration of both monocot and dicot plants (Giridhar et al. 2003). The use of $AgNO_3$ in the tissue culture medium to enhance the growth and development of various plants is showed in Table 4.1. Actually the exact mechanism of $AgNO_3$ action on plants is unclear. However, few existing evidences recommend its interference in ethylene perception mechanism. In recent years, $AgNO_3$ has been employed in tissue culture studies for inhibiting ethylene action because of its water solubility and lack of phytotoxicity at effective concentrations (Kumar et al. 2009). Giridhar et al. (2004) established direct somatic embryogenesis from hypocotyl explants of in vitro regenerated plantlets of *C. arabica* and *C. canephora* on modified MS medium containing 1.7–11.9 mg/l silver nitrate supplemented with 0.2 mg/l BA and 0.5 mg/l IAA. Maximum embryos per explants were produced at 6.8 mg/l silver nitrate in *C. canephora* and *C. arabica*, respectively. The addition of $AgNO_3$ enhanced the frequency of the shoot formation from 61.3 to 87.6% of *Phaseolus vulgaris* (Dang and Wei 2009). Gutiérrez-Miceli et al. (2010) reported the optimum concentrations of NAA and BA to stimulate callus growth and the optimum concentrations of NAA, KIN, and $AgNO_3$ for callus redifferentiation in *Dianthus caryophyllus* L.

Recently, Abdollahi and Rashidi (2018) reported the silver nitrate produced the highest number of embryos and regenerated plants were obtained in chickpea (*Cicer arietinum*) anther culture, when 10 mg/l 2,4-D was combined with 15 and 50 mg/l silver nitrate in the anther culture medium, respectively. On the other hand, the medium fortified with of $AgNO_3$ in the combinations of BA and kinetin medium resulted in a significant positive effect on the development of multiple shoots and proliferation in *Catharanthus roseus* (Panigrahi et al. 2018). Addition of $AgNO_3$ to the culture medium significantly enhanced shoot proliferation in the in vitro *Capsicum chinense* plantlets. In vitro flower induction and fruit formation were successfully carried out in the presence of GA_3 and $AgNO_3$ also maximum root induction was observed in the medium containing IBA, GA_3, and $AgNO_3$ (Bora et al. 2018).

Table 4.1 Beneficiary concentration(s) of silver nitrate for plants

Plant	Beneficiary silver nitrate concentrations	Part of plant	References
Brassica juncea	20.0 μM/l	Shoot regeneration	Chi et al. (1990)
Garcinia mangostana	20.0 μM/l	Callus induction	Chongjin et al. (1997)
Albizia procera	15.0 μM/l	Shoot regeneration	Kumar et al. (1998)
Coffea canephora	2.0 mg/l	Shoot regeneration	Fuentes et al. (2000)
Vanilla planifolia	40.0 μM/l	Shoot multiplication and root formation	Giridhar et al. (2001)
Manihot esculenta	12.0 mg/l	Shoot organogenesis	Zhang et al. (2001)
Coffea arabica	20.0 μM/l	Shoot growth	Giridhar et al. (2003)
C. canephora and *C. arabica*	40.0 μM/l	Direct somatic embryogenesis	Giridhar et al. (2004)
Salvia miltiorrhiza	300 μmol/l	Hairy root culture	Yan et al. (2006)
Azadirachta excelsa	10.0 mg/l	Defoliation	Nor et al. (2008)
Phaseolus vulgaris	2.0 mg/dm	Shoot development	Dang and Wei (2009)
Salvia miltiorrhiza	15.0 μM/l	Hairy root culture	Xiao et al. (2010)
Dianthus caryophyllus	47.08 μM/l	Shoot multiplication and callus induction	Gutiérrez-Miceli et al. (2010)
Oroxylum indicum	2.0 mg/l	Shoot multiplication	Bansal and Gokhale (2012)
Momordica cymbalaria	3.0 mg/l	Shoot regeneration	Balkhande et al. (2013)
Hordeum vulgare	1.5 mg/l	Callus induction	Haque et al. (2015)
Arachis hypogaea	2.0 mg/l	Shoot multiplication	Rajinikanth et al. (2015)
Polygonum tinctorium	7.0 mg/l	Shoot regeneration	Kim et al. (2016)
Gossypium hirsutum	2.0 mg/l	Shoot regeneration and inhibiting phenol secretion	Prem Kumar et al. (2016)
Solanum nigrum	0.4 and 6.0 mg/l	Shoot multiplication and in vitro flowering	Geetha et al. (2016)
Silybum marianum	$5.887.10^{-4}$ mol/l	Suspension culture	Vildová et al. (2016)
Ophiorrhiza mungos	2.5 μM/l	Suspension culture	Deepthi and Satheeshkumar (2016)

Table 4.1 (continued)

Plant	Beneficiary silver nitrate concentrations	Part of plant	References
Prosopis cineraria	0.59 μM/l	Inhibition of shoot defoliation and shoot bud multiplication	Venkatachalam et al. (2017b)
Gloriosa superba	200 and 500 mg/l	Suspension culture	Mahendran et al. (2018a, b)
Cicer arietinum	15 and 50 mg/l	Anther culture	Abdollahi and Rashidi (2018)
Catharanthus roseus	0.1 μM/l	Shoot multiplication	Panigrahi et al. (2018)
Capsicum chinense	3 mg/l	Shoot proliferation	Bora et al. (2018)

Venkatachalam et al. (2017b) proved the effect of $AgNO_3$ on tissue culture medium to avoid leaf and shoot defoliation as well as multiple shoot induction of *Prosopis cineraria*. The highest percent of shoot multiplication was observed on the medium containing BAP and KIN. However, leaf and shoot defoliation happened in the culture medium. Therefore, to overcome the defoliation occurrence, various concentrations of $AgNO_3$ were tested. Among the various concentrations tested, $AgNO_3$ at 0.59 μM/l dose was found to be best for inhibition of leaf and shoot defoliation. Addition of silver nitrate ($AgNO_3$) in combination with BAP had significantly increased the multiple shoot bud development as well as shoot elongation in *P. cineraria*. Defoliation of shoot lets has been noted during in vitro organogenesis of *Azadirachta excelsa*. The addition of $AgNO_3$ in to the medium the shoot bud defoliation was inhibited and resulting high percentage of shoot bud regeneration was observed (AbdShukor et al. 2008). Geetha et al. (2016) tested the role of $AgNO_3$ on in vitro flowering and shoot regeneration of *Solanum nigrum*. Leaf and axillary bud explants were responded effectively in all the combinations of $AgNO_3$ in the medium compared to control. Maximum number of multiple shoots was found in leaf explants cultured on MS media supplemented with BAP (2.0 mg/l), NAA (0.5 mg/l), and $AgNO_3$ (0.4 mg/l). The regenerated shoots were subcultured; the maximum number of in vitro *flowers* was obtained from axillary bud explants in BAP (2.0 mg/l), KIN (1.0 mg/l), IAA (0.5 mg/l), and $AgNO_3$ (6.0 mg/l) supplemented media when compared to leaf and control.

Giridhar et al. (2003) revealed $AgNO_3$ holds the growth rate which was significantly increased in coffee. $AgNO_3$ at 40 μM/l nearly doubled the growth of arabica shoots, whereas leaf number was more with 10 μM/l, and the chlorophyll content was higher with 20 μM/l $AgNO_3$. In robusta, the highest percent of shoot length and most leaves were observed on 20 μM/l silver nitrate, but 10 μM induced more leaf area and chlorophyll content. Silver nitrate has produced positive effect on all the shoot regeneration parameters for direct and indirect regeneration, and rooting parameters are tested for regeneration of *Oroxylum indicum* (Bansal and Gokhale

2012). Previously, $AgNO_3$ has been used to prevent callus formation in tree and woody species, viz., *Garcinia mangostana* (Chongjin et al. 1997), *Albizia procera* (Kumar et al. 1998), and *Manihot esculenta* (Zhang et al. 2001), as well as in other plants, viz., *Vanilla planifolia* (Giridhar et al. 2001). Such silver nitrate supported multiple shoot formation has been reported in different plants, viz., *Brassica* sp. (Chi et al. 1990) and *Coffea* sp. (Fuentes et al. 2000). Supplementation of $AgNO_3$ in culture media caused significant positive effect on shoot number with its best response being observed at 2 mg/l concentration. Balkhande et al. (2013) suggested the $AgNO_3$ highly influenced the shoot bud formation and their subsequent proliferation. The best medium composition for multiple shoot induction was observed on MS medium supplemented with 2 mg/l BAP and 3 mg/l $AgNO_3$.

Haque et al. (2015) reported that the $AgNO_3$ induced more shoots from embryo-derived organogenic callus in *Hordeum vulgare* using five different concentrations of $AgNO_3$ alone and/or combination with five different doses of L-proline or L-glutamine were tested, and the highest frequency of callusing was recorded in 3.0 mg/l $AgNO_3$ alone and 1.0 mg/l BAP + 1.5 mg/l $AgNO_3$ + 150 mg/l glutamine. Rajinikanth et al. (2015) established the effect of $AgNO_3$ on in vitro direct regeneration; efficiency of leaflet explants was studied in two varieties of *A. hypogaea*. The highest multiple shoot induction was observed in 2.0 mg/l $AgNO_3$ + 10 mg/lBAP + 1.0 mg/l NAA combination. Similarly, Kim et al. (2016) reported the shoot regeneration from internode explants of *Polygonum tinctorium* was improved using the ethylene inhibitor $AgNO_3$ and the polyamine putrescine. Maximum number of shoot and shoot length was obtained in MS medium supplemented with 7 mg/l $AgNO_3$ and further increasing the $AgNO_3$ concentration, shoot and shoot length production was decreased.

Prem Kumar et al. (2016) reported the $AgNO_3$ promotes highest frequency of shoot induction and also inhibits the ethylene production and phenolic secretion in cotton (*Gossypium hirsutum*). There are different concentrations of $AgNO_3$ (0.5–3.0 mg/l) were tested, the greatest number of shoots produced at 2.0 mg/l $AgNO_3$ with no phenolic secretion. The regenerated shoots were healthy with intense green color and fast multiply on the same medium. Silver nitrate as an elicitor to enhanced the flavonolignans and flavonoid taxifolin production in *Silybum marianum* cell suspension culture. The highest content of taxifolin production in *S. marianum* cell suspension culture was obtained after silver nitrate (5.887.10–4 mol/l) treatment and 72 h exposure (Vildová et al. 2016). Lithospermic acid B is more responsive to silver ions (Ag+) in hairy root cultures of *Salvia miltiorrhiza* (Xiao et al. 2010). Synergistic effects of $AgNO_3$ on *Salvia miltiorrhiza* hariy root culture produce increased production of tanshinones. The combination of yeast extract and $AgNO_3$ (300 μmol/l) yielded the highest tanshinone I content, which was 14-fold increased when compared to control (Yan et al. 2006). Recently, Mahendran et al. (2018b) investigated the $AgNO_3$-treated cell suspension cultures of *Gloriosa superba* showed enhanced biosynthesis of colchicine and thiocolchicoside content levels for 15 and 30 days of exposure, respectively. Among the five concentrations of $AgNO_3$ tested, the maximum production of colchicine and thiocolchicoside contents was obtained from cells developed at 500 and 200 mg/l $AgNO_3$, respectively, for 15 days period of time compared to without

$AgNO_3$-treated cells. Similarly, Deepthi and Satheeshkumar (2016) reported that the 2.5 μM/l concentration of $AgNO_3$ produced a significant increase in biomass as well as camptothecin production level on the 10th day of incubation of *Ophiorrhiza mungos* cell suspension cultures when compared to control cultures.

4.3 Effect of Silver Nanoparticles on Tissue Culture Medium

4.3.1 Using Ag NPs to Avoid Microbial Contamination

The antibacterial property of NPs is one of the most established and well-studied achievements of NPs, with the most renowned being Ag NPs, TiO_2 NPs, and ZnO NPs (Applerot et al. 2012; Guzman et al. 2012). The effectiveness of NPs in the elimination of microbial contaminants in plant tissue cultures depends on their dimensions, size, distribution, and type. Several researchers have reported the survival and subsequent regeneration of explants that were affected negatively by NPs during application periods. Therefore, the effects of different types and levels of NPs on explants obtained from various plant species should be investigated in order to determine the best dose with minimal or no phytotoxicity. The effectiveness of NPs can also be improved by synergistic effects when they are combined with sterilizing agents or antibiotics.

In plant tissue culture technology, microbial contamination is a major problem. The contamination can extinguish the whole process and efficacy of the micropropagation method even earlier induction of explants. The contaminant sources mainly are explants and laboratory condition (Leifert et al. 1994). Typically, the many explants collected from field- and greenhouse-grown plants are surface sterilized with different solution before the establishment of in vitro cultures. Surface disinfection of the explants is a major step prior to in vitro culture initiation because microorganisms grow faster than explants in the tissue culture medium. Several sterilizing agents such as bromine water (BW), calcium hypochlorite (CaOCl), ethanol (EtOH), hydrogen peroxide (H_2O_2), sodium hypochlorite (NaOC1), mercuric chloride ($HgCl_2$), silver nitrate ($AgNO_3$), antibiotics, and fungicides are used. Among them, some disinfecting agents have failed to eliminate contaminants in explants, and instead they profoundly affected organogenesis due to their phytotoxicity. Metal and metal oxide NPs have been proven to be useful for the elimination of various microorganisms (Wang et al. 2017). A varied range of nanoparticles such as silver (Ag), aluminum oxide (Al_2O_3), CuO, iron oxide (Fe_3O_4), gold (Au), magnesium oxide (MgO), nickel (Ni), silicon (Si), SiO_2, titanium dioxide (TiO_2), and ZnO have been reported to possess antimicrobial activities against various microorganisms (Beyth et al. 2015). For the first time, Abdi et al. (2008) tested the silver nanoparticles (Ag NPs) to control the microbial contamination in *Valeriana officinalis*. Moreover, the Ag NPs treatment in the tissue culture medium did not affect shoot multiplication and subsequent rooting. Similarly, the immature capitulum of *Gerbera jamesonii* treated with 1.5% NaOCl and 200 mg/l Ag NPs had positively eliminated all microbial contaminants and showed no contrary effects on micropropagation (Fakhrfeshani et al. 2012). Gouran et al. (2014) employed the leaf explants of *Vitis vinifera* were successfully decontaminated with 95% EtOH and 100 mg/l Ag NPs.

Mahna et al. (2013) examined the effect of Ag NPs treatment on surface decontamination of *Arabidopsis* seeds, potato leaves, and tomato cotyledons. The explants treated with 100 mg/l Ag NPs for 1–5 min were found to be best for disinfecting (100%) the seeds, leaves, or cotyledons and had no side effects on explant viability. However, seed germination and leaf and cotyledon survival were reduced when they are treated with highest amount of Ag NPs. Removal of microbial contamination from in vitro cultures is a massive challenge when it comes to woody plants. The explants treated with 100 mg/l Ag NPs for 60 min after EtOH and Clorox exposure completely stopped microbial contamination, but only limited explants survived. However, supplementation of 4 mg/l Ag NPs to the medium sufficiently controlled internal contaminants in the olive explants (Rostami and Shahsavar 2009). Various explants of *Araucaria excelsa* immersion in Ag NPs (200 mg/l) for 180 min reduced the contamination rate from 61.5 to 11.3% (Sarmast et al. 2011). Potato and tobacco shoot buds were cultured on MS medium containing Ag NPs or TiO_2 NPs observed contamination-free cultures (Safavi et al. 2011; Safavi 2012). *Rosa hybrida* explants were treated with 200 mg/l Ag NPs solution for 20 min; the bacterial contamination was reduced (Shokri et al. 2014). On the other hand, Kalsaitkar et al. (2014) reported that internal bacterial contamination was significantly reduced in callus cultures of *Bacopa monnieri* by addition of 160 mg/l Ag NPs to the MS medium. Combination of Ag NPs and thymol inhibited microbial growth in *Cynodon dactylon*. Contamination-free cultures were initiated from nodal explants treated with 200 mg/l Ag NPs solution and 100 mg/l thymol for 60 min (Taghizadeh and Solgi 2014). Similarly, Arab et al. (2014) tested the Ag NP (100 and 150 mg/l) immersion and addition to the MS medium; internal and external contamination was reduced in G × N15 (hybrid of almond × peach) rootstocks. Recently, 0% contamination was detected in *Vanilla planifolia* shoot cultures when supplementation of 50–200 mg/l Ag NPs to the MS medium (Spinoso-Castillo et al. 2017).

4.3.2 Effect of Ag NPs on Micropropagation

The positive response of Ag NPs confirmed in many plant species showed in Table 4.2, for instance, the effects of Ag NPs on shoot regeneration of *Vanilla planifolia* using a temporary immersion bioreactor system. The maximum number of shoots was achieved on the medium containing 25 or 50 mg/l Ag NPs, whereas the lowest number of shoots was observed on MS medium supplemented with 200 mg/l of Ag NPs (Spinoso-Castillo et al. 2017). In vitro *Chrysanthemum* growth improved at 10 mg/l of Ag NPs dose (Duong Tan Nhut et al. 2014). Several studies have shown positive effects of NPs on callus induction, shoot regeneration, and growth. In *Tecomella undulata*, the shoot induction rate, number of shoots, and callus formation were increased when the stem explants were cultured on MS medium containing with 10 mg/l Ag NPs, 2.5 mg/l BAP, and 0.1 mg/l IAA (Aghdaei et al. 2012).

Table 4.2 Beneficiary effect of silver nanoparticles concentration(s) for plants

Plant	Beneficiary silver nanoparticles (Ag NPs) concentrations	Part of plant	References
Tecomella undulata	10.0 mg/l	Multiple shoots and callus formation	Aghdaei et al. (2012)
Brassica juncea	50.0 mg/l	Shoot regeneration	Sharma et al. (2012)
Bacopa monnieri	100.0 ppm	Seed germination	Krishnaraj et al. (2012)
Boswellia ovalifoliolata	4.0 μg/ml	Seed germination	Savithramma et al. (2012)
Brassica juncea	50.0 ppm	Seedling growth	Sharma et al. (2012)
Phaseolus vulgaris and *Zea mays*	16.0 ppm	Plant growth profile and biochemical attributes	Salama (2012)
Lactuca sativa	10.0 mg/l	Shoot and root growth	Gruyer et al. (2013)
Artemisia annua	900 mg/l	Suspension culture	Zhang et al. (2013)
Chrysanthemum	10.0 mg/l	Regeneration of shoots	Duong Tan Nhut et al. (2014)
Bacopa monnieri	16×10^3	Callus formation	Kalsaitkar et al. (2014)
Alternanthera sessilis	2.5 mg/l	Multiple shoot bud induction	Venkatachalam et al. (2017a)
Chrysanthemum morifolium	1.5 ppm	Shoot and root regeneration	Luong Thien Nghia et al. (2017)
Arabidopsis thaliana	30 mg/l	Suspension culture	Selivanov et al. (2017)
Vanilla planifolia	25 mg/l	Shoot multiplication	Spinoso-Castillo et al. (2017)
Gloriosa superba	0.3 mg/l	Somatic embryogenesis and germination	Mahendran et al. (2018a, b)

The positive effects of Ag NPs on micropropagation may be due to inhibition of ethylene production in the medium. Number of shoots, shoot length, and percentage of initiated shoots were significantly higher when nodal explants of *T. undulata* were cultured on MS medium fortified with 60 mg/l Ag NPs, 2.5 mg/l BAP, and 0.1 mg/l IAA (Sarmast et al. 2015). However, at higher concentrations (above 60 mg/l), a harmful effect on shoot regeneration was evident. Sharma et al. (2012) reported that the Ag NPs at 50 mg/l into MS medium improved the growth characteristics of *Brassica juncea* seedlings. On the other hand, hydrogen peroxide, malondialdehyde (MDA), and proline contents were decreased through activating antioxidant enzymes.

Recently, Venkatachalam et al. (2017a) investigated the effect of phycochemical-coated silver nanocomplexes as novel growth-stimulating hormone for plant regeneration of *Alternanthera sessilis* L. The medium supplemented with *Gracilaria foliifera* silver nanoparticles (GFAg NPs) + Adenine sulfide (AdS) combinations exhibited maximum percent of multiple shoot proliferation as well as mean number of shoots per culture. The highest percent of multiple shoot development (100%) with maximum mean number of shoots was observed on MS medium containing 2.5 mg/l GFAg NPs +50 mg/l AdS combination followed by 3.0 mg/l GFAg NPs +50 mg/l AdS combination, whereas the 0.5 mg/l GFAg NPs +50 mg/l AdS combination gave 51.4% of multiple shoot bud proliferation with 38.5 shoots/culture. The rate of multiple shoot bud induction and mean number of shoots were significantly enhanced with increasing concentrations of growth regulators in the medium but, beyond the optimum level, lead to a decrease. Out of three growth regulator combinations tested, the GFAg NPs + AdS combination significantly increased multiple shoot bud development as well as number of shoots compared to the BAP + AdS and KIN + AdS combinations. Selivanov et al. (2017) reported the addition of silver nanoparticles (30 mg/l) to a culture medium has a positive effect on the enhanced growth of biomass in cell suspension culture of *Arabidopsis thaliana.*

Similarly, Mahendran et al. (2018a) studied the phycomolecule-coated silver nanoparticles induced high-frequency somatic embryogenesis and plant regeneration from *Gloriosa superba* L. *Gloriosa superba* rhizome explants were placed on the MS medium supplemented with different doses of *Ulva lactuca* silver nanoparticles (ULAg NPs) along with 2,4-D and NAA for embryogenic callus formation. Interestingly, an enhanced rate of embryogenic callus formation was noticed on the medium containing ULAg NPs with 2,4-D and NAA combinations. Among the different combinations tested, the highest percentage of embryogenic callus initiation (96%) was on the MS medium supplemented with 0.1 mg/l ULAg NPs along with 2.0 mg/l 2,4-D and 0.5 mg/l NAA combination. To assess the effect of ULAg NPs on embryo development, among the various ULAg NP combinations examined, the highest percentage of somatic embryo production was 95.7% on the MS medium containing 0.4 mg/l ULAg NPs along with 20% ULE, BAP (2.0 mg/l), KIN (1.0 mg/l), and NAA (0.5 mg/l) combinations. On the other hand, high-frequency embryo germination (98.1%) was observed on the medium containing 0.3 mg/l ULAg NPs in combination with 5.0 mg/l GA_3, 2.0 mg/l BAP, 0.5 mg/l AdS, and 20% ULE. This suggests that the addition of phycomolecule-coated ULAg NPs along with ULE and PGRs could enhance the embryo conversion rate significantly. It is presumed that seaweed extract might act as a plant growth promoting substance along with other growth hormones under in vitro conditions. Interestingly, the addition of ULE and ULAg NPs along with plant growth regulators in the medium exhibited a pivotal role in the enhanced production of somatic embryos in *G. superba* over PGRs alone.

Luong Thien Nghia et al. (2017) investigated the effects of Ag NPs on *Chrysanthemum morifolium* ramat. cv. "JIMBA" growth and its ability in three culture systems: in vitro solid medium system, in vitro liquid medium system, and microponic (combined of micropropagation and hydroponic). The 3 cm *Chrysanthemum* shoots and silver nanoparticles of diameter smaller than 20 nm were used in the experiments. Resultant, the SNP concentration was suitable for growth and development of chrysanthemum in solid, liquid medium and microponic system. Microponic system not only improved plant growth but also reduced the succulent phenomenon. SNP uptake was likely dependent on concentrations and culture systems. At low concentrations (1, 5, and 10 ppm), SNPs were completely absorbed after 4 weeks of culture in all systems, but not at high concentrations (20 ppm). The amount of absorbed silver nanoparticles was directly proportional to the culture period and inversely proportional to the concentration of SNPs supplemented to the medium. In all of our investigated systems, the hydroponic system showed the highest capability of SNPs absorption.

Ewais et al. (2015) studied the effect of biosynthesized silver nanoparticles (Ag NPs) on growth, anatomy, and protein, and DNA of *Solanum nigrum* callus was investigated in vitro. Three concentrations of aqueous silver nanoparticles conjugated with camel milk were used in the present study. Murashige and Skoog (MS) nutrient media supplemented with different concentrations of Ag NPs, 0, 2, 4, and 8 mg/l were evaluated for their effects on the callus induction from the leaf explants of *Solanum nigrum*. Compact calli with white and greenish color were obtained after 10 days upon control culture (no Ag NPs), whereas friable watery calli with white, greenish, or yellowish color were observed after 10–13 days upon culture supplemented with Ag NPs. Interestingly, it was found that exposure to Ag NPs increased callus fresh weight and the intensity of callus formation. These were associated with the deformity of callus morphology. Light microscopy observations showed that the nanoparticles damaged cell wall. This was clear with the high concentration of Ag NPs. Variation in protein pattern was recorded between callus control (no Ag NPs) and callus exposure to Ag NPs. Genetic stability between callus control and callus under Ag NPs exposure was interrelated.

Kalsaitkar et al. (2014) studied the induced effect of silver nanoparticles (Ag NPs) on in vitro callus induction from *Bacopa monnieri*. Distinct morphological changes on callogenesis were observed on stress application. The effect of Ag NPs on callus induction and autoclaving was studied within MS media and without MS media. The MS medium supplemented with lower concentration of Ag NPs capable for medium growth of callus as that control and simultaneously reducing the rate of contamination. Biosynthesis of Ag NPs was carried out by using *Fusarium oxysporum* fungus and characterized by UV-Vis spectrophotometer and LM 20. Krishnaraj et al. (2012) studied the effect of biologically synthesized Ag NPs on hydroponically grown *Bacopa monnieri* growth metabolism and found that biosynthesized Ag NPs showed a significant effect on seed germination and induced the synthesis of protein and carbohydrate and decreased the total phenol contents and catalase and peroxidase activities. Also, biologically synthesized Ag NPs enhanced seed germination and seedling growth of trees *Boswellia ovalifoliolata* (Savithramma et al. 2012). Ag NPs increased plant growth profile (shoot and root length, leaf area) and biochemical attributes (chlorophyll, carbohydrate and protein contents, antioxidant enzymes) of *Brassica juncea*, common bean, and corn (Salama 2012; Sharma et al. 2012). However, Gruyer et al. (2013) reported Ag NPs have both positive and negative effect on root elongation depending on the plant species. They reported that root length was increased in barley but was inhibited in lettuce. Also, Yin et al. (2012) studied the effects of Ag NPs on germination of 11 wetland plant species (*Lolium multiflorum, Panicum virgatum, Carex lurida, C. scoparia, C. vulpinoidea, C. crinita, Eupatorium fistulosum, Phytolacca americana, Scirpus cyperinus, Lobelia cardinalis, Juncus effusus*) and found Ag NPs enhanced the germination rate of one species (*E. fistulosum*). Ag NPs induce root growth by blocking ethylene signaling in *Crocus sativus* (Rezvani et al. 2012). Zhang et al. (2013) explored the potential of Ag NPs on increasing the artemisinin content in hairy root cultures of *Artemisia annua*. The enhanced production of artemisinin content (3.9-fold) was obtained in hairy root cultures of *A. annua* for 3 days treatment with 900 mg/l Ag NPs.

4.4 Conclusion

In this review, the role of silver nitrate and their nanoparticles are potent regulators of plant growth and development. The influence of exogenously applied silver ions in the form of $AgNO_3$ in plant tissue culture media significantly regulates the ethylene activity and increase the plant growth in most of the plants. We have clearly conveyed the major physiological effects of $AgNO_3$ in plant systems, viz., direct or indirect organogenesis, somatic embryogenesis, in vitro rooting, flowering, and cell suspension culture. Nanotechnology is an evolutionary science and has presented many novel applications in the field of life science and medicine. However, due to their unique properties, a number of researches have been done on the toxicological effect of Ag NPs on plants, yet research focusing on the realization of the positive effects of Ag NPs on plant remains incomplete. Various studies have shown positive effect of Ag NPs on plant growth and development. It is evident from collected information that effect of Ag NPs varies from plant to plant and depends on their mode of application, size, and concentrations. This chapter revealed that the influence of $AgNO_3$ and Ag NPs on plant growth and development in various plants. In the beginning, more difficult works are needed to understand physiological, biochemical, and molecular mechanisms of plants in relation to Ag NPs. Also, more studies are needed to explore the mode of action of Ag NPs, their interaction with biomolecules, and their impact on the regulation of gene expressions in plants. Hence, it is necessary to explain the physiological mechanisms at the gene regulation level to find out the actual role of silver ions in signaling and to see how they influence regulation of ethylene action in plants.

References

Abdi G, Salehi H, Khosh-Khui M (2008) Nano silver: a novel nanomaterial for removal of bacterial contaminants in valerian (*Valeriana officinalis* L.) tissue culture. Acta Physiol Plant 30:709–714

Abdollahi MR, Rashidi S (2018) Production and conversion of haploid embryos in chickpea (*Cicer arietinum* L.) anther cultures using high 2,4-D and silver nitrate containing media. Plant Cell Tissue Organ Cult 133:39–49

Aghdaei M, Sarmast MK, Salehi H (2012) Effects of silver nanoparticles on *Tecomella undulata* (Roxb.) Seem. micropropagation. Adv Hortic Sci 26:21–24

Applerot G, Lellouche J, Perkas N, Nitzan Y, Gedanken A, Banin E (2012) ZnO nanoparticle-coated surfaces inhibit bacterial biofilm formation and increase antibiotic susceptibility. RSC Adv 2:2314–2321

Arab MM, Yadollahi A, Hosseini-Mazinani M, Bagheri S (2014) Effects of antimicrobial activity of silver nanoparticles on in vitro establishment of G × N15 (hybrid of almond×peach) rootstock. J Genet Eng Biotechnol 12:103–110

Bais HP, Sudha GS, Ravishankar GA (2001) Influence of putrescine $AgNO_3$ and polyamine inhibitors on the morphogenetic response in untransformed and transformed tissues of *Cichorium intybus* and their regenerants. Plant Cell Rep 20:547–555

Balkhande S, Kure SR, Surwase BS (2013) Influence of silver nitrate on shoot regeneration from excised meristems of *Momordica cymbalaria* Hook.: a diminishing species. Res J Biotech 8:42–47

Bansal YK, Gokhale M (2012) Effect of additives on micropropagation of an endangered medicinal tree *Oroxylum indicum* L. Vent. In: Recent advances in plant in vitro culture. Intech, Rijeka, pp 183–196

Beyth N, Houri-Haddad Y, Domb A, Khan W, Hazan R (2015) Alternative antimicrobial approach: nano-antimicrobial materials. Evid Based Complement Alternat Med 2015:1–16

Bora G, Gogoi HK, Handique PJ (2018) Callus mediated *in vitro* regeneration of Naga chilli (*Capsicum chinense* Jacq.): the fiery pepper from North East India. Int J Curr Microbiol App Sci 7:1312–1324

Chi GL, Barfield DG, Sim GE, Pua EC (1990) Effect of $AgNO_3$ and Aminoethoxyvinylglycine on *in vitro* shoot organogenesis from seedling explants of recalcitrant brassica genotypes. Plant Cell Rep 9:195–198

Chongjin JG, Siewkeng N, Prakash L, Chiangshong Goh CJ, Ng SK, Loh CS (1997) The role of ethylene on direct shoot bud regeneration from mangosteen (*Garcinia mangostana* L.) leaves cultured *in vitro*. Plant Sci 124:193–202

Dang W, Wei ZM (2009) High frequency plant regeneration from the cotyledonary node of common bean. Biol Plant 53:312–316

Deepthi S, Satheeshkumar K (2016) Enhanced camptothecin production induced by elicitors in the cell suspension cultures of *Ophiorrhiza mungos* Linn. Plant Cell Tissue Organ Cult 124:483–493

Eapen S, George L (1997) Plant regeneration from peduncle segments of oil seed brassica species: influence of $AgNO_3$ and silver thiosulphate. Plant Cell Tissue Organ Cult 51:229–232

Ewais EA, Desouky SA, Elshazly EH (2015) Evaluation of callus responses of *Solanum nigrum* L. exposed to biologically synthesized silver nanoparticles. Nanosci Nanotechnol 5:45–56

Fakhrfeshani M, Bagheri A, Sharifi A (2012) Disinfecting effects of nano silver fluids in Gerbera (*Gerbera jamesonii*) capitulum tissue culture. J Biol Environ Sci 6:121–127

Fuentes SR, Calheiros MB, Manettiflho J, Vieira LG (2000) The effects of silver nitrate and different carbohydrate sources on somatic embryogenesis in *Coffea canephora*. Plant Cell Tissue Organ Cult 60:5–13

Geetha G, Harathi K, Naidu CV (2016) Role of silver nitrate on in vitro flowering and shoot regeneration of *Solanum nigrum* (L.)—an important multipurpose medicinal plant. Am J. Plant Sci 7:1021

Giridhar P, Reddy OB, Ravishankar GA (2001) Silver nitrate influences in vitro shoot multiplication and root formation in *Vanilla planifolia* Andr. Curr Sci 81:1166–1170

Giridhar P, Indu EP, Vijaya Ramu D, Ravishankar GA (2003) Effect of silver nitrate on *in vitro* shoot growth of *Coffee*. Trop Sci 43:144–146

Giridhar P, Indu EP, Vinod K, Chandrashekar A, Ravishankar GA (2004) Direct somatic embryogenesis from *Coffea arabica* L and *Coffea canephora* P ex Fr. under the influence of ethylene action inhibitor-silver nitrate. Acta Physiol Plant 26:299–305

Gouran A, Jirani M, Mozafari AA, Saba MK, Ghaderi N, Zaheri S (2014) Effect of silver nanoparticles on grapevine leaf explants sterilization at *in vitro* conditions. In: 2nd National conference on nanotechnology from theory to application, Isfahan, Iran, pp 1–6

Griffitt RJ, Luo J, Gao J, Bonzongo JC, Barber DS (2008) Effects of particle composition and species on toxicity of metallic nanomaterials in aquatic organisms. Environ Toxicol Chem 27:1972–1978

Gruyer N, Dorais M, Bastien C, Dassylva N, Triffault-Bouchet G (2013) Interaction between sliver nanoparticles and plant growth. In: International symposium on new technologies for environment control, energy-saving and crop production in greenhouse and plant factory– greensys, Jeju, Korea, pp 6–11

Gutiérrez-Miceli FA, Arias L, Juarez-Rodríguez N, Abud-Archila M, Amaro-Reyes A, Dendooven L (2010) Optimization of growth regulators and silver nitrate for micropropagation of *Dianthus caryophyllus* L. with the aid of a response surface experimental design. In Vitro Cell Dev Biol Plant 46:57–63

Guzman M, Dille J, Godet S (2012) Synthesis and antibacterial activity of silver nanoparticles against gram-positive and gram-negative bacteria. Nanomed Nanotechnol Biol Med 8:37–45

Haque M, Siddique AB, Shahinul Islam SM (2015) Effect of silver nitrate and amino acids on high frequency plants regeneration in barley (*Hordeum vulgare* L.). Plant Tissue Cult Biotechnol 25:37–50

Jasim B, Thomas R, Mathew J, Radhakrishnan EK (2017) Plant growth and diosgenin enhancement effect of silver nanoparticles in Fenugreek (*Trigonella foenum-graecum* L.). Saudi Pharm J 25:443–447

Kalsaitkar P, Tanna J, Kumbhare A, Akre S, Warade C, Gandhare N (2014) Silver nanoparticles induced effect on *in-vitro* callus production in *Bacopa monnieri*. Asian J Biol Life Sci 3:167–172

Khodakovskaya MV, De Silva K, Biris AS, Dervishi E, Villagarcia H (2012) Carbon nanotubes induce growth enhancement of tobacco cells. ACS Nano 6:2128–2135

Kim JK, Baskar TB, Park SU (2016) Silver nitrate and putrescine enhance *in vitro* shoot organogenesis in *Polygonum tinctorium*. Biosci Biotechnol Res Asia 13:53–58

Krishnaraj C, Jagan EG, Ramachandran R, Abirami SM, Mohan N, Kalaichelvan PT (2012) Effect of biologically synthesized silver nanoparticles on *Bacopa monnieri* (Linn.) Wettst. Plant growth metabolism. Process Biochem 47:651–658

Kumar S, Sarkar AK, Kuhikannan C (1998) Regeneration of plants from leaflet explants of tissue culture raised Safed Siris (*Albizia procera*) Plant Cell Tissue Organ Cult 54:137–143

Kumar V, Parvatam G, Ravishankar GA (2009) $AgNO_3$: a potential regulator of ethylene activity and plant growth modulator. Electron J Biotechnol 12:8–9

Kumari M, Ernest V, Mukherjee A, Chandrasekaran N (2012) *In vivo* nanotoxicity assays in plant models. In: Nanotoxicity. Humana Press, Totowa, pp 399–410

Leifert C, Morris CE, Waites WM (1994) Ecology of microbial saprophytes and pathogens in tissue culture and field-grown plants: reasons for contamination problems *in vitro*. Crit Rev Plant Sci 13:139–183

Ma X, Geiser-Lee J, Deng Y, Kolmakov A (2010) Interactions between engineered nanoparticles (ENPs) and plants: phytotoxicity, uptake and accumulation. Sci Total Environ 408:3053–3061

Mahendran D, Kavi Kishor PB, Geetha N, Venkatachalam P (2018a) Phycomolecule-coated silver nanoparticles and seaweed extracts induced high-frequency somatic embryogenesis and plant regeneration from *Gloriosa superba* L. J Appl Phycol 30:1425–1436

Mahendran D, Kavi Kishor PB, Sreeramanan S, Venkatachalam P (2018b) Enhanced biosynthesis of colchicine and thiocolchicoside contents in cell suspension cultures of *Gloriosa superba* L. exposed to ethylene inhibitor and elicitors. Ind Crop Prod 120:123–130

Mahna N, Vahed SZ, Khani S (2013) Plant in vitro culture goes nano: nanosilver-mediated decontamination of ex vitro explants. J Nanomed Nanotechnol 4:1

Nhut DT, Tam HT, HIen NTT, Cuong LK, Luan VQ, Nam NB et al (2014) Effects of nanosilver on growth of *Chrysanthemum* sp., *Fragaria* sp. and *Gerbera* sp. cultured *in vitro*. J Biotechnol 12:103–111

Nghia LT, Tung HT, Huy NP, Luan VQ, Nhut DT (2017) The effects of silver nanoparticles on growth of *Chrysanthemum morifolium* Ramat. cv. "Jimba" in different cultural systems. Vietnam J Sci Technol 55:503
Nor AA, Jainol JE, Aziah MY, Mihdzar AK (2008) Defoliation of *in vitro* shootlets of *Azadirachta excelsa* (Jack) M. Jacobs-a possible solution. Malaysian Forester 71:37–44
Panigrahi J, Dholu P, Shah TJ, Gantait S (2018) Silver nitrate-induced *in vitro* shoot multiplication and precocious flowering in *Catharanthus roseus* (L.) G. Don, a rich source of terpenoid indole alkaloids. Plant Cell Tissue Organ Cult 132:579–584
Park EJ, Yi J, Kim Y, Choi K, Park K (2010) Silver nanoparticles induce cytotoxicity by a Trojan-horse type mechanism. Toxicol In Vitro 24:872–878
Prem Kumar G, Sivakumar S, Siva G, Vigneswaran M, Senthil Kumar T, Jayabalan N (2016) Silver nitrate promotes high-frequency multiple shoot regeneration in cotton (*Gossypium hirsutum* L.) by inhibiting ethylene production and phenolic secretion. In Vitro Cellular & Developmental Biology - Plant 52(4):408–418
Rajinikanth M, Mahitha B, Rama Swamy N (2015) Influence of silver nitrate on leaflet based direct regeneration in *Arachis hypogaea* L. Vegetos 28:63–69
Rezvani N, Sorooshzadeh A, Farhadi N (2012) Effect of nano-silver on growth of saffron in flooding stress. World Acad Sci Eng Technol 1:517–522
Rostami AA, Shahsavar A (2009) Olive "Mission" explants. Asian J Plant Sci 8:505–509
Safavi K (2012) Evaluation of using nanomaterial in tissue culture media and biological activity. In: 2nd International conference on ecological, environmental and biological sciences (EEBS' 2012), Bali, Indonesia, pp 5–8
Safavi K, Mortazaeinezahad F, Esfahanizadeh M, Asgari MJ (2011) *In vitro* antibacterial activity of nanomaterial for using in tobacco plants tissue culture. World Acad Sci Eng Technol 79:372–373
Salama HMH (2012) Effects of silver nanoparticles in some crop plants, common bean (*Phaseolus vulgaris* L.) and corn (*Zea mays* L.). Int Res J Biotechnol 3:190–197
Sarmast M, Salehi H, Khosh-Khui M (2011) Nano silver treatment is effective in reducing bacterial contaminations of *Araucaria excelsa* R. Br. var. glauca explants. Acta Biol Hung 62:477–484
Sarmast MK, Niazi A, Salehi H, Abolimoghadam A (2015) Silver nanoparticles affect ACS expression in *Tecomella undulata in vitro* culture. Plant Cell Tissue Organ Cult 121:227–236
Savithramma N, Ankanna S, Bhumi G (2012) Effect of nanoparticles on seed germination and seedling growth of *Boswellia ovalifoliolata* an endemic and endangered medicinal tree taxon. Nano Vis 2:2
Selivanov NY, Selivanova OG, Sokolov OI, Sokolova MK, Sokolov AO, Bogatyrev VA, Dykman LA (2017) Effect of gold and silver nanoparticles on the growth of the *Arabidopsis thaliana* cell suspension culture. Nanotechnol Russ 12:116–124
Sharma P, Bhatt D, Zaidi MGH, Saradhi PP, Khanna PK, Arora S (2012) Silver nanoparticle-mediated enhancement in growth and antioxidant status of *Brassica juncea*. Appl Biochem Biotechnol 167:2225–2233
Shokri S, Babaei A, Ahmadian M, Hessami S, Arab MM (2014) The effects of different concentrations of nano silver on elimination of bacterial contaminations and phenolic exudation of Rose (*Rosa hybrida* L.) *in vitro* culture. Int J Farm Allied Sci 3:50–54
Spinoso-Castillo JL, Chavez-Santoscoy RA, Bogdanchikova N, Pérez-Sato JA, Morales-Ramos V, Bello-Bello JJ (2017) Antimicrobial and hormetic effects of silver nanoparticles on in vitro regeneration of vanilla (*Vanilla planifolia* Jacks. ex Andrews) using a temporary immersion system. Plant Cell Tissue Organ Cult 129:195–207
Taghizadeh M, Solgi M (2014) The application of essential oils and silver nanoparticles for sterilization of Bermuda grass explants in *in vitro* culture. Int J Horticult Sci Technol 1:131–140
Venkatachalam P, Malar S, Thiyagarajan M, Indiraarulselvi P, Geetha N (2017a) Effect of phycochemical coated silver nanocomplexes as novel growth-stimulating compounds for plant regeneration of *Alternanthera sessilis* L. J Appl Phycol 29:1095–1106
Venkatachalam P, Jinu U, Gomathi M, Mahendran D, Ahmad N, Geetha N, Sahi SV (2017b) Role of silver nitrate in plant regeneration from cotyledonary nodal segment explants of *Prosopis*

cineraria (L.) Druce.: a recalcitrant medicinal leguminous tree. Biocatal Agric Biotechnol 12:286–291

Vildová A, Hendrychová H, Kubeš J, Tůmová L (2016) Influence of $AgNO_3$ treatment on the flavonolignan production in cell suspension culture of *Silybum marianum* (L.) Gaertn. World Acad Sci Eng Technol Int J Biol Biomol Agric Food Biotechnol Eng 8:959–962

Wang L, Hu C, Shao L (2017) The antimicrobial activity of nanoparticles: present situation and prospects for the future. Int J Nanomedicine 12:1227

Xiao Y, Gao S, Di P, Chen J, Chen W, Zhang L (2010) Lithospermic acid B is more responsive to silver ion (Ag+) than rosmarinic acid in *Salvia miltiorrhiza* hairy root cultures. Biosci Rep 30:33–40

Yan Q, Hu ZD, Wu JY (2006) Synergistic effects of biotic and abiotic elicitors on the production of tanshinones in *Salvia miltiorrhiza* hairy root culture. Chinese Tradit Herbal Drugs 31:188–191

Yin L, Colman BP, McGill BM, Wright JP, Bernhardt ES (2012) Effects of silver nanoparticle exposure on germination and early growth of eleven wetland plants. PLoS One 7:1–7

Zhang P, Phansiri S, Puanti-Kaerlas J (2001) Improvement of cassava shoot organogenesis by the use of silver nitrate *in vitro*. Plant Cell Tissue Organ Cult 67:47–54

Zhang B, Zheng LP, Yi Li W, Wen Wang J (2013) Stimulation of artemisinin production in *Artemisia annua* hairy roots by Ag-SiO_2 core-shell nanoparticles. Curr Nanosci 9:363–370

Exploring the Phytoremediation Potential of *Calotropis gigantea* L. Using a Combined FTIR and Principal Component Analysis

5

P. Sangeetha, Perumal Venkatachalam,
and Natesan Geetha

Abstract

Chromium (Cr) is a main concern pollutant and has been recognized to be harmful to fauna, flora, and human beings. The intent of the present study was to investigate the phytoremediation potential of *Calotropis gigantea* L. plant using a combined FTIR and principal component analysis (PCA) for effective Cr removal. Plant leaves were collected from unpolluted and polluted areas (near tanneries). After authentication, the composite leaf sample of each site was thoroughly washed and allowed to dry. The dried samples were finely powdered and used for FTIR analysis. The FTIR spectra of both the samples showed totally 8 peak values and corresponding 12 functional groups such as alcohols, phenols, alkanes, carboxylic acids, alkynes, alkenes, aromatics, amines, esters, ethers, aliphatic amines, and alkyl chlorides. The peak value's difference was found to be very low between samples. To confirm this, peak values were subjected to PCA which revealed that no functional groups of both the samples contributing variation were identified. This observation suggests that this plant has a protective mechanism to detoxify heavy metal Cr and could be used for phytoremediation purpose in the future.

Keywords

Calotropis gigantea · Phytoremediation · Chromium · FTIR · PCA

P. Sangeetha
Mother Teresa Women's University, Kodaikanal, Tamil Nadu, India

Periyar University, Salem, Tamil Nadu, India

P. Venkatachalam
Department of Biotechnology, School of Biosciences, Periyar University, Salem, Tamil Nadu, India

N. Geetha (✉)
Department of Biotechnology, School of Biosciences, Periyar University, Salem, Tamil Nadu, India

Department of Botany, Bharathiar University, Coimbatore, Tamil Nadu, India

M. Kumar et al. (eds.), *In vitro Plant Breeding towards Novel Agronomic Traits*,
https://doi.org/10.1007/978-981-32-9824-8_5

5.1 Introduction

Natural waters have been found to be contaminated with numerous heavy metals arising mostly from mining wastes and industrial discharges (Dönmez and Aksu 2001). The presence of heavy metals in the environment is of chief concern because of their toxicity, bioaccumulating capability, and risk to human life and the environment (Volesky 1994; Volesky and Holan 1995). Heavy metals are among the usual pollutants that are not subjected to bacterial attack or other breakdown or degradation process and are stable accumulation to the marine environment (El-Nady and Atta 1996). As a result of this, their concentrations often exceed the permissible levels than normally found which can greatly disturb biological processes. In accordance with World Health Organization (WHO), the metals of most urgent concern are chromium, copper, zinc, iron, cadmium, and lead (WHO 2011). Tannery is one of the industries, which continuously polluting soil (Vankar and Dwivedi 2009). Most of the tanneries in India are located near the riverbanks, i.e., Ganges River in North India and Palar River in South India (Mondal and Singh 2010), because it requires a large amount of fresh water to process hides and skins. A variety of chemicals are used to process the raw materials in tannery industries, which cause the surface water as well as groundwater contamination. Such type of problem is also faced in and around Dindigul district, Tamil Nadu, India (Paul Basker 2000). The discharge of chromium (VI) into aquatic ecosystems has become a matter of concern in all the tannery industries situated areas in India over the last few decades. Chromium's interaction with biological systems is very diverse and complex. According to review of available literature, about 10% of ingested Cr hexavalent (VI) is absorbed in the gastrointestinal tract. Even though Cr remains as a highly effective tanning agent for producing leather with suitable characteristic features essential for many end uses, acute exposure to Cr (VI) causes various health problems including kidney damage, dermatitis, asthma, etc. (Saha et al. 2011; Mohan and Pittman 2006; Gong et al. 2010; Nasr et al. 2011). Fourier transform infrared spectroscopy (FTIR) is a nondestructive analytical technique that is less time-consuming and cheaper than the present "gold standard" methods, providing exclusive molecular chemical information based on the FTIR spectrum (McCann et al. 1992; Yu 2004). Principal component analysis works based on transformation of the original variables into new axes or principal components. It is possible to express maximum total variation in data set in only a few principal components and represent the decrease of the variation (Martini et al. 2015). FTIR spectroscopy has been used in conjunction with PCA to detect and classify a broad range of cell wall mutants from a large mutagenized *Arabidopsis* population (Chen et al. 1998).

Calotropis gigantea belongs to family Asclepiadaceae and is one of the drought-resistant and salt-tolerant medicinal plants (Sharma and Tripathi 2009) which is used in different ways to treat the infectious diseases (Hemalatha et al. 2011). The medicinal properties in *C. gigantea* are due to the presence of various phytochemical constituents reported from various parts, e.g., plants contain the cardenolides, root bark contains benzoylinesolone and benzoylisolinelone, leaves and stalk of the plant contain calotropin and calotropagenin, while the flower and latex contain uzarigenin and terpenol ester (Yoganarasimhan 2011). The uptake of radiostrontium and radiocesium by this plant was studied and suggested that it could be used as potential candidate plant for phytoremediation of these radionuclides from soils and solutions (Eapen et al. 2006). FTIR has been used to analyze the chemical constituents of this plant leaves which were collected from SIPCOT area, Cuddalore district, Tamil Nadu, India. For this, leaf samples were collected in both winter and summer seasons at different industrially polluted locations. One sample was collected from an unpolluted area which is used as control sample. Only minor changes in their FTIR peak values were noticed and suggested that this plant has a protective mechanism to detoxify various pollutants (Ramamurthy and Kannan 2007). The aim of the present study was to assess the phytoremediation potential of this plant for efficient chromium elimination using a combined FTIR and PCA analysis.

5.2 Materials and Methods

Calotropis gigantea plant leaves were collected from unpolluted area (near Kattu Road Batlagundu, Dindigul district, Tamil Nadu, India) and from polluted area (near tanneries, Dindigul). After collection, the leaves were thoroughly washed with water to remove dust particles. The cleaned samples were oven-dried at 60 °C for 4 h to remove moisture content. Dried leaves were then ground into fine powder and were subjected to FTIR analysis.

For FTIR analysis, the leaf powder of both the samples were carefully pressed into pellets using KBr. The samples were then examined in a PerkinElmer FITR spectrometer (Model 2000) under the transmission mode, i.e., 4000–400 cm^{-1}. FTIR spectra of both the samples were compared. The data of functional groups were subjected to PCA to find out the variation between functional groups of both the samples using Past3 software.

5.3 Results and Discussion

The ever-increasing use of wide variety of heavy metals in industries and agriculture has caused a serious environmental pollution. Phytoremediation is a developing skill that can potentially address the problems of contaminated agricultural land or water by providing effective and affordable technological solution for removal of heavy metals from polluted areas. Soil is a very precise component of the biosphere because it is not only a geochemical sink for contaminations but also acts as a natural buffer controlling the transportation of chemical elements and substances to the ecosystem. Heavy elements which originated from diverse sources may finally reach the surface soil, and their further destiny depends upon soil chemical and physical properties (Hashem and Al-Farraj 1995). The level of chromium was 0.0102 ppm/g in the unpolluted soil, whereas it was 6.46 ppm/g in the polluted soil of Dindigul district (unpublished data). The maximum permissible limit of Cr for normal water or soil sample is 0.006 μg/g as per WHO standards.

The FTIR spectra of plant samples collected from two different areas, i.e., unpolluted (control) and polluted sites, are shown in Figs. 5.1 and 5.2. The absorption peak values and their functional groups for the samples collected from polluted and unpolluted sites are given in Table 5.1. In both the samples, totally 8 peak values and 12 functional groups were identified such as alcohols, phenols, alkanes, carboxylic acids, alkynes, alkenes, aromatics, amines, esters, ethers, aliphatic amines, and alkyl chlorides. The peak value difference was found to be very low between samples collected from polluted and unpolluted sites. The strong and broad absorption

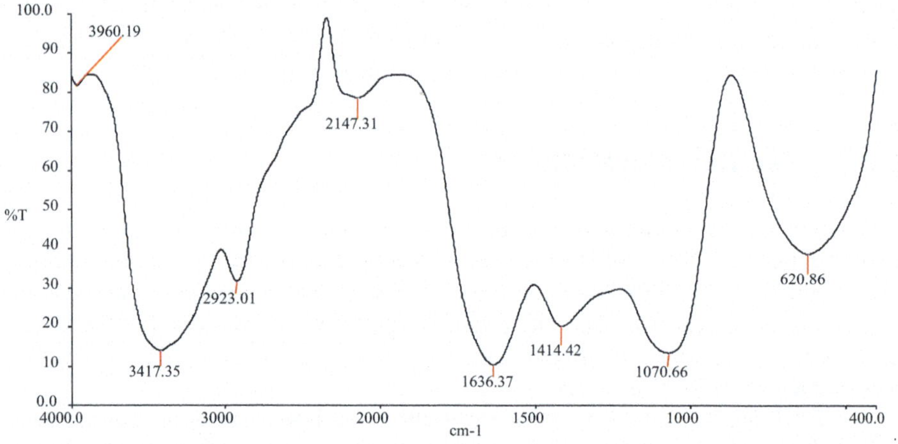

Fig. 5.1 FTIR spectrum of control leaf sample

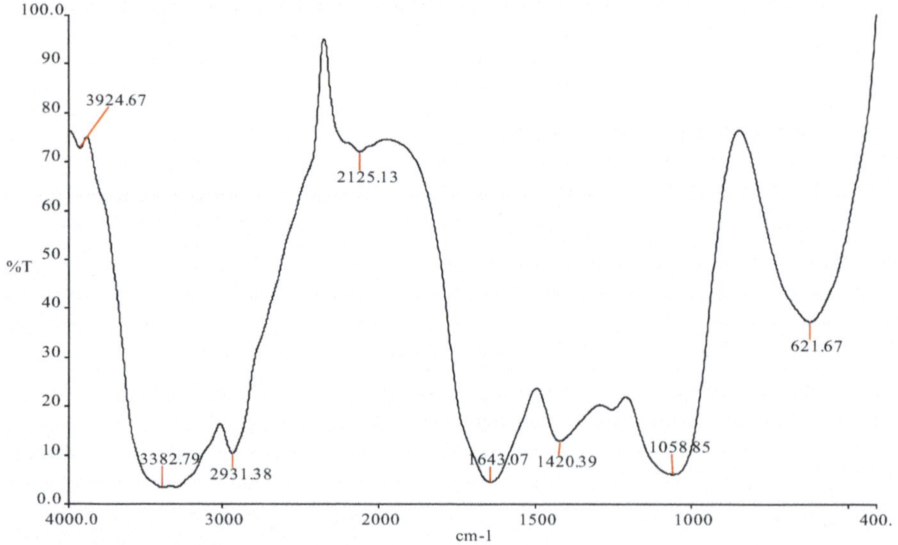

Fig. 5.2 FTIR spectrum of polluted leaf sample

Table 5.1 IR Peaks assignments for samples collected from unpolluted and polluted sites

Control sample				Polluted sample		
Sl. no.	Peak values	Chemical assignment	Functional group	Peak values	Chemical assignment	Functional group
1.	3960.19			3924.67		
2.	3417.35	Strong, broad stretch (O-H)	Alcohols and phenols	3382.79	Strong, broad stretch (O-H)	Alcohols and phenols
3.	2923.01	Medium (C-H) variable broad stretch (O-H)	Alkanes carboxylic acids	2931.38	Medium (C-H) variable broad stretch (O-H)	Alkanes carboxylic acids
4.	2147.31	Week stretch C≡C	Alkynes	2125.13	Week stretch C≡C	Alkynes
5.	1636.37	Medium stretch C=C	Alkenes	1643.07	Medium stretch C=C	Alkenes
		Medium stretch C-C	Aromatics		Medium stretch C-C	Aromatics
		Bend N-H	Amines		Bend N-H	Amines
6.	1414.42	Bend back medium C-H	Alkanes	1420.39	Bend back medium C-H	Alkanes
		Medium stretch C-C	Aromatics		Medium stretch C-C	Aromatics
7.	1070.66	Strong stretch C-O	Esters, ethers, alcohols, carboxylic acids	1058.85	Strong stretch C-O	Esters, ethers, alcohols, carboxylic acids
		Variable stretch C-N	Aliphatic amines		Variable stretch C-N	Aliphatic amines
8.	620.86	Medium stretch C-X	Alkyl halides	621.67	Medium stretch C-X	Alkyl halides

bands at 3417.35 cm^{-1} for control and 3382.79 cm^{-1} for polluted samples correspond to the presence of functional groups of alcohols and phenols. These are related to stretching (ν) vibration of hydroxyl groups (Barud et al. 2013). The peaks at 2923.01 cm^{-1} and 2931.38 cm^{-1} are representatives of functional groups of alkanes and carboxylic acids present in both the samples. The peaks at 2147.31 cm^{-1} and at 2125.13 cm^{-1} correspond to alkyne ($C \equiv C$) group of both the samples. Bands at 1636.37 cm^{-1} and 1643.07 cm^{-1} denote the presence of alkenes, aromatics, and amines in both the samples. The involvement of amino groups for the uptake of chromium has been discussed by Siva and Vidhya (2015) who studied the use of *Allium cepa* to assess the phytoremediation capacity of chromium. Absorption spectra for both the samples at 1414.42 cm^{-1} and 1420.39 cm^{-1} are due to medium stretching vibrations of C-H and C-C groups. The presence of peaks at 1070.66 cm^{-1} and 1058.85 cm^{-1} in both the samples is due to strong stretching vibrations of C-O and variable stretch vibrations of C-N groups within the peak range of 1070–1030 cm^{-1} which predict the presence of esters, ethers, alcohols, carboxylic acids, and aliphatic amines. The peaks at 620.86 cm^{-1} for control sample and 621.67 cm^{-1} for polluted sample are assigned to the presence of alkyl halides.

When analyzing the FTIR peak values, it is understood that the plants collected from both the sites did not show any structural changes in their organic constituents except slight peak value difference between samples. This is similar to the report of Ramamurthy and Kannan (2007) who found out some minor differences in peak values but no changes in organic constituents of *Calotropis gigantea* when they applied FTIR technique in the leaf samples collected from polluted (SIPCOT) and unpolluted area of Cuddalore district, Tamil Nadu, India even at different seasons and different localities. They collected leaf sample from SIPCOT area where many industries are located and more chances for contamination with different type of pollutants. In the present study, leaf samples were collected from in and around tannery industries only, where chromium is the main pollutant. Thus, these studies clearly indicate this plant may have some kind of exclusion mechanism for any heavy metal/s as mentioned earlier (Brooks 1972).

The data of functional groups were subjected to PCA analysis which is one of the most common multivariate statistical techniques to detect minute variations between two or more samples. In biplot of PC1 against PC2, all the peak values of both the samples are found in the central axis of the plot except few which may be due to slight differences in peak values. Combining FTIR with PCA can enable to distinguish spectral fingerprint that can be related to the biochemical changes in a plant (Manrique and Lajolo 2004; Dokken and Davis 2007). This is an emerging technique for studying biochemical changes of plant tissue (Yu 2005). In addition to PCA analysis, using Past3 software, eigenvalues and % variance for both the samples were obtained. The eigenvalues give a measure of the variance accounted by the corresponding eigenvectors or components (Hammer et al. 2001). From these values, it is understood that there is only 0.003% variation between samples collected from unpolluted and polluted sites (data not available). This small % variation is also confirmed by scree plot graph (data unavailable).

5.4 Conclusion

Heavy metals pollution generated by industrial people put pressure on human health all over the world. Finding a solution for this situation is a permanent task of researchers. In our study, application of combined FTIR–PCA for exploring phytoremediation potential of *Calotropis gigantea* clearly indicates that there is no difference in the chemical constituents of leaf samples collected from control and polluted areas. This observation suggests that this plant has a protective mechanism to detoxify heavy metal Cr and could be used as phytoremediation purpose in the future. And also this study proves over again the capability of this plant to phytotransform any heavy metal/s as investigated by earlier researchers.

Acknowledgment The author thank DST-CURIE funding agency for providing infrastructure facilities to carry out this experimental work at Mother Teresa Women's University, Kodaikanal, Tamil Nadu, India.

References

Barud H, Júnior A, Saska S, Mestieri L et al (2013) Antimicrobial Brazilian propolis (EPPAF) containing biocellulose membranes as promising biomaterial for skin wound healing. Evid Based Complement Alternat Med 2013:1–10. Article ID 703024

Brooks RR (1972) Geobotany and geochemistry in mineral exploration. Harper and Row, New York, p 292

Chen L, Carpita NC, Reiter WD, Wilson RH, Jeffries C, McCann MC (1998) A rapid method to screen for cell-wall mutants using discriminant analysis of Fourier transform infrared spectra. Plant J 16:385–392

Dokken KM, Davis LC (2007) Infrared imaging of sunflower and maize root anatomy. J Agric Food Chem 55:10517–10530

Dönmez G, Aksu Z (2001) Bioaccumulation of copper (II) and nickel (II) by the non-adapted and adapted growing Candida sp. Water Res 35:1425–1434

Eapen S, Shraddha S, Throat V, Kaushik CP, Raj K, D'Souza SF (2006) Phytoremediation of radiostrontium (^{90}Sr) and radiocesium (^{137}Cs) using giant milky weed (*Calotropis gigantea* R.Br.) plants. Chemosphere 65:2071–2073

El-Nady FE, Atta MM (1996) Toxicity and bioaccumulation of heavy metals to some marine biota from the Egyptian coastal waters. J Environ Sci Health Part A 31:1529–1545

Gong Y, Liu X, Huang L, Chen W (2010) Stabilization of chromium: an alternative to make safe leathers. J Hazard Mater 179:540–544

Hammer O, Harper DAT, Ryan PD (2001) PAST: paleontological statistics software package for education and data analysis. Palaeontol Electron 4:9

Hashem AR, Al-Farraj MM (1995) Mineral analysis of soil, *Euphorbia hirta* L. and mycoflora from the industrial Yanbu city, Saudi Arabia. Qatar Univ Sci J 15:83–89

Hemalatha M, Arirudran B, Thenmozhi A, Rao USM (2011) Antimicrobial effect of separate extract of acetone, ethyl acetate, methanol and aqueous from leaf of Milkweed (*Calotropis gigantea* L.). Asian J Pharm 1:102–107

Manrique GD, Lajolo FM (2004) Cell wall polysaccharide modifications during post harvest ripening of papaya fruit (*Carica papaya*). Postharvest Biol Technol 33:11–26

Martini D, Taddei F, Ciccoritti R, Pasquini M, Nicoletti I, Corradini D, D'Egidio MG (2015) Variation of total antioxidant activity and of phenolic acid, total phenolics and yellow coloured pigments in durum wheat (*Triticum turgidum* L. var. *durum*) as a function of genotype, crop year and growing area. J Cereal Sci 65:175–185

McCann MC, Hammouri MK, Wilson RH, Belton PS, Roberts K (1992) Fourier transform infrared microspectroscopy is a new way to look at plant cell walls. Plant Physiol 100:1940–1947

Mohan D, Pittman CU Jr (2006) Activated carbons and low cost adsorbents for remediation of tri- and hexavalent chromium from water. J Hazard Mater 137:762–811

Mondal NC, Singh VP (2010) Need of groundwater management in tannery belt: a scenario about Dindigul town, Tamil Nadu. J Geol Soc India 76:303–309

Nasr MM, Gondal MA, Seddigi ZS (2011) Detection of hazardous pollutants in chrome tanned leather using locally developed laser-induced breakdown spectrometer. Environ Monit Assess 175:387–395

Paul Basker J (2000) Tanneries in Dindigul district. Dossier on Tannery pollution in Tamil nadu. Peace Trust, Dindigul, Tamil Nadu, India 208–210 p

Ramamurthy N, Kannan S (2007) Fourier transform infrared spectroscopic analysis of a plant (*Calotropis gigantea* Linn) from an industrial village, Cuddalore dt, Tamilnadu, India. Romanian J Biophys 17:269–276

Saha R, Nandi R, Saha B (2011) Sources and toxicity of hexavalent chromium. J Coord Chem 64:1782–1806

Sharma AP, Tripathi BD (2009) Assessment of atmospheric PAHs profile through *Calotropis gigantea* R.Br. leaves in the vicinity of an Indian coal-fired power plant. Environ Monit Assess 149:477–482

Siva GV, Vidhya U (2015) Phytoremediation of heavy metal chromium using *Allium cepa* L. as model system. Int J Curr Biotechnol 3:1–5

Vankar PS, Dwivedi AK (2009) Sulphates for skin preservation—a novel approach to reduce tannery effluent salinity hazards. J Hazard Mater 163:207–212

Volesky B (1994) Advances in biosorption of metals: selection of biomass types. FEMS Microbiol Rev 14:291–302

Volesky B, Holan ZR (1995) Biosorption of heavy metals. Biotechnol Prog 11:235–250

WHO G (2011) Guidelines for drinking-water quality. World Health Org 216:303–304

Yoganarasimhan SN (2011) Medicinal plants of India. Regional Research Institute (Ay.) Bangalore, Tamil Nadu Ayurvedic uses and pharmacological activities of *Calotropis procera* Linn. Asian J Tradit Med 6:97

Yu P (2004) Application of advanced synchrotron radiation-based fourier transform infrared (SR-FTIR) microspectroscopy to animal nutrition and feed science; a novel approach. Br J Nutr 92:869–885

Yu P (2005) Molecular chemistry imaging to reveal structural features of various plant feed tissues. J Struct Biol 150:81–89

Role of Plant-Microbe Interaction in Phytoremediation

6

Geeta Prakash, Renu Soni, Reema Mishra, and Shweta Sharma

Abstract

As a result of rapidly increasing population and the related anthropogenic activities, our natural resources are becoming severely contaminated with various inorganic and organic compounds. Among the several remediation techniques available for removal of pollutants, phytoremediation is considered to be the most efficient and environment friendly. Phytoremediation is a process where plants and associated microbes are utilized for removal of harmful chemicals from the polluted sites. The rhizospheric microbiome is comprised of several bacteria, endophytes and AMF that plays significant role in plant growth and development. The rhizospheric microbiota facilitates nutrient availability, releases growth-promoting phytohormones and provides protection against abiotic and biotic stresses to plants. The interaction of these soilborne microbes with plants helps in uptake, sequestration and detoxification of contaminants from polluted sites. This chapter aims at highlighting the role of plant-microbe interactions in the process of phytoremediation. A detailed understanding of the plant-microbe interaction in phytoremediation will help in developing more efficient methods for detoxification of the polluted sites leading to environmental cleanup.

Keywords

AMF · Bacteria · Fungi · Heavy metals · Phytoremediation · Plant-microbe interaction

Authors Geeta Prakash, Renu Soni, Reema Mishra and Shweta Sharma have equally contributed to this chapter.

G. Prakash · R. Soni · R. Mishra · S. Sharma (✉)
Department of Botany, Gargi College, University of Delhi, New Delhi, Delhi, India

M. Kumar et al. (eds.), *In vitro Plant Breeding towards Novel Agronomic Traits*,
https://doi.org/10.1007/978-981-32-9824-8_6

6.1 Introduction

A rapid increase in world population has led to manifold increase in global energy demand which in turn has resulted in overexploitation of resources, urbanization, industrialization, land use changes, etc. All these anthropogenic activities have culminated into a drastic increase in release of toxic pollutants into the environment (Glick 2010). The major classes of pollutants detected in environment are polycyclic aromatic hydrocarbons (PAHs), heavy metals, pesticides, petroleum hydrocarbons, furans, dioxins, halogenated hydrocarbons, etc. (Kaya 2006; Singh and Jain 2003; Gerhardt et al. 2009). These pollutants, which are released due to plethora of human activities such as crude oil extraction, accidental oil spills, improper waste management practices, excessive use of pesticides, etc. are the principal causes of soil, air and water pollution (Conte et al. 2005; Paria 2008; Gerhardt et al. 2009; James and Strand 2009; Mc-Guinness and Dowling 2009). Out of all these, heavy metals are considered as one of the most harmful contaminants. Heavy metals are extremely toxic as they are mutagens and carcinogens which lead to damages to central nervous system, mental functions and various vital organs such as lungs, kidney, liver, etc. Long-term exposure to heavy metals might result in Alzheimer's disease, Parkinson's disease, muscular dystrophy, sclerosis, etc. (Chaney et al. 1999; IOSHIC 1999; Ferner 2001). Heavy metals such as cadmium (Cd), chromium (Cr), copper (Cu), lead (Pb), mercury (Hg), nickel (Ni), arsenic (As) and zinc (Zn) are released into the environment due to activities such as mining, landfill activities, industrial solid waste disposal, industrial effluent disposal, airborne industrial wastes and use of chemical-based pesticides and herbicides (Griffin et al. 1997; Adriano 1986; Oleksyn and Innes 2000). In addition to this use of leaded petrol for land transports, burning of nonrenewable energy resources, nuclear wastes have been the major contributors of heavy metal-based atmospheric pollution. Apart from the adverse effects on human health, presence of these toxic pollutants in environment has led to severe ecological damages as well such as habitat destruction, loss of biodiversity, deterioration of air quality, poor vegetation development and imbalance in ecosystem (Conte et al. 2005; Escalante-Espinosa et al. 2005; Scullion 2006; Paria 2008; Glick 2010).

In view of such widespread effects, various physical and chemical remediation technologies were developed for proper management of the pollutants. These remediation techniques included incineration, landfill sites, excavation, etc. But these technologies proved to be of limited use due to their low efficiency, high cost and possibility of secondary air and groundwater contamination (Mc-Guinness and Dowling 2009; Singh and Jain 2003; Scullion 2006; Gao et al. 2007).These limitations led the emphasis towards the development of more efficient and environment friendly techniques. Phytoremediation is one such approach that has been a method of choice for remediation of polluted sites for past few years. Phytoremediation is a process where plants and plant-microbe interactions is utilized to detoxify the contaminated sites (Gratão and Braz 2005; Kraomer 2005). Phytoremediation is an environment friendly, cost effective and efficient method as compared to the existing methods (Hao et al. 2011). The underlying mechanism includes extraction, volatilization, accumulation and degradation of harmful chemicals by different plant species (Chaney et al. 1997; Glick 2003; Morikawa and Erkin 2003; Singh and Jain 2003). Various plant species have been utilized successfully for remediation of harmful chemicals from the polluted sites. A phytoremediator plant species must possess certain properties such as fast growth, large biomass, hardiness and tolerance to toxic chemicals (Cherian and Oliveira 2005). Phytoremediator plant species are capable of accumulating and tolerating much higher concentrations of toxic pollutants without any visible phytotoxic symptoms as compared to other plants (Brooks 1977; Salt et al. 1998; Reeves and Baker 2000; Vassilev et al. 2005). The different phytoremediator plant species differ in their ability to absorb contaminants from the soil depending on their biomass, rate of transpiration, abundance of enzymes and the associated microbes. The microbes residing in the rhizosphere of plants play a significant role in promoting growth and development in plants. The rhizospheric microbiome helps in absorption of nutrients and release of growth-promoting plant hormones and provides resistance to abiotic as well as biotic stress (Bulgarelli et al. 2013; Arshad et al. 2007; Weyens et al. 2009a). In addition, the interaction between plants and microbes has been reported to play a significant role in the process of phytoremediation by sequestration, degradation and detoxification of harmful chemicals (Weyens et al. 2009a; Weyens et al. 2009b). In this chapter, we review the role of plant-microbe interaction in the process of phytoremediation, the underlying mechanisms and its applications in detoxification of environment.

6.2 Role of Rhizospheric Microbiome in Plant Growth and Development

Soil is considered to be a reservoir of diverse forms of microscopic organisms which includes bacteria, fungi, protozoa and algae (Glick 2012). All these microorganisms play diverse roles in maintaining the ecological balance. Among these different life forms, bacteria are found to be the most common (Schoenborn et al. 2004). Most of the bacteria inhabiting soil have been found to be concentrated in areas around the roots of the plant called as the rhizosphere. The presence of dense population of bacteria in rhizosphere is due to the presence of nutrients such as amino acids, sugars, organic acid, etc. in the exudates secreted by the roots (Whipps 1990; Bais et al. 2006; Badri et al. 2009; Badri and Vivanco 2009).The rhizosphere thus has a vast potential for maintaining plant health and nutrition (Berendsen et al. 2012). These rhizospheric bacteria have been reported to play a significant role in promoting growth and development in plants; therefore they are also known as plant growth-promoting bacteria (PGPBs) or plant growth-promoting rhizobacteria (PGPRs; Kloepper and Schroth 1978). The PGPBs include free-living, symbiotic as well as the endophytic bacteria which inhabit the internal tissues of the plants.

There are several mechanisms by which PGPBs promote growth and development in plants. One of the most direct mechanisms by which PGPBs promote plant growth is by facilitating the absorption of nutrients such as N, Fe, P, etc. Several PGPBs such as *Rhizobium* sp., *Azospirillum* sp., *Pantoea agglomerans*, *Frankia* sp.

and *Azoarcus* spp. are well known for fixing atmospheric nitrogen and facilitating its absorption by plants (Hill and Crossman 1983; Verma et al. 2001; Hurek et al. 2002). Phosphate-solubilizing bacteria such as *Achromobacter xylosoxidans* and *Bacillus pumilus* increase the availability of inorganic phosphorus to plants by solubilization and mineralization (Rodríguez and Fraga 1999; Richardson 2001). PGPBs also facilitate the uptake of iron by synthesizing low molecular weight molecules called siderophores, which has exceptionally high affinity for Fe^{3+} ions (Neilands 1981; Hider and Kong 2010). Inoculation of mung bean plants growing under iron-limiting conditions with siderophore producing *Pseudomonas* strain GRB3 showed reduced chlorotic symptoms and enhanced chlorophyll levels as compared to uninoculated plants (Sharma et al. 2003). In addition PGPBs have also been reported to produce plant hormones, thereby affecting plants' hormonal balance and stress response. Production of cytokinin and gibberellins is reported in case of several PSPBs such as *Azotobacter* spp., *Rhizobium* spp., *Pantoea agglomerans*, *Rhodospirillum rubrum* and *Pseudomonas fluorescens* (Tien et al. 1979; Williams and de Mallorca 1982; Atzorn et al. 1988; Taller and Wong 1989; Timmusk et al. 1999). Many strains of *Rhizobia* sp. have been reported to produce IAA which helps in nodule formation in plants (Mathesius et al. 1998; Badenoch-Jones et al. 1984).

PGPBs also promote growth of plants indirectly by acting as biocontrol agents and providing protection against pathogenic microbes. The underlying mechanism involves secretion of a range of antibiotics to prevent the proliferation of pathogens Mazurier et al. 2009; Raaijmakers et al. 2002; Haas and Keel 2003). Some PGPBs secrete lytic enzymes such as chitinases, cellulases, proteases and lipases which digests the cell walls of many pathogenic microorganisms mostly fungi such as *Botrytis cinerea*, *Sclerotium rolfsii*, *Fusarium oxysporum*, *Phytophthora* sp. (Ordentlich et al. 1988; Singh et al. 1999; Frankowski et al. 2001; Kim et al. 2008). PGPBs are also known to trigger a response called induced systemic resistance (ISR) in plants. ISR priming helps in activating defence mechanism via jasmonate and ethylene signalling in plants in response to infection by pathogenic microbes (Verhagen et al. 2004; Pieterse et al. 2009).

Many genera of rhizobacteria like *Bacillus subtilis*, *Burkholderia cepacia* and *Staphylococcus epidermidis* have been reported to promote plant growth by releasing volatile organic compounds (VOCs) (Ryu et al. 2003; Kai et al. 2007; Effmert et al. 2012; Bitas et al. 2013). Volatile organic compounds are low molecular weight compounds (<300 Da) that are capable to move over long distances through water-filled pores and porous soil structures by diffusion (Kai et al. 2007; Insam and Seewald 2010). VOCs are reported to be involved in inducing ISR in response of the non–pathogenic rhizobacteria (Van Loon et al. 1998; Lugtenberg and Kamilova 2009).

Bacterial endophytes have also been found to play a significant role in promoting plant growth and development. The plant growth-promoting bacterial endophytes (PGPBEs) are capable of influencing plant growth directly by biofertilization, i.e. facilitating nutrient absorption, and by phytostimulation, i.e. modulating the levels of growth-promoting plant hormones. They might also act indirectly by acting as biocontrol agents thereby protecting plants against various pathogenic organisms. PGPBEs thus act by three mechanisms, i.e. phytostimulation, biofertilization and biocontrol (Bloemberg and Lugtenberg 2001). Several PGPBEs that produce ACC deaminase have been known to enhance plant growth like *Arthrobacter* sp. and *Bacillus* sp. in pepper plants and *Pseudomonas putida* and *Rhodococcus* spp. in peas (Belimov et al. 2001; Sziderics et al. 2007).

Apart from bacteria, plant growth-promoting fungi (PGPFs) are also very beneficial for plant growth and development. PGPFs facilitate nutrient availability to plants, produce growth-promoting phytohormones such as auxin and also provide protection against several abiotic and biotic stresses (Pozo and Azcon-Aguilar 2007; Berendsen et al. 2012). PGPF also elicits induced systemic resistance (ISR) and mycorrhizal-induced resistance (MIR) by regulating various phytohormonal pathways to provide protection against pathogenic microbes (Pieterse et al. 1998; Pozo and Azcon-Aguilar 2007; Pozo et al. 2008; Van der Ent et al. 2009; Zamioudis and Pieterse 2012). Mycorrhiza, which is a symbiotic association of fungi with the roots of higher plants, has also been reported to significantly promote plant growth. Several genera of AMF like *Scutellospora*, *Gigaspora*, *Glomus*, *Acaulospora* and *Entrophospora* have been found to play important roles in plant growth and development. AMF facilitated plant growth by promoting the absorption of nutrients such as phosphorus (P), copper (Cu), zinc (Zn), etc. In addition, AMF also increases the water holding capacity of soil and water absorbing capacity of the root system by improving root architecture (Marulanda et al. 2003; Wu et al. 2012). The AMF association also acts as biocontrol agents and provides protection against root pathogens (Bagyaraj 2014; Bagyaraj et al. 2015).

These studies clearly highlight the significant role of rhizospheric microbes in facilitating growth and development of plants as summarized in Fig. 6.1.

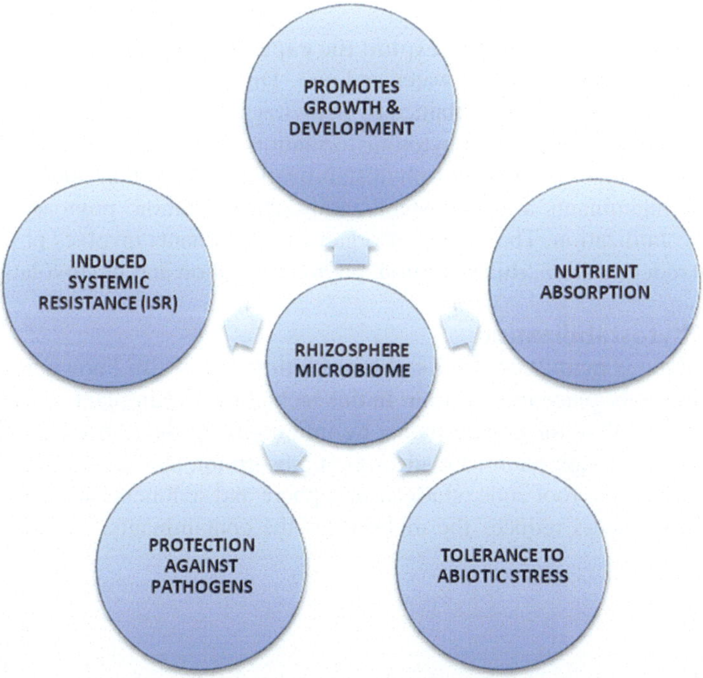

Fig. 6.1 Role of rhizospheric microbiome in plant growth and development

6.3 Phytoremediation Process

Phytoremediation is defined as an in situ remediation technology which reduces mobility, concentration, or toxicity of contaminants in contaminated environments (air, water and soil) through the use of living plants (USEPA 2000a; Cunningham and Ow 1996).This process is based on the combined action of selected or genetically modified plants, along with their associated microbes. Both plants as well as their associated microbes have the capability of direct uptake of contaminants (Macek et al. 2000). Phytoremediation can be used to remove both organic pollutants like petroleum, pesticides, polycyclic aromatic hydrocarbons (PAHs) and polychlorinated biphenyls (PCBs) and for inorganic pollutants like heavy metals and radionuclides, present in soil and water (Salt et al. 1998). This technique is most effective when concentration of contaminants ranges from low to medium levels, as high concentration levels of contaminant can inhibit activity and growth of microbes and plants (USEPA 2000a). Basic processes of this technique involve uptake, translocation, transformation, compartmentalization and mineralization (Schnoor et al. 1995).The chemical and physical properties of the compound (e.g. molecular weight, water solubility, etc.), abiotic conditions (e.g. pH, temperature, soil moisture content and organic matter) as well as plant characteristics (e.g. enzymes, root system, etc.) affect the uptake and translocation process (Suresh and Ravishankar 2004; Susarla et al. 2002).

6.3.1 Mechanisms of Phytoremediation

Different processes are used to exploit the capability of plants for the removal of contaminants from polluted water and soil. Different technological subsets of phytoremediation have been explained by Schwitzguebel (2000); these are phytoextraction (phytoaccumulation), phytotransformation, phytovolatilization, phytodegradation, phytorhizofiltration and phytostabilization. The uptake mechanisms for inorganic contaminants are phytostabilization, rhizofiltration, phytoaccumulation and phytovolatilization. The uptake of organic contaminants involves phytostabilization, rhizodegradation, rhizofiltration, phytodegradation and phytovolatilization.

6.3.1.1 Phytostabilization

This technique is mainly used for soil remediation and is also known as phytoimmobilization or in-place inactivation. In this technique certain plant species having tendency to stabilize the contaminants by immobilizing them in the groundwater and soil either through plant root adsorption, absorption and accumulation or precipitation within the root zone (sludge rhizosphere and sediment) are used (USEPA 2000a). This process reduces the mobility of the contaminants, minimizing their leaching into the groundwater which in turn reduces the bioavailability of these contaminants into the food chain, but they are not really removed from the soil (Erakhrumen 2007; Vangronsveld et al. 2009). Plant species induced changes in soil environment and chemistry, may result in contaminant's adsorption onto the root surface or soil or cause precipitation of metals onto the root. This process has been successful in remediation of soil and sediments from metals and other inorganic pollutants (EPA 2000). This process is useful for the treatment of heavy metals like cadmium (Cd), lead (Pb), arsenic (As), chromium (Cr), zinc (Zn) and copper (Cu). Absorption of these onto the root reduces the valence number which results in the complex formation further making them immobile. Several plants convert hazardous metals into a less toxic state by releasing redox enzymes, which reduces the metal stress and damage caused by it. For example, Chromium (VI) is converted into less toxic Chromium (III), which has less mobility (Wuana and Okieimen 2011; Ghosh 2010; Yoon et al. 2006).This technology has several advantages such as contaminants/biomass disposal is not required and is effective in preserving the ground- and surface water where rapid immobilization is needed (Zhang et al. 2009). Soil erosion is reduced due to plant cover, and it also reduces the amount of available water (USEPA 2000a).

6.3.1.2 Phytoextraction

This process is also known as phytoaccumulation, phytoabsorption or phytosequestration. Phytoextraction involves absorption of contaminants present in the surrounding environment (soil and water) by the roots of plant (soil and water). This is followed by translocation and accumulation of the contaminants in the above-ground parts (shoots, leaves, etc.) of the plant (USEPA 2000a; Rafati et al. 2011; Sekara et al. 2005). These above-ground parts can later on be harvested which removes the contaminants and gain energy by burning. It also helps in recycling of the metal from the ash (Erdei et al. 2005; USEPA 2000a). Phytoextraction is more suitable for removal of metals like zinc, nickel and copper as these are absorbed by majority of plants approximately 400 plants such as *Thlaspi caerulescens*, *Ipomea alpina*, *Astragalus racemosus* (Reeves and Baker 2000), most of which are Ni hyperaccumulators, while Zn, Cd, Co, Pb and Cu hyperaccumulators are less in number (McGrath et al. 2001). Phytoaccumulator plant must have the ability to extract large amount of contaminants (heavy metal) into their roots and translocate these into above-ground parts and should also be able to produce large amount of biomass. These plants are adapted to tolerate high concentrations of metal in their shoots and also have mechanisms to detoxify these metals. (Brennan and Shelley 1999).This process can also be applied for commercial extraction of metals by cropping in mineral industry (Sheoran et al. 2009).

6.3.1.3 Phytofiltration

This process uses plants for filtration purpose which inhibit the organic contaminants present in surface and wastewater from mixing into groundwater or the water streams, by absorption or adsorption of the contaminants (Padmavathiamma and Li 2007; Khan et al. 2007). Phytofiltration minimizes movement of contaminant movement in the soil (Memon et al. 2001; Sakai et al. 2012). This process can also be used for rhizofiltration in which plant roots are used to remediate contaminated groundwater, wastewater and surface water having low concentrations of contaminants (Mesjasz-Przybylowicz et al. 2004; Ensley 2000). It can be used for metals which mainly accumulate within the roots like Pb, Zn, Cd, Cu, Ni and Cr. Roots are harvested once they are saturated with contaminants (USEPA 2000a).

6.3.1.4 Phytovolatilization

In Phytodegradation plants and microorganisms are used for removal of organic contaminants like insecticides, herbicides, chlorinated solvents and inorganic contaminants from soil. This process detoxifies soil from organic contaminants by uptake of contaminants, metabolization followed by their degradation (Pivetz 2001; Garbisu and Alkorta 2001). These volatile forms of the contaminants are transpired into the air along with water vapour from the foliage (USEPA 2000a). Diffusion of pollutants also occurs from other parts of plant like stem, before these travels to the leaves (Raskin and Ensley 2000). It can be used for volatile organic contaminants like trichloroethene, as well as inorganic chemicals having volatile forms, such as arsenic and selenium (EPA 2000). This process does not eliminate the contaminants permanently as these pollutants are only transferred from one media to another which after sometime again gets deposited into soil (Padmavathiamma and Li 2007).

6.3.1.5 Phytodegradation

This process is also referred as phytotransformation. In this process plants and microorganisms are used to for removal of organic contaminants like insecticides, herbicides, chlorinated solvents and inorganic contaminants from soil detoxify soil from organic contaminants like chlorinated solvents, insecticides, herbicides and inorganic contaminants involves uptake of contaminants, metabolization followed by their degradation (Pivetz 2001; Garbisu and Alkorta 2001). The breakdown of organic contaminants occurs within plant tissue by enzymes, such as oxygenase and dehalogenase that help in their degradation. This process is used to decontaminate soil, sediment, sludge as well as ground and surface water (EPA 2000). Both plants as well as associated microbial community play an important role in attenuating pollutants. This process involves the degradation of contaminants internally or externally through metabolic processes of the plant (Prasad and Freitas 2003). In the rhizosphere (within plant root zone) the contaminants are also degraded with the help of plant associated microbes (Vidali 2001; Yang et al. 2005), and this is also known as rhizodegradation or phytostimulation. Rhizodegradation is effective in decontaminating the contaminated soil, mostly with organic chemicals like polycyclic aromatic hydrocarbons (PAHs), petroleum hydrocarbons, chlorinated solvents, pesticides, benzene, polychlorinated biphenyls (PCBs), toluene, xylenes and ethylbenzene (EPA 2000).

6.3.1.6 Vegetative Caps

In this process, plant cover is maintained over the polluted site which utilizes the evapotranspiration mechanism, resulting in prevention of leaching of pollutants from the disposal. Plants used for this process should have an ability of evapotranspiring the required amount of water. Soil should also have maximum water storage capacity with less capability of water drainage (Pivetz 2001).

6.3.1.7 Hydraulic Control

In this method, plants are used to control soil field capacity and water table by transpiring large amount of water which reduces infiltration. This in turn reduces leaching of contaminants therefore influences the movement of a groundwater plume. It therefore, enhances the upward water flow from the water table through the vadose zone. It is useful for a wide range of contaminants in sediment, soil or groundwater (EPA 2000; Pivetz 2001).

6.3.1.8 Advantages of Phytoremediation over Conventional Techniques

Phytoremediation is more cost-effective, environment-friendly clean-up technology driven by solar energy. Recycling of metal further provides economic advantages. It is effective for large areas where other approaches would be ineffective and expensive. It has both ex situ and in situ application. In situ application reduces the chances of disturbance of the surrounding system and soil. It also avoids transport of contaminated media, therefore reducing the spread of contamination. It is therefore a green technology having potential to detoxify variety of organic and inorganic contaminants. Using trees and creating green cover, this also provides aesthetic appeal. Implementation of this technology is relatively easy as it does not require specialized personnel or expensive equipment. It is effective in treating polluted sites contaminated with more than one pollutant, and disposal sites are not needed. It has potential to remediate bioavailable fraction of contaminants and prevent leaching of pollutants and soil erosion therefore reduces the secondary air- and waterborne contaminants.

However phytoremediation process has certain limitations which need to be taken into consideration. Phytoremediation is a lengthy process, thus may require several years for effective remediation. As this process depends on the rooting depth of detoxifying plants, therefore it is applicable only to surface soil. Also it is only applicable to sites with low pollutant concentrations. Due to the production of hazardous waste, it requires proper handling and disposal. Efficiency of this process can be reduced due to unfavourable climatic conditions which limit plant growth and biomass production. The use of non-native and invasive species affects the biodiversity of site. In extensively contaminated areas, vegetation conservation is complicated and the food chain also gets contaminated which threatens human health (Ghosh and Singh 2005; Hegedusova et al. 2009; Raskin and Ensley 2000; Lewis

2006; Prasad and Freitas 2003; USEPA 2000a, b; Rajakaruna et al. 2006; Pivetz 2001).

6.4 Bacterial Endophyte-Aided Phytoremediation

Endophytes are organisms that occupy plant organs, and their infection is generally not noticeable. Endophytes generally include endophytic bacteria, endophytic fungi and endophytic actinomycetes (Raghukumar 2008). Bacterial endophytes can provide several advantages to the host plant, mainly growth promotion and pathogen protection.

6.5 Diversity of Heavy Metal-Resistant Endophytes

Even though heavy metals are toxic to plants, it has been explained that many plants are metal resistant and that few of them are metal hyperaccumulators (Robinson et al. 2003; Rosa et al. 2004; Pulford and Watson 2003). The hyperaccumulator endophytes and many metal-resistant endophytes were extracted from various hyperaccumulating plants, such as *Alyssum bertolonii*, *Alnus firma*, *Brassica napus*, *Nicotiana tabacum*, *Thlaspi caerulescens*, *T. goesingense* and *Solanum nigrum* (Table 6.1). The reported metal-resistant endophytes include *Arthrobacter*, *Bacillus*, *Clostridium*, *Curtobacterium*, *Enterobacter*, *Leifsonia*, *Microbacterium*, *Paenibacillus*, *Pseudomonas*, *Xanthomonadaceae*, *Staphylococcus*, *Stenotrophomonas*, and *Sanguibacter* (Table 6.1).

The endophytes play an important role in host plant adjustment to polluted environments and that they can improve phytoremediation by mobilizing/degrading or immobilizing contaminants in the soil, enhancing plant growth, reducing phytotoxicity and improving plants' metal tolerance as well as in other ways (Zhang et al. 2011a; Weyens et al. 2010; Chen et al. 2010; Germaine et al. 2009).

Plant growth-promoting bacteria belonging to genera like *Bacillus*, *Beijerinckia*, *Burkholderia*, *Erwinia*, *Enterbacter*, *Flavobacterium*, *Gluconacetobacter*, *Klebsiella*, *Pseudomonas* and *Serratia* are important for improving plant growth (Delvasto et al. 2009; Dardanelli et al. 2010; Nadeem et al. 2010). PGP bacteria may enhance plant growth through several ways like phytohormone production, synthesis of chelating agents and siderophores, organic acid, fixation of atmospheric nitrogen and specific enzyme activity (Jing et al. 2007; Kamnev and Lelie 2000). In addition, some PGP bacteria also have the potential to cause reduction in ethylene production that causes improved plant growth. For example, endophytic bacteria *Serratia nematodiphila* LRE07, *Enterobacter* sp. LSE04, *Enterobacter aerogenes* LRE17 and *Acinetobacter* sp. LSE06 from *Solanum nigrum*, endophytic bacteria *Enterobacter* sp. 12 J1 from *Allium macrostemon* and endophytic bacteria *Acinetobacter*, *Agrobacterium tumefaciens*, *Bacillus* sp., *B. subtilis* and *B. megaterium* from *Commelina communis* can synthesize IAA to enhance plant growth and improve phytoremediation (Chen et al. 2010; Sheng et al. 2008a; Zhang et al.

Table 6.1 Bacterial endophytes involved in phytoremediation

Endophytic bacteria	Host plant	Metal	References
Ochrobactrum sp., *Bacillus* sp.	*Oryza sativa*	Cd, Pb and As	Pandey et al. (2013)
SAN 1 and SAN 2 (rhizobacterial isolates)	*Triticum aestivum* L.	Cd	Hassan et al. (2016)
Pseudomonas putida	*Eruca sativa* L.	Cd	Kamran et al. (2015)
*Pseudomonas marginalis*Dp1, *Pseudomonas brassicacearum* Am3, and *Rhodococcus* ssp.	*Pisum sativum* L.	Cd	Safronova et al. (2006)
Enterobacter (CIK-521R), *Leifsonia* (CIK-521), Klebsiella (CIK- 518) and *Bacillus* (CIK-517, CIK-519)	*Zea mays* L.	Cd	Ahmad et al. (2016)
Rhodococcus erythropolis EC 34, *Achromobacter* sp. 1AP2 and *Microbacterium* sp. 3ZP2	*Trifolium repens* L.	Zn and Cd	Pereira et al. (2015)
Pseudomonad strains (MKRh1, MKRh3, and MKRh4) *Pseudomonad* strains, namely, MKRh1, MKRh3 and MKRh4	Black gram	Cd	Ganesan (2008)
Rhizobacterial isolates, namely, (ACC8, ACC5, AN8, AN12, *Azotobacter* and RN5)	*Zea mays* L.	Pb	Hassan et al. (2014)
Variovorax paradoxus, *Rhodococcus* sp. and *Flavobacterium* sp.	*Brassica juncea* L.	Cd	Belimov et al. (2015)
Bacillus mucilaginosus (HKK-1), *Bacillus megaterium* (HKP-1), *Azotobacter chroococcum* (HKN-5)	*Brassica juncea* L.	Pb, Zn	Wu et al. (2006)
Xanthomonas ssp., *Azomonas* sp., *Pseudomonas* sp., *Bacillus* sp. and *Bacillus* sp.	*Brassica napus* L.	Cd	Sheng and Xia (2006)
Bacillus subtilis	*Brassica juncea L.*	Ni	Zaidi et al. (2006)
Pseudomonas sp., *Bacillus* sp.	*Brassica juncea L.*	Cr	Rajkumar et al. (2006)
Pseudomonas putida	*Brassica napus* L.	Ni	Farwell et al. (2006)
Brevibacillus	*Trifolium repens* L.	Zn	Vivas et al. (2006)
Microbacterium oxydans, *Rhizobium galegae*, *Microbacterium oxydans*, *Clavibacter xyli*, *Acidovorax avenae*, *Microbacterium arabinogalactanolyticum*, *M. oxydans*, *M. arabinogalactanolyticum* and *M. oxydans*	*Alyssum murale* L.	Ni	Abou-Shanab et al. (2006)
Pseudomonas putida KNP9	*Phaseolus vulgaris*	Cd and Pb	Tripathi et al. (2005)
Pseudomonas	Chickpea	Ni	Tank and Saraf (2009)
Azotobacter spp.	*Triticum aestivum*	Cd and Cr	Joshi and Juwarkar (2009)

(continued)

Table 6.1 (continued)

Endophytic bacteria	Host plant	Metal	References
Glomus mosseae	Sorghum	Cu, Pb and Cd	Gonzalez-Chavez et al. (2004)
Cellulosimicrobium cellulans KUCr3	Chilli plants	Cr	Chatterjee et al. (2009)
Stenotrophomonas maltophilia	*Astragalus bisulcatus*	Se	Di Gregorio et al. (2005)
Serratia sp.MSMC541	*Lupinus uteus*	As, Cd and Cu	Aafi et al. (2012)

2011b). Several endophytes produce cytokinins and/or gibberellins, which can enhance growth of plants and modify plant morphology or plant phenotype under both stressed and non-stressed conditions (Feng et al. 2006; Oelmüller et al. 2009; Hamayun et al. 2010). Bacterial 1-aminocyclopropane-1-carboxylic (ACC) deaminase can reduce the amount of ACC, an ethylene precursor, by hydrolyzation, ensuing in a reduction of ethylene secretion (Hardoim et al. 2008). Some endophytes can also enhance plant growth through nitrogen (N) fixation. A good example is the nitrogen-fixing endophytes of sugarcane, which provide abundant N to the plant and improve plant growth (Muthukumarasamy et al. 2002). Many other plants also like rice, maize, wheat, poplar and grass have been occupied by nitrogen-fixating endophytes (Rajkumar et al. 2009; Dory et al. 2009; Hurek and Reinhold-Hurek 2003).

In addition to plant growth-promoting potential, some endophytes can also reduce metal toxicity through several ways including, production of iron chelators, siderophores, organic acids and various degrading enzymes (Saravanan et al. 2007; Sheng et al. 2008b; Soleimani et al. 2010; Yousaf et al. 2010). Some endophytic bacteria also solubilize mineral phosphate. For example, Kuklinsky-Sobral et al. (2004) established that 52% of the endophytic bacteria extracted from soybean plants can solubilize mineral phosphate. Similarly, Puente et al. (2009) stated that endophytic bacteria in cacti seeds can enhance the development of cactus seedlings on barren rock by increasing the removal of P_2O_5, Fe_2O_3, K_2O and MgO from the substrate. Some endophytes have been involved in heavy metal mobilization through the production of low-molecular-mass organic acids. Another example is the secretion of 5-ketogluconic acid by endophytic diazotroph *Gluconacetobacter diazotrophicus*, which helps in solubilization of various Zn sources of Zn such as $Zn_3(PO_4)_2$, ZnO or $ZnCO_3$.This process thus make Zn readily available for uptake by plants (Saravanan et al. 2007). Kuffner et al. (2010) also demonstrated that some *Actinobacteria* endophytes could synthesize metal-mobilizing metabolites into the contaminated soil and increase the metal accumulation in the leaves of *Salix caprea*. Chen et al. (2010) explained that four heavy metal-resistant endophytic bacteria enhanced Cd accumulation in root, stem and leaf tissues of *Solanum nigrum* L. growing in three different levels of Cd-contaminated soil; they also explained that the accumulation potential vary with the Cd concentration in the soil. Similarly, Mastretta et al. (2009) found that the inoculation of *Nicotiana tabacum* with

Cd-resistant endophyte *Sanguibacter* sp. improved the concentration of Cd in shoot tissues. Sheng et al. (2008a) also demonstrated that the inoculation of *Brassica napus* with Pb-resistant endophytic bacteria improved Pb uptake into the shoot. An increased uptake of Cd in aerial parts of plants was observed, when *Brassica napus* plants were inoculated by *Pseudomonas tolaasii* ACC23, *Pseudomonas fluorescence* ACC9 and *Mycobacterium* sp. ACC14 (Dell'Amico et al. 2008). *Microbacterium arborescens* extracted from rhizosphere of *Prosopis juliflora* when inoculated to rye grass conferred tolerance against Cr, Cd, Zn, Cu and Pb. Plants inoculated with *Microbacterium arborescens* strains led to higher amount of metals accumulating in root and shoot. In another example, inoculation of *Frankia* sp. reduced accumulation of Ni and enhanced nodulation in *Alnus glutinosa* (Wheeler et al. 2001). Isolation of *Streptomyces mirabilis* P16B-1 from soil of uranium mining area showed tolerance against several metals including Zn, Co, Cd, Ni and Cu. When *Sorghum bicolor* was inoculated with this strain, there was limited bioavailability of metal. Similarly, *Rhodococcus erythropolis* isolated from psychrotrophic metal-contaminated soil considerably decreased Cr uptake and enhanced plant growth in *Pisum sativum* plants. The enhancement in plant growth was due to P solubilization and IAA production potential of *Rhodococcus erythropolis* (Trivedi et al. 2007).

The production of siderophores pyoverdine and pyochelin by rhizosphere bacteria *Pseudomonas aeruginosa* raised the Cr and Pb concentration in the rhizosphere, thus making them accessible for maize uptake (Braud et al. 2009a). Braud et al. (2009b) reported that addition of heavy metals, Al, Cu, Ga, Mn and Ni in iron-limited succinate medium causing pyoverdine production in the *Pseudomonas aeruginosa*. For example, production of the siderophores pyoverdine and pyochelin reduced the toxicity of Al, Co, Cu, Ni, Pb and Zn in *P. aeruginosa* (Braud et al. 2010), but pyochelin enhanced the toxicity of vanadium in *P. aeruginosa* (Baysse et al. 2000). Organic acids secreted by plant-associated microbes play a key role in the complexation of toxic and essential ions and enhance their mobility for plant uptake. *Pseudomonas fluorescens* G10 and *Microbacterium* sp. G16, which are metal-resistant endophytic bacteria, have also been reported to increase the Pb accumulation of Pb in rape by secretion of organic acid (Sheng et al. 2008a). In addition, maize roots were able to dissociate Cd from Cd-organic acid complex by a process mediated by root surface and thereby led to enhanced Cd uptake. Their results specify that free Cd ions were significantly available to maize roots than intact Cd-organic ligand complexes. Panfili and co-workers (2009) also reported a similar effect on the Cd uptake and availability to durum wheat roots in the presence of citric acid. Another significant metabolite that has capacity to increased metal mobilization and phytoremediation is endophytically produced surfactants called biosurfactants. Juwarkar et al. (2007) analysed the metal mobilization capacity of *P. aeruginosa* BS2 under in vitro column experiments and demonstrated the production of a biosurfactant and rhamnolipid that helps in the solubilization of Cd and Pb compounds from artificially metal-contaminated soil. Similarly Venkatesh and Vedaraman (2012) calculated the potential of rhamnolipids produced by *P. aeruginosa* to mobilize Cu in contaminated soils. Sheng et al. (2008b) assessed the potential of the biosurfactant-producing bacterial strain *Bacillus* sp. J119 in enhancing the Cd uptake by sudangrass, rape, maize and tomato in soil artificially contaminated with several different levels of Cd. They found that the inoculation of live bacterium *Bacillus* sp. J119 to soils considerably enhanced the plant Cd uptake when compared to dead bacterium-inoculated control.

The secretion of extracellular polymeric substances (EPS), mucopolysaccharides and proteins by endophytic microbes can also play a key role in complexing toxic metals and in reducing their mobility in the soils. Joshi and Juwarkar (2009) demonstrated the immobilization of Cd and Cr after inoculation of EPS producing *Azotobacter* spp. and explained that these isolates were able to bind 15.2 mg g^{-1} of Cd and 21.9 mg g^{-1} of Cr. Besides they demonstrated that the inoculation of *Azotobacter* to the metal-contaminated soils decreased Cd (−0.5) and Cr (−0.4) uptake by *Triticum aestivum*.

6.6 Fungi-Assisted Phytoremediation

Roots of majority of land plants are inhabited by arbuscular mycorrhizal fungi (AMF; fungal hyphae enter the cortical cells of roots and form arbuscules), ectomycorrhizal fungi (EMF; fungal hyphae do not enter the cell but grow around the roots and form Hartig net) and fungal endophytes. AMF (mostly belonging to *Glomeromycota*) helps plant in acquiring nutrients from the soil and also protects plant from different abiotic and biotic stresses and in exchange gets photoassimilates from the host plant (Smith and Smith 2011; Bücking and Kafle 2015; Buscot 2015). It is believed that EMF largely belongs to *Basidiomycetes* and *Ascomycetes* with few members belonging to *Zygomycetes* and has been found to be associated with the roots of woody plants and exchange soil nutrients with carbohydrates (Tedersoo et al. 2010; Buscot 2015). These mycorrhizal associations have also been shown to assist in phytoremediation (Hildebrandt et al. 2007; Upadhyaya et al. 2010; Colpaert et al. 2011; Luo et al. 2014). The fungal hyphae have large surface area and act as sinks of heavy metals. Reports have also suggested that fungal vesicles or spores store the toxic compounds and detoxify them (Gohre and Paszkowski 2006; Cornejo et al. 2013). It is also believed that AMF releases a glycoprotein, glomalin, which helps sequester the heavy metals and hence confers tolerance to the host (Gonzalez-Chavez et al. 2004; Singh 2012).

In a report by Fomina et al. (2005), the EMF has been shown to solubilize Cd, Cu, Pb and Zn metals. Some of the EMF species are considered as ‘rock eating’ fungi and conferred Al tolerance by excreting oxalate (van Scholl et al. 2008).

Glomus intraradices assisted *Helianthus annuus* showed enhanced chromium (Cr) accumulation and Cr tolerance (Davies et al. 2001). *Pisum sativum* plants associated with AMF *Glomus intraradices* were found to be tolerant to Cd stress (Rivera-Becerril et al. 2002). Increased biomass and nutrient uptake was observed in *Zea mays* plants (grown in various depth of soil covered with coal fly ash) inoculated with AMF *Glomus mosseae* and *Glomus versiforme* (Bi et al. 2003). Dual inoculation of *Glomus deserticola* (AMF) and *Trichoderma koningii* (saprobe fungi) promoted the accumulation of Cd in *Eucalyptus globulus* (Arriagada et al. 2004). *Glomus mosseae* AMF inoculated *Phaseolus vulgaris* and *Triticum aestivum* displayed enhanced metal accumulation on soil artificially contaminated with Cu, Zn, Cd and Pb (Rabie 2005). *Glomus mosseae* AMF infected *Pteris vittata* plants displayed increased As uptake in As containing soil or solution (Trotta et al. 2006). Dual inoculation of AMF and nitrogen fixing *Rhizobia* on the host plant *Vigna sinensis* (cow pea) lead to increased dry weight, root/shoot ratio, leaf number, leaf pigments and N and P levels under heavy metal stress conditions (Al-Garni 2006).

On diesel contaminated soil, *Melilotus albus* associated with *Glomus* sp. AMF increased plant biomass, antioxidant activity and nutrient uptake (Hernández-Ortega et al. 2012). Inoculation of *Glomus etunicatum* to *Calopogonium mucunoides* increased the biomass and nutrient uptake (P, S and Fe) and Pb tolerance (De Souza et al. 2012). The association of AMF *Glomus mosseae* with two genotypes of *Cajanus cajan* exposed to Cd and Zn, promoted biomass production and combated oxidative stress by enhancing the antioxidant enzymes (superoxide dismutase, glutathione reductase, peroxidase and catalase) activity (Garg and Kaur 2013).

Sunflower plants infected with AMF *Rhizophagus irregularis* exhibited higher shoot Cd and Zn levels on soil polluted with low Cd concentration (Hassan et al. 2013). Three AMF species *Glomus mosseae*, *Glomus geosporum* and *Glomus versiforme* when inoculated on *Oryza sativa* increased the uptake of arsenic (Chan et al. 2013). AMF inoculation alleviated oxidative stress induced due to trace metal toxicity in *Miscanthus* grown in highly polluted soil (Firmin et al. 2015).

Fungal endophytes are microorganisms that live inside the plant organs for completing their life cycle and do not cause any harm to the plant (Wilson 1993). They gain entry into a particular host plant through flower, leaf, root and stem. This mutualistic interaction between plants and fungi can be found in all the ecosystems. Fungal endophytes have been categorized in to two classes: C- endophytes (C for clavicipitaceous) and NC-endophytes (NC for non-clavicipitaceous targets). C-endophytes mostly infect grass species while NC-endophyte target nonvascular plants (Arnold 2007). Rodriguez et al. (2009) further divided endophytic fungi into four classes based on colonization, host range, tissue specificity and transmission patterns.

The plants in this association have many advantages like enhanced fitness, ecological adaptation and tolerance against abiotic and biotic stresses (Arnold 2007; Rodriguez et al. 2009; Bourdel et al. 2016; Meena et al. 2017). These endophytes have been shown to up-/downregulate various phytohormonal pathways for promoting plant growth (Glick 2014; Deng et al. 2017).Endophytic fungi can also synthesize various metabolites which have a thermo-protective function (Redman et al. 2002). The endophytic fungi also help plants to grow on soils contaminated with different pollutants like heavy metals, radionuclides, pesticides and different organic compounds and thus aid in phytoremediation. They are equipped with different mechanism for detoxification, sequestration, chelation or degradation of contaminants and also regulate plant growth by synthesis of different plant growth regulators. It is today the most cost-effective and environment-friendly approach for phytoremediation (Aly et al. 2011; Khan and Doty 2011; Deng and Cao 2017).

Many AMF and endophytic fungal species (*Glomus*, *Alternaria*, *Phoma*, *Fusarium*, *Aspergillus*, *Mucor*, etc.) have been identified at the metal-contaminated sites suggesting their role in assisting phytoremediation (Ortega-Larrocea et al. 2010; Li et al. 2012; Deng et al. 2014a, b). These endophytes have been shown to up-/downregulate various phytohormonal pathways thereby promoting plant growth (Glick 2014).

Glycine max plants assisted with *Penicillium funiculosum* LHL06 under Cu stress showed increased biomass, nutrient uptake, proline and glutamine levels and metal extraction efficiency. The association also reduced the electrolyte leakage and lipid peroxidation and ABA levels (Khan and Lee 2013). *Glycine max* plants inoculated with *Paecilomyces formosus* showed improved growth in Ni-contaminated soils. Under Ni stress, the plants inoculated with the fungus showed decreased lipid peroxidation, increased accumulation of glutathione, peroxidase, catalase, polyphenol oxidase, superoxide dismutase and linolenic acid accumulation and downregulation of stress-induced hormones, jasmonic acid and abscisic acid (Bilal et al. 2017).

Infection of *Lolium perenne* with the endophytic fungus *Neotyphodium lolii* conferred Zn stress tolerance (Monnet et al. 2001). *Trichoderma atroviride* mitigated Cd and Ni toxicity when inoculated on *Brassica juncea* (L.) Coss.var. *foliosa* grown on metal-contaminated soil (Cao et al. 2008). *Brassica napus* plants inoculated with the metal (Cd, Pb, Zn and Cu)-resistant *Rhodotorula* sp. CBSB79 displayed increased biomass improved metal phytoextraction capacity on multi-metal-contaminated soils (Wang et al. 2013). *Penicillium* sp. CBRF65 and *Fusarium* sp. CBRF44 conferred metal tolerance to *Brassica napus* plants grown on metal-contaminated soil by synthesizing auxin and siderophores, solubilizing phosphate, increasing biomass and extracting Cd and Pb (Shi et al. 2017). Khan et al. (2017a, b) tested different endophytic fungi on *Solanum nigrum* and observed improved growth parameters in terms of root and shoot length, chlorophyll content and biomass under cadmium stress. Heavy metal-tolerant fungi, *Trichoderma* sp. PDR-416 and *Penicillium aculeatum* PDR-4, when applied to *Sorghum bicolor* × *sudanense* (Sorghum-sudangrass) increased its biomass, chlorophyll content and concentration of As, Cu, Pb and Zn in roots and As, Pb and Zn in shoots (Babu et al. 2014a). The same group also reported increased concentration of Cu and Pb in maize shoots and roots, increased biomass and increased chlorophyll content when inoculated with *Trichoderma* sp. PDR-28 (Babu et al. 2014b).

A metal (Cd, Pb, Zn and Cu)-tolerant yeast *Cryptococcus* sp. CBSB78 increased the biomass extraction quantity of Pb, Cu and Zn and survival frequency of *Brassica alboglabra* on multi-metal-polluted soils (Deng et al. 2012). On Pb- and Cd-contaminated soils, another metal (Zn, Cd and Pb)-resistant fungus, *Lasiodiplodia* sp. MXSF31, extracted from the stem of *Portulaca oleracea* increased the extraction levels of Cd by *Brassica napus* and its biomass (Deng et al. 2014b). Engineered self-fusion product of *Mucor* sp. CBRF59 increased the biomass and Cd extraction by *Brassica napus* plants in Pb- and Cd-contaminated soil (Deng et al. 2013). Interspecific fusion product of *Mucor* sp. (Cd-resistant) CBRF59 and *Fusarium* sp. CBRF14 (Zn-resistant) enhanced growth and metal extraction, concentration and metal movement from root to shoot in *Brassica napus* plants grown on multi-metal-polluted soil (Deng et al. 2014a).

Acacia auriculiformis associated with fungi, *Trichoderma* H8 and *Aspergillus* G16, when applied to *Brassica juncea* (L.) Coss.var. *foliosa* Bailey in Ni- and Cd-contaminated soil lead to increased yield, translocation factor, phytoextraction and bioconcentration factor of the mustard (Jiang et al. 2008).

Festuca pratensis and *Festuca arundinacea* grasses when inoculated with *Neotyphodium uncinatum* and *Neotyphodium coenophialum* fungi exhibited increased biomass and synthesized higher amount water soluble phenols and showed enhanced degradation of total petroleum hydrocarbons in petroleum-contaminated soil (Soleimani et al. 2010). Ren et al. (2011) studied the effect of *Neotyphodium coenophialum* fungus on *Lolium arundinaceum* grass grown in Cd containing solution. The infection increased the Cd buildup, biomass and transport from root to shoot under Cd stress (Ren et al. 2011).

Penicillium janthinellum LK5 increased the aluminium phytoextraction potential and salicylic concentration and reduced oxidative stress of *Solanum lycopersicum* under Al stress (Khan et al. 2015).

A broad host range fungus *Phomopsis liquidambari* promoted plant seedling growth, accelerated litter decomposition for the release of nutrients (Chen et al. 2013). The same fungus also exhibits polycyclic aromatic hydrocarbon degradation capacity suggesting its role in bioremediation (Dai et al. 2010). *Fusarium oxysporum*, a non–pathogenic fungus, increased the biomass, metal uptake and translocation and improved the phytoextraction potential of heavy metals (Zn, Pb, Cd and Cu) of *Sedum alfredii* (Zhang et al. 2012).

All these reports suggest to the fact that fungal-assisted phytoremediation delivers an alternative way to improve bioremediation of metal-polluted soils. Few of the endophytic fungi involved in alleviating phytoremediation have been listed in Table 6.2.

6.7 Role of Phytoremediation in Removal of Pollutants

Phytoremediation method plays very important role in removal of various toxic contaminants and pollutants from the environment. It is one of the most preferred green technologies which very efficiently remove harmful contaminants from soil, air and water resources. Some of the reports of successful remediation of toxic pollutants using plant-microbe interaction have been discussed below.

6.7.1 Phytoremediation of Soil Pollutants

Major pollutants of soil are urban wastes, industrial wastes, agricultural practices, biological agents, heavy metals and radioactive pollutants. According to Duffus (2002), any metal whose atomic number is e > 20 is considered as heavy metal. Also the metallic elements of relatively high molecular weight are known to be heavy

Table 6.2 Fungal endophytes involved in phytoremediation

Endophytic fungus	Host plant	Metal	References
Neotyphodium lolii	*Lolium perenne*	Zn	Monnet et al. (2001)
Trichoderma H8 and *Aspergillus G16*	*Brassica juncea*	Ni and Cd	Jiang et al. (2008)
Neotyphodium uncinatum and *Neotyphodium coenophialum*	*Festuca pratensis* and *Festuca arundinacea*	Petroleum	Soleimani et al. (2010)
Neotyphodium coenophialum	*Lolium arundinaceum*	Cd	Ren et al. (2011)
Cryptococcus sp. *CBSB78*	*Brassica alboglabra*	Pb, Cu and Zn	Deng et al. (2012)
Fusarium oxysporum	*Sedum alfredii*	Zn, Pb, Cd and Cu	Zhang et al. (2012)
Penicillium funiculosum LHL06	*Glycine max*	Cu	Khan and Lee (2013)
Penicillium aculeatum PDR-4 and *Trichoderma* sp. *PDR-*	*Sorghum bicolor* × *sudanense*	As, Cu, Pb and Zn	Babu et al. (2014a)
Trichoderma sp. *PDR-28*	*Zea mays*	Cu and Pb	Babu et al. (2014b)
Lasiodiplodia sp. *MXSF31*	*Brassica napus*	Cd	Deng et al. (2014b)
Penicillium janthinellum LK5	*Solanum lycopersicum*	Al	Khan et al. (2015)
Paecilomyces formosus	*Glycine max*	Ni	Bilal et al. (2017)
Penicillium sp. *CBRF65* and *Fusarium* sp.	*Brassica napus*	Cd and Pb	Shi et al. (2017)

metal, such as Cu, Cd, Cr, Pb, Hg and Zn. These are one of the components in soil (Lasat 2000). Some of these are important micronutrient for plant growth, like Cu, Zn, Ni, Mn and Co, while Pb, Cd and Hg have biological function (Gaur and Adholeya 2004). Higher concentration of these metals has detrimental effects on the environment. Presence of heavy metals deprives the soil of microbial diversity and nutrients. Plant growth and development is affected due to the presence of high concentration of heavy metals (Carlson et al. 1991; White et al. 2006).

As these are not chemically degraded and are either converted into nontoxic compounds or physically removed (Gaur and Adholeya 2004).The soil problems aggravates due to disposal of these metals (Vidali 2001). Heavy metals also have threatening effects on the human health (Khan et al. 2010). Therefore, phytoremediation is critical for removal of the heavy metals from the environment. Around 400 plant species belonging to 45 families such as Lamiaceae, Asteraceae, Caryophyllaceae, Brassicaceae, Cyperaceae, Fabaceae, Cunouniaceae, Flacourtiaceae, Poaceae, Euphobiaceae and Violaceae have been reported to hyperaccumulate heavy metal such as Co, Cd, Cu, Ni, Mn, Pb, Sb, Zn, Se and Tl (Lasat 2000; Ghosh and Singh 2005). Therefore, by using hyperaccumulator plant species in phytoremediation techniques, efficiency for removal of the heavy metals from contaminated sites increases (Padmavathiamma and Li 2007). This technique is effective in remediation of water, soil and sediments from heavy metal by using naturally occurring or genetically modified plants (Flathman and Lanza 1998). Yoon et al. (2006) reported that *Phyla nodiflora was* very efficient in accumulating Zn and Cu in its shoots and therefore was appropriate for phytoextraction, whereas *Gentiana pennelliana* was most suitable for phytostabilization of sites contaminated with Cu, Pb and Zn. The first field trial was conducted on phytoextraction of cadmium (Cd) and zinc (Zn) by Baker and Brooks (1989). Caçador and Duarte (2015) reported the phytotransformation of toxic form of Cr (VI) to Cr (III) which is less toxic by halophytes. Microorganisms can also transform toxic metals to their less toxic states. Blake et al. (1993) reported that *Pseudomonas maltophilia* facilitates the transformation and precipitation of various toxic oxyanions and metal cations at a toxic waste site in Oak Ridge in Anderson County, Tennessee. Lin et al. (2015) recently reported *Youngia erythrocarpa* (a farmland weed) as a new Cd hyperaccumulator plant. Most of the Ni hyperaccumulator plants belong to the family Brassicaceae and particularly *Alyssum* and *Thlaspi* (Rascio and Navari-Izzo 2011).Symbiotic bacteria also play important role in facilitating plant growth in poor soil with metal accumulation. For example, Grison et al. (2014, 2015) recently isolated symbiotic heavy metal-resistant bacterium *Rhizobium metallidurans* sp. nova, from a Zn-hyperaccumulating legume *Anthyllis vulneraria.* On inoculation of these bacteria, there was an increase of up to 36% of Zn concentration in the shoots of in this legume. The metal accumulation capacity is improved by adding acidifying agents or chelaters in the soil when contaminants are in unavailable form. Ethylene diamine disuccinate (EDDS) increases the efficiency of metal accumulation in shoots of *Brachiaria decumbens* by solubilizing metals from the soil (Yeh and Pan 2012; Santos et al. 2006).It was reported that the cereal crops have high ability of heavy metal accumulation, for example, sorghum (*Sorghum bicolor*), maize (*Zea mays* L.), and alfalfa (*Medicago sativa* L.) (Vijayarengan 2005). These plants therefore can be successfully used to detoxify the heavy metal-polluted soils (Ebbs et al. 1998). Ornamental plants *Tagetes patula* and *Syngonia* sp. are reported to have arsenic accumulating property and therefore can be exploited to detoxify arsenic-contaminated soil (Huq et al. 2005). Bacterial genes, such as merA (for mercuric reductase) and merB (for organomercurial lyase), were transferred to *Arabidopsis thaliana*. This transgenic plant has an ability to detoxify organic Hg in the soil (Bizily et al. 2000). In a study of mine soils in France, Frérot et al. and coworkers (2006) reported an increased biomass of plants when a mixture of nonleguminous and leguminous plants like *Anthyllis vulneraria* were used.

6.8 Phytoremediation of Air Pollutants

Particulate matters (PMs), volatile organic compounds (VOCs), ground-level ozone (O_3), nitrogen dioxides (NO_2) and sulphur dioxide (SO_2) represent the major air pollutants posing serious threat. Particulate matter (PM) is a mixture of liquid and solid substances having different origins, chemical compositions and shapes (Pastuszka

2007). Particulate matter consists of a relatively nonreactive part such as calcium or carbon, to which chemicals like organic compounds (e.g. PAHs), environmentally persistent free radicals (EPFRs) and toxic metals are adsorbed making them more toxic. As a result particulate matter represents one of the most hazardous pollutants for human health (Masiol et al. 2012; Gehling and Dellinger 2013; Louwies et al. 2015). Significant amounts of PM are scavenged by plants as their foliage adsorb PM (sPM) or stabilize them in waxes (wPM) (Saebo et al. 2012; Dzierzanowski et al. 2011).Trees and shrubs act as biofilter for PM flow, by accumulation of PM on the foliage, thus reducing its amount by 50% (Popek et al. 2012). Plant-specific characters which affects PM scavenging capacity includes leaf structure and size, pubescence, surface roughness, wax content, thickness and ultrastructure, along with the climatic conditions such as air and precipitation, composition of PM and its quantity (Nowak et al. 2006; Wang et al. 2015). Increased biomass results in increased surface area for PM absorbance. Indoor air is also polluted with PM (WHO 2000). Gawrońska and Bakera (2015) reported that indoor plant *Chlorophytum comosum* is capable of accumulating PM.

Air quality both indoor and outdoor is negatively affected by the presence of volatile compounds (VOCs) such as alcohols, aldehydes, aliphatic, alicyclic and aromatic hydrocarbons, chlorinated hydrocarbons, esters, glycol ethers, ketones phenolics, terpenes, etc. (Hess-Kosa 2002). Although VOCs have role in O_3 formation, they also have adverse effects on human health (Bluyssen et al. 2011). Ugrekhelidze et al. (1997) reported that plants can remove toluene and benzene from air and can assimilate it within different tissues. The diffusion of the VOC into the foliage is influenced by several factors such as stomatal characteristics, hair growth and wax layer. A lipophilic VOC like benzene can easily diffuse through the cuticle, whereas a hydrophilic VOC like formaldehyde has difficulty in penetration. It was reported that enhanced catalase activity and higher stomatal density in *Melissa officinalis* as compared to *Hedera helix* resulted in higher efficiency of removal of formaldehyde (Jin et al. 2013). Microbes associated with the above- and below-ground plant parts as well as with the root growth can also facilitate the VOCs phytoremediation by synthesis of biosurfactants or biofilm formation, therefore enhancing the bioavailability of VOCs for the plant (Ramos et al. 2009). De Kempeneer et al. (2004) reported that the time for removal of VOCs decreased when *Azalea indica* leaves were inoculated with *Pseudomonas* TVA8, as compared to control plants without inoculums. Chun et al. (2010) reported that there was high removal rate of toluene, benzene, o-xylene and m/p-xylene by plants when inoculated with bacteria isolated from nutrient-rich growing media. Weyens et al. (2015) reported an overview of research on phyllosphere microorganisms in the framework of VOC. Similarly, Dela Cruz et al. (2014) reported more than 100 indoor plants having the capability of removing VOCs. Genetic modification of plants also resulted in increased efficiency and removal rate of VOCs. Genetically modified *Nicotiana tabacum* plants with the mammalian Cytochrome P450 2E1 gene resulted in enhanced removal rate of VOC (James et al. 2008). Sawada and co-workers (2007) reported that genetically modified tobacco plants with the bacterial genes for two key enzymes in the monophosphate pathway showed 20% higher formaldehyde removal capacity as compared to wild-type plants.

6.9 Phytoremediation of Water Pollutants

The sources of water include surface water and groundwater. Misuse and overuse of these resources lead to danger. These water resources can be contaminated by POPs (persistent organic pollutants), pesticides (eg. DDT), heavy metals, inadequately or untreated municipal sewage and nutrients (agricultural runoff, domestic wastewater, industrial effluents etc) resulting in increased level of nutrients which in turn leads to eutrophication. All these contaminants have negative effect on health (Mann et al. 2014). Crude oil spillage has negative impacts on soil, water and ecosystems as well as to humans (Liang et al. 2009). Petroleum pollutants cause environmental degradation due to inherent properties like solubility, volatility, biodegradability and affects humans as well as aquatic life forms. Petroleum pollutants include hydrocarbon chains, aromatics benzene, toluene, ethyl benzene and xylene (abbreviated as BTEX). These pollutants at low concentration can be detoxified by remediated plants (Ndimele 2010). These plants are associated with the petroleum-degrading bacteria in the rhizosphere or in the roots (Siciliano et al. 2003). Moore and co-workers (2006) reported vast endophytic bacterial diversity in poplar trees growing on a BTEX-contaminated site. These bacteria increased the pollutant-degrading capabilities thus improving phytoremediation. Genetically engineered endophytic bacteria having toluene-degrading genes, inoculated with lupine, enhanced the phytoremediation of volatile, water-soluble organic pollutants (Barac et al. 2004). Ndimele (2010) listed plant species having ability to remediate petroleum hydrocarbons as well as plants species that can tolerate petroleum hydrocarbons. Different plants species in aquatic ecosystem have the capability to remediate oil hydrocarbon due to their fast-growing ability (Bhatia and Goyal 2014). According to White et al. (2006), fibrous roots of these aquatic plants increase the surface area for microbial colonization.

Sanusi and co-workers (2012) reported that *Paspalum vaginatum* Sw. (Biscuit grasses) is a species having potential to remediate petroleum hydrocarbons. Fibrous root systems of water hyacinths (*Eichhornia crassipes*) were reported to remediate surface waters from the floating petroleum hydrocarbons (Ndimele 2010). Huesemann et al. (2009) reported that a marine macrophyte *Zostera marina* (eelgrass) significantly remediates submerged marine sediments by removing polychlorinated biphenyls and polynuclear aromatic hydrocarbons. Reeds, dominant plants in coastal wetlands are beneficial for remediating the petroleum-contaminated wetlands due to their strong vitality and great root surface area (Wang et al. 2011). On the other hand, floating species phytoremediate pollutants through their ability for biosorption and bioaccumulation of contaminants from the liquid medium (Rahman and Hasegawa 2011). Due to anoxic and hypoxic conditions of aquatic ecosystem, degradation of crude oil is very slow and incomplete. Moreira et al. (2011) reported some macrophytes which have the ability to transport atmospheric oxygen to roots from the shoots and enhance the aerobic respiration of rhizosphere microbial community. Phreatophytes are deep-rooted plants having

Table 6.3 Applications of phytoremediation in alleviation of soil, water and air pollution

Type of pollutants	Pollutants remediated	Plant species/microbes	Mechanism	References
Soil pollutants	Zn, Cu	*Phyla nodiflora*	Phytoextraction	Yoon et al. (2006)
	Cu, Pb, Zn	*Gentiana pennelliana*	Phytostabilization	Yoon et al. (2006)
	Cr (VI)	Halophytes	Phytotransformation	Caçador and Duarte (2015)
	Toxic oxyanions and metal cations	*Pseudomonas maltophilia*	Transformation	Blake et al. (1993)
	Cd	*Youngia erythrocarpa*	Phytoextraction (hyperaccumulation)	Lin et al. (2015)
	Ni	*Alyssum* and *Thlaspi*	Phytoextraction (hyperaccumulation)	Rascio and Navari-Izzo (2011)
	Zn	*Anthyllis vulneraria* associated with symbiotic bacterium *Rhizobium metallidurans*	Phytoextraction (hyperaccumulation)	Grison et al. (2014, 2015)
	Mine soil	*Anthyllis vulneraria*	Phytostabilization	Frérot et al. (2006)
Water pollutants	BTEX	Endophytic bacterial diversity in poplar tree	Phytoremediation.	Moore et al. (2006)
	Petroleum hydrocarbons	*Paspalum vaginatum*	Phytoremediation	Sanusi et al. (2012)
	Surface floating petroleum hydrocarbons	*Eichhornia crassipes*	Phytoremediation	Ndimele (2010)
	Polychlorinated biphenyls and polynuclear aromatic hydrocarbons	*Zostera marina*	Phytoremediation	Huesemann et al. (2009)
	Groundwater polluted with pesticides, nutrients or heavy metals	Poplar and cottonwood trees	Phytoremediation	Gatliff (1994)
Air pollutants	Particulate matter	*Chlorophytum comosum*	Accumulation	Gawrońska and Bakera (2015)
	Volatile compounds (VOCs)	*Azalea indica* leaves inoculated with *Pseudomonas* sp.	Phytovolatilization	De Kempeneer et al. (2004)
	2-formaldehyde	*Melissa officinalis*	Enhanced catalase activity and higher stomatal density	Jin et al. (2013)

the capability of remediating polluted groundwater through hydraulic control. For example, eucalyptus and tamarisk are deep-rooted and can withdraw huge amount of water from a deep water table. Poplar and cottonwood trees were used to treat groundwater polluted with pesticides, nutrients or heavy metals (Gatliff 1994) (Table 6.3).

6.10 Conclusions

In this chapter the role of plant-microbe interaction in phytoremediation has been discussed in detail. Several reports have firmly established the significance of plant-microbe interactions in phytoremediation of harmful chemicals. A comprehensive understanding of the role of microbes in phytoremediation process is necessary for developing more efficient management practices for detoxification of contaminants. In view of the ever-increasing population and the associated environment-degrading activities such as mining, transport, agrochemicals, etc., development of more efficient plants with better capability for removal of contaminants is the only sustainable way for protection of environment. In conclusion, we firmly believe that plant-microbe interaction plays a significant role in phytoremediation of harmful pollutants leading to an efficient and cost-effective method for detoxification of contaminated sites.

References

Aafi NE, Brhada F, Dary M, Maltouf AF, Pajuelo E (2012) Rhizostabilization of metals in soils using *Lupinus luteus* inoculated with the metal resistant rhizobacterium *Serratia* sp. MSMC 541. Int J Phytoremediation 14:261–274

Abou-Shanab RAI, Angle JS, Chaney RL (2006) Bacterial inoculants affecting nickel uptake by *Alyssum murale* from low, moderate and high Ni soils. Soil Biol Biochem 38:2882–2889

Adriano DC (1986) Trace elements in the terrestrial environment. Springer, New York, NY, pp 1–533

Ahmad I, Akhtar MJ, Asghar HN, Ghafoor U, Shahid M (2016) Differential effects of plant growth-promoting rhizobacteria on maize growth and cadmium uptake. J Plant Growth Regul 35:303–315

Al-Garni SMS (2006) Increased heavy metal tolerance of cowpea plants by dual inoculation of an arbuscular mycorrhizal fungi and nitrogen-fixer *Rhizobium* bacterium. Afr J Biotechnol 5:133–142

Aly AH, Debbab A, Proksch P (2011) Fungal endophytes: unique plant inhabitants with great promises. Appl Microbiol Biotechnol 90:1829e1845

Arnold AE (2007) Understanding the diversity of foliar fungal endophytes: progress, challenges, and frontiers. Fungal Biol Rev 21:51–66

Arriagada CA, Herrera MA, García-Romera I, Ocampo JA (2004) Tolerance to Cd of soybean (*Glycine max*) and eucalyptus (*Eucalyptus globulus*) inoculated with arbuscular mycorrhizal and saprobe fungi. Symbiosis 36:285–299

Arshad M, Saleem M, Hussain S (2007) Perspectives of bacterial ACC deaminase in phytoremediation. Trends Biotchnol 25:356–362

Atzorn R, Crozier A, Wheeler CT, Sandberg G (1988) Production of gibberellins and indole-3-acetic acid by Rhizobium phaseoli in relation to nodulation of *Phaseolus vulgaris* roots. Planta 175:532–538

Babu AG, Shim J, Shea PJ, Oh BT (2014a) *Penicillium aculeatum* PDR-4 and *Trichoderma* sp. PDR-16 promote phytoremediation of mine tailing soil and bioenergy production with *Sorghum* Sudan grass. Ecol Eng 69:186e191

Babu AG, Shim J, Bang KS, Shea PJ, Oh BT (2014b) *Trichoderma virens* PDR-28: a heavy metal-tolerant and plant growth-promoting fungus for remediation and bioenergy crop production on mine tailing soil. J Environ Manag 132:129e134

Badenoch-Jones J, Summons RE, Rolfe BG, Letham DS (1984) Phytohormones, rhizobium mutants and nodulation in legumes. IV. Auxin metabolites in pea root nodules. J Plant Growth Regul 3:23–39

Badri DV, Vivanco JM (2009) Regulation and function of root exudates. Plant Cell Environ 32:666–681

Badri DV, Weir TL, van der Lelie D, Vivanco JM (2009) Rhizosphere chemical dialogues: plant-microbe interactions. Curr Opin Biotechnol 20:642–650

Bagyaraj DJ (2014) Mycorrhizal fungi. Proc Indian Natl Sci Acad 80:415–428

Bagyaraj DJ, Sharma MP, Maiti D (2015) Phosphorus nutrition of crops through arbuscular mycorrhizal fungi. Curr Sci 108:1288–1212

Bais HP, Weir TL, Perry LG, Gilroy S, Vivanco JM (2006) The role of root exudates in rhizosphere interactions with plants and other organisms. Annu Rev Plant Biol 57:233–266

Baker AJM, Brooks RR (1989) Terrestrial higher plants which hyper accumulate metallic elements – review of their distribution, ecology and phytochemistry. Biorecovery 1:81–126

Barac T, Taghavi S, Borremans B, Provoost A, Oeyen L et al (2004) Engineered endophytic bacteria improve phytoremediation of water-soluble, volatile, organic pollutants. Nat Biotechnol 22:583–588

Baysse C, Vos DD, Naudet Y, Vandermonde A, Ochsner U, Meyer JM et al (2000) Vanadium interferes with siderophore mediated iron uptake in *Pseudomonas aeruginosa*. Microbiol-SGM 146:2425–2434

Belimov AA, Safronova VI, Sergeyeva TA, Egorova TN, Matveyeva VA, Tsyganov VE et al (2001) Characterization of plant growth promoting rhizobacteria isolated from polluted soils and containing 1–aminocyclopropane–1–carboxylate deaminase. Can J Microbiol 47:642–652

Belimov AA, Puhalsky IV, Safronova VI, Shaposhnikov AI, Vishnyakova MA, Semenova EV, Zinovkina NY, Makarova NM, Wenzel W, Tikhonovich IA (2015) Role of plant genotype and soil conditions in symbiotic plant-microbe interactions for adaptation of plants to cadmium-polluted soils. Water Air Soil Pollut 226:264

Berendsen RL, Pieterse CMJ, Bakker P (2012) The rhizosphere microbiome and plant health. Trends Plant Sci 17:478–486

Bhatia M, Goyal D (2014) Analyzing remediation potential of waste water through wetland plants: a review. Environ Prog Sustain Energy 33:9–27

Bi YL, Li XL, Christie P, Hu ZQ, Wong MH (2003) Growth and nutrient uptake of arbuscular mycorrhizal maize in different depths of soil overlying coal fly ash. Chemosphere 50:863–869

Bilal S, Khan AL, Shahzad R, Asaf S, Kang S-M, Lee I-J (2017) Endophytic *Paecilomyces formosus* LHL10 augments *glycine max* L. adaptation to Ni-contamination through affecting endogenous phytohormones and oxidative stress. Front Plant Sci 8:870

Bitas V, Kim HS, Bennett JW, Kang S (2013) Sniffing on microbes: diverse roles of microbial volatile organic compounds in plant health. Mol Plant-Microbe Interact 26:835–843

Bizily SP, Rugh CL, Meagher RB (2000) Phytodetoxification of hazardous organomercurials by genetically engineered plants. Nat Biotechnol 18:213–217

Blake RC, Choate DM, Bardhan S, Revis N, Barton LL et al (1993) Chemical transformation of toxic metals by a *Pseudomonas* strain from a toxic waste site. Environ Toxicol Chem 12:1365–1376

Bloemberg GV, Lugtenberg BJJ (2001) Molecular basis of plant growth promotion and biocontrol by rhizobacteria. Curr Opin Plant Biol 4:343–350

Bluyssen PM, Janssen S, van den Brink LH, de Kluizenaar Y (2011) Assessment of wellbeing in an indoor office environment. Build Environ 46:2632–2640

Bourdel G, Roy-Bolduc A, St-Arnaud M, Hijri M (2016) Concentration of petroleum-hydrocarbon contamination shapes fungal endophytic community structure in plant roots. Front Microbiol 7:685
Braud A, Jézéquel K, Bazot S, Lebeau T (2009a) Enhanced phytoextraction of an agricultural Cr, Hg- and Pb-contaminated soil by bioaugmentation with siderophore producing bacteria. Chemosphere 74:280–286
Braud A, Hannauer M, Milsin GLA, Schalk IJ (2009b) The *Pseudomonas aeruginosa* pyochelin-iron uptake pathway and its metal specificity. J Bacteriol 191:5317–5325
Braud A, Geoffroy V, Hoegy F, Mislin GLA, Schalk IJ (2010) The siderophores pyoverdine and pyochelin are involved in *Pseudomonas aeruginosa* resistance against metals: another biological function of these two siderophores. Environ Microbiol Rep 2:419–425
Brennan MA, Shelley ML (1999) A model of the uptake, translocation, and accumulation of lead (Pb) by maize for the purpose of phytoextraction. Ecol Eng 12:271–297
Brooks RR (1977) Copper and cobalt uptake by Haumaniastrum species. Plant Soil 48:541–544
Bücking H, Kafle A (2015) Role of arbuscular mycorrhizal fungi in the nitrogen uptake of plants: current knowledge and research gaps. Agronomy 5:587–612
Bulgarelli D, Schlaeppi K, Spaepen S, Ver Loren van Themaat E, Schulze-Lefert P (2013) Structure and functions of the bacterial microbiota of plants. Annu Rev Plant Biol 64:807–838
Buscot F (2015) Implication of evolution and diversity in arbuscular and ectomycorrhizal symbioses. J Plant Physiol 172:55–61
Caçador I, Duarte B (2015) Chromium phyto-transformation in salt marshes: the role of halophytes. Phytoremediation:211–217
Cao L, Jiang M, Zeng Z, Liu Y (2008) *Trichoderma atroviride* F6 improves phytoextraction efficiency of mustard (*Brassica juncea* (L) Coss. var. *foliosa* Bailey) in Cd, Ni contaminated soils. Chemosphere 71:1769e1773
Carlson CL, Adriano DC, Sajwan KS, Abels SL, Thoma DP et al (1991) Effects of selected trace metals on germinating seeds of six plant species. Water Air Soil Pollut 59:231–240
Chan WF, Li H, Wu FY, Wu SC, Wong MH (2013) Arsenic uptake in upland rice inoculated with a combination or single arbuscular mycorrhizal fungi. J Hazard Mater 262:1116–1122
Chaney R, Malik M, Li Y, Brown SL, Brewer EP, Angle JS, Bake AJ (1997) Phytoremediation of soil metals. Curr Opin Biotechnol 8:279–284
Chaney RL, Li YM, Angel JS, Baker AJM, Reeves RD, Brown SL, Homer FA, Malik M, Chin M (1999) Improving metal hyperaccumulators wild plants to develop commercial phytoextraction systems: approaches and progress. In: Terry Nand Banuelos GS (ed) Phytoremediation of contaminated soil and water. CRC Press, Boca Raton, FL, pp 129–158
Chatterjee S, Sau GB, Mukherjee SK (2009) Plant growth promotion by a hexavalent chromium reducing bacterial strain, Cellulosimicrobium cellulans KUCr3. World J Microbiol Biotechnol 25:1829–1836
Chen L, Luo S, Xiao X, Guo H, Chen J, WanY Li B, Xu T, Xi Q, Rao C, Liu C, Zeng G (2010) Application of plant growth-promoting endophytes (PGPE) isolated from Solanum nigrum L. for phytoextraction of Cd-polluted soils. Appl Soil Ecol 46:383–389
Chen Y, Wang H, Li L, Dai C (2013) The potential application of the endophyte *Phomopsis liquidambari* to the ecological remediation of long-term cropping soil. Appl Soil Ecol 67:20e26
Cherian S, Oliveira MM (2005) Transgenic plants in phytoremediation: recent advances and new possibilities. Environ Sci Technol 39:9377–9390
Chun SC, Yoo MH, Moon YS, Shin MH, Son KC, Chung IM, Kays SJ (2010) Effect of bacterial population from rhizosphere of various foliage plants on removal of indoor volatile organic compounds. Korean J Hortic Sci 28:476–483
Colpaert JV, Wevers JHL, Krznaric E, Adriaensen K (2011) How metal-tolerant ecotypes of ectomycorrhizal fungi protect plants from heavy metal pollution. Ann For Sci 68:17–24
Conte P, Agrettoa A, Spaccinia RA, Piccoloa A (2005) Soil remediation: humic acids as natural surfactants in the washings of highly contaminated soils. Environ Pollut 135:515–522

Cornejo P, Pérez-Tienda J, Meier S, Valderas A, Borie F, Azcón-Aguilar C, Ferrol N (2013) Copper compartmentalization in spores as a survival strategy of arbuscular mycorrhizal fungi in Cu-polluted environments. Soil Biol Biochem 57:925–928

Cunningham SD, Ow DW (1996) Promises and prospect of phytoremediation. Plant Physiol 110:715–719

Dai C, Tian L, Zhao Y, Chen Y, Xie H (2010) Degradation of phenanthrene by the endophytic fungus *Ceratobasidum stevensii* found in *Bischofia polycarpa*. Biodegradation 21:245e255

Dardanelli MS, Manyani H, González-Barroso S, Rodríguez-Carvajal MA, Gil-Serrano AM, Espuny MR, López-Baena FJ, Bellogín RA, Megías M, Ollero FJ (2010) Effect of the presence of the plant growth promoting rhizobacterium (PGPR) *Chryseobacterium balustinum* Aur9 and salt stress in the pattern of flavonoids exuded by soybean roots. Plant Soil 328:483–493

Davies FT, Puryear JD, Newton RJ, Egilla JN, Saraiva Grossi JA (2001) Mycorrhizal fungi enhance accumulation and tolerance of chromium in sunflower (*Helianthus annuus*). J Plant Physiol 158:777–786

De Kempeneer L, Sercu B, Vanbrabant W, Van Langenhove H, Verstraete W (2004) Bioaugmentation of the phyllosphere for the removal of toluene from indoor air. Appl Microbiol Biotechnol 64:284–288. https://doi.org/10.1007/s00253-003-1415-3

De Souza LA, De Andrade SAL, De Souza SCR, Schiavinato MA (2012) Arbuscular mycorrhiza confers Pb tolerance in *Calopogonium mucunoides*. Acta Physiol Plant 34:523–531

Dela Cruz M, Christensen JH, Thomsen JD, Müller R (2014) Can ornamental potted plants remove volatile organic compounds from indoor air?—a review. Envrion Sci Pollut Res 21:13909–13928

Dell'Amico E, Cavalca L, Andreoni V (2008) Improvement of Brassica napus growth under cadmium stress by cadmium-resistant rhizobacteria. Soil Biol Biochem 40:74–84

Delvasto P, Ballester A, Muñoz JA, González F, Blázquez ML, Igual JM et al (2009) Mobilization of phosphorus from iron ore by the bacterium *Burkholderia caribensis* FeGL03. Miner Eng 22:1–9

Deng Z, Cao L (2017) Fungal endophytes and their interactions with plants in phytoremediation: a review. Chemosphere 168:1100–1106

Deng Z, Wang W, Tan H, Cao L (2012) Characterization of heavy metal-resistant endophytic yeast *Cryptococcus* sp. CBSB78 from rapes (*Brassica chinensis)* and its potential in promoting the growth of *Brassica* spp. in metal-contaminated soils. Water Air Soil Pollut 223:5321e5329

Deng Z, Zhang R, Shi Y, Hu L, Tan H, Cao L (2013) Self-fusion of protoplast improves phytoremediation of Cd and Pb contaminants in endophytic *Mucor* sp. CBRF59. Chemosphere 91:41e47

Deng Z, Cao L, Zhang R, Shi Y, Hu L, Tan H, Cao L (2014a) Enhanced phytoremediation of multimetal contaminated soils by interspecific fusion between the protoplasts of endophytic *Mucor* sp.CBRF59 and *Fusarium* sp. CBRF14. Soil Biol Biochem 77:31e40

Deng Z, Zhang R, Shi Y, Hu L, Tan H, Cao L (2014b) Characterization of Cd-, Pb-, Zn-resistant endophytic *Lasiodiplodia* sp. MXSF31 from metal accumulating *Portulaca oleracea* and its potential in promoting the growth of rape in metal contaminated soils. Environ Sci Pollut Res 21:2346e2357

Di Gregorio S, Lampis S, Vallini G (2005) Selenite precipitation by a rhizospheric strain of *Stenotrophomonas* sp isolated from the root system of *Astragalus bisulcatus*: a biotechnological perspective. Environ Int 31:233–241

Dory SL, Oakley B, Xin G, Kang JW, Singleton G, Khan Z, Vajzovic A, Staley JT (2009) Diazotrophic endophytes of native black cottonwood and willow. Symbiosis 47:23–33

Duffus JH (2002) 'Heavy metals'—a meaningless term? Pure Appl Chem 74:793–807

Dzierzanowski K, Popek R, Gawronska H, Saebo A, Gawronski SW (2011) Deposition of particulate matter of different size fractions on leaf surface and waxes of urban forest species. Int J Phytoremediation 13:1037–1046

Ebbs SD, Brady DJ, Kochian LV (1998) Role of uranium speciation in the uptake and translocation by plants. J Exp Bot 49:1183–1190

Effmert U, Kalderás J, Warnke R, Piechulla B (2012) Volatile mediated interactions between bacteria and fungi in the soil. J Chem Ecol 38:665–703

Ensley BD (2000) "Rationale for the use of phytoremediation." phytoremediation of toxic metals: using plants to clean-up the environment. John Wiley Publishers, New York
EPA (2000) A citizen's guide to phytoremediation. EPA 542-F-98-011. United States Environmental Protection Agency, pp 6. Available at: http://www.bugsatwork.com/XYCLONYX/EPA_GUIDES/PHYTO.PDF
Erakhrumen AA (2007) Phytoremediation: an environmentally sound technology for pollution prevention, control and remediation in developing countries. Educ Res Rev 2:151–156
Erdei L, Mezôsi G, Mécs I, Vass I, Fôglein F, Bulik L (2005) Phytoremediation as a program for decontamination of heavy-metal polluted environment, in Proceedings of the 8th Hungarian Congress on plant physiology and the 6th Hungarian Conference on photosynthesis
Escalante-Espinosa E, Gallegos-Martínez ME, Favela-Torres E, Gutiérrez-Rojas M (2005) Improvement of the hydrocarbon phytoremediation rate by *Cyperus Laxus* lam. Inoculated with a microbial consortium in a model system. Chemosphere 59:405–413
Farwell AJ, Vesely S, Nero V, Rodriguez H, Shah S, Dixon DG, Glick BR (2006) The use of transgenic canola (*Brassica napus*) and plant growth-promoting bacteria to enhance plant biomass at a nickel-contaminated field site. Plant Soil 288:309–318
Feng Y, Shen D, Song W (2006) Rice endophyte *Pantoea agglomerans* YS19 promotes host plant growth and affects allocations of host photosynthates. J Appl Microbiol 100:938–945
Ferner DJ (2001) Toxicity heavy metals. Emed J 2:1
Firmin S, Labidi S, Fontaine J, Laruelle F, Tisserant B, Nsanganwimana F et al (2015) Arbuscular mycorrhizal fungal inoculation protects *Miscanthus* × *giganteus* against trace element toxicity in a highly metal-contaminated site. Sci Total Environ 527–528:91–99
Flathman PE, Lanza GR (1998) Phytoremediation: current views on an emerging green technology. J Soil Contam 7:415–432
Fomina MA, Alexander IJ, Colpaert JV, Gadd GM (2005) Solubilization of toxic metal minerals and metal tolerance of mycorrhizal fungi. Soil Biol Biochem 37:851e866
Frankowski J, Lorito M, Scala F, Schmid R, Berg G, Bahl H (2001) Purification and properties of two chitinolytic enzymes of *Serratia plymuthica* HRO-C48. Arch Microbiol 176:421–426
Frérot H, Lefebvre C, Gruber W, Collin C, Dos Santos A et al (2006) Specific interactions between local metallicolous plants improve the phytostabilization of mine soils. Plant Soil 282:53–65
Ganesan V (2008) Rhizoremediation of cadmium soil using a cadmium-resistant plant growth-promoting rhizopseudomonad. Curr Microbiol 56:403–407
Gao YZ, Ling WT, Zhu LZ, Zhao BW, Zheng QS (2007) Surfactant-enhanced phytoremediation of soils contaminated with hydrophobic organic contaminants: potential and assessment. Pedosphere 17:409–418
Garbisu C, Alkorta I (2001) Phytoextraction: a cost-effective plant-based technology for the removal of metals from the environment. Bioresour Technol 77:229–236
Garg N, Kaur H (2013) Response of antioxidant enzymes, phytochelatins and glutathione production towards Cd and Zn stresses in *Cajanus cajan* (L.) mill sp. genotypes colonized by arbuscular mycorrhizal fungi. J Agron Crop Sci 199:118–133
Gatliff EG (1994) Vegetative remediation process offers advantages over traditional pump-and-treat technologies. Remediation 4:343–352
Gaur A, Adholeya A (2004) Prospects of arbuscular mycorrhizal fungi in phytoremediation of heavy metal contaminated soils. Curr Sci 86:528–534
Gawrońska H, Bakera B (2015) Phytoremediation of particulate matter from indoor air by *Chlorophytum comosum* L. plants. Air Qual Atmos Health 8:265–272.
Gehling W, Dellinger B (2013) Environmentally persistent free radicals and their lifetimes in PM2.5. Environ Sci Technol 47:8172–8178
Gerhardt KE, Huang XD, Glick BR, Greenberg BM (2009) Phytoremediation and rhizoremediation of organic soil contaminants: potential and challenges. Plant Sci 176:20–30
Germaine KJ, Keogh E, Ryan D, Dowling D (2009) Bacterial endophyte-mediated naphthalene phytoprotection and phytoremediation. FEMS Microbiol Lett 296:226–234
Ghosh S (2010) Wet land macrophytes as toxic metal accumulators. Int J Environ Sci 1:523–528

Ghosh M, Singh SP (2005) A review on phytoremediation of heavy metals and utilization of it's by products. Appl Ecol Environ Res 3:1–18
Glick BR (2003) Phytoremediation: synergistic use of plants and bacteria to clean up the environment. Biotechnol Adv 21:383–393
Glick BR (2010) Using soil bacteria to facilitate phytoremediation. Biotechnol Adv 28:367–374
Glick BR (2012) Plant growth-promoting bacteria: mechanisms and applications. Scientifica. Hindawi Publishing Corporation, Article ID 963401 2012:1–15
Glick BR (2014) Bacteria with ACC deaminase can promote plant growth and help to feed the world. Microbiol Res 169:30–39
Gohre V, Paszkowski U (2006) Contribution of arbuscular mycorrhizal symbiosis to heavy metal phytoremediation. Planta 223:1115–1122
Gonzalez-Chavez MC, Carrillo-Gonzalez R, Wright SF, Nichols KA (2004) The role of glomalin, a protein produced by arbuscular mycorrhizal fungi, in sequestering potentially toxic elements. Environ Pollut 130:317–323
Gratão LP, Braz J (2005) Phytoremediation: green technology for the clean-up of toxic metals in the environment. Plant Physiol 17:53–64
Griffin M, Antikainen M, Hon WC, Pihakaski-Maunsbach K, Yu XM, Chun JU, Yang DSC (1997) Antifreeze proteins in winter rye. Plant Physiol 100:327–332
Grison CM, Mazel M, Sellini A, Escande V, Biton J et al (2014) The leguminous species *Anthyllis vulneraria* as a Zn-hyperaccumulator and eco- Zn catalyst resources. Environ Sci Pollut Res Int 22:5667–5676
Grison C, Jackson S, Merlot S, Dobson A, Grison C (2015) *Rhizobium metallidurans* sp. nov, a symbiotic heavy metal resistant bacterium isolated from the *Anthyllis vulneraria* Zn-hyperaccumulator. Int J Syst Evol Microbiol 65:1525–1530
Haas D, Keel C (2003) Regulation of antibiotic production in root-colonizing *Pseudomonas* spp. and relevance for biological control of plant disease. Annu Rev Phytopathol 41:117–153
Hamayun M, Sumera AK, Iqbal I, Ahmad B, Lee I (2010) Isolation of a gibberellin-producing fungus (Penicillium sp. MH7) and growth promotion of crown daisy (Chrysanthemum coronarium). J Microbiol Biotechnol 20:202–207
Hao HZ, Chen TB, Jin MG, Lei M, Liu CW, Zu WP, Huang LM (2011) Recent advance in solidification/stabilization technology for the remediation of heavy metals-contaminated soil. J Appl Ecol 22:816–824
Hardoim PR, Overbeek LS, Elsas JD (2008) Properties of bacterial endophytes and their proposed role in plant growth. Trends Microbiol 16:467–471
Hassan SE, Hijri M, St-Arnaud M (2013) Effect of arbuscular mycorrhizal fungi on trace metal uptake by sunflower plants grown on cadmium contaminated soil. New Biotechnol 30:780–787
Hassan W, Bano R, Bashir F, David J (2014) Comparative effectiveness of ACC-deaminase and/or nitrogen-fixing rhizobacteria in promotion of maize *(Zea mays* L.) growth under lead pollution. Environ Sci Pollut Res 21:10983–10996
Hassan W, Bashir S, Ali F, Ijaz M, Hussain M, David J (2016) Role of ACC deaminase and/or nitrogen fixing rhizobacteria in growth promotion of wheat (*Triticum aestivum* L.) under cadmium pollution. Environ Earth Sci 75:1–14
Hegedusova A, Jakabova S, Simon L (2009) Induced phytoextraction of lead from contaminated soil. Acta Universitatis Sapientiae Agric Environ 1:116–122
Hernández-Ortega HA, Alarcón A, Ferrera-Cerrato R, Zavaleta-Mancera HA, López-Delgado HA, Mendoza-López MR (2012) Arbuscular mycorrhizal fungi on growth, nutrient status, and total antioxidant activity of *Melilotus albus* during phytoremediation of a diesel-contaminated substrate. J Environ Manag 95:319–324
Hess-Kosa K (2002) Indoor air quality sampling methodologies. Lewis Publishers, Roca Raton-London-New York-Washington
Hider RC, Kong X (2010) Chemistry and biology of siderophores. Nat Prod Rep 27:637–657
Hildebrandt U, Regvar M, Bothe H (2007) Arbuscular mycorrhiza and heavy metal tolerance. Phytochemistry 68:139–146

Hill A, Crossman SM (1983) Characterization of N2 –fixing bacteria associated with sweet potato roots. Can J Microbiol 29:860–862

Huesemann MH, Hausmann TS, Fortman TJ, Thom RM, Cullinan V (2009) *In situ* phytoremediation of PAH and PCB contaminated marine sediments with eelgrass (*Zostera marina*). Ecol Eng 35:1395–1404

Huq SI, Joardar J, Parvin S (2005) Marigold (*Tagetes patula*) and ornamental arum (*Syngonia* sp.) as phytoremediators for arsenic in pot soil. Bangladesh J Bot 34:65–70

Hurek T, Reinhold-Hurek B (2003) Azoarcus sp. strain BH72 as a model for nitrogen-fixing grass endophytes. J Biotechnol 106:169–178

Hurek T, Handley LL, Reinhold-Hurek B, Piché Y (2002) Azoarcus grass endophytes contribute fixed nitrogen to the plant in an unculturable state. Mol Plant-Microbe Interact 15:233–242

Insam H, Seewald MS (2010) Volatile organic compounds (VOCs) in soils. Biol Fertil Soils 46:199–213

IOSHIC International Occupational Safety and Health Information Centre (1999) Basics of chemical safety. International Labour Organization, Geneva

James AC, Strand SE (2009) Phytoremediation of small organic contaminants using transgenic plants. Curr Opin Biotechnol 20:237–241

James CA, Xin G, Doty SL, Strand SE (2008) Degradation of low molecular weight volatile organic compounds by plants genetically modified with mammalian Cytochrome P450 2E1. Environ Sci Technol 42:289–293

Jiang M, Cao L, Zhang R (2008) Effects of *Acacia* (*Acacia auriculaeformis* A. Cunn)-associated fungi on mustard (*Brassica juncea* (L.) Coss.var. foliosa Bailey) growth in Cd- and Ni-contaminated soils. Lett Appl Microbiol 47:561e565

Jin CL, Zhou XJ, Zhao HT, Liu XM, Feng K (2013) Comparison of removal of formaldehyde capacity between *Hedera helix* and *Melissa officinalis*. Asian J Chem 25:3823–3826

Jing Y, He Z, Yang X (2007) Role of soil rhizobacteria in phytoremediation of heavy metal contaminated soils. J Zhejiang Univ (Sci) 8:192–207

Joshi PM, Juwarkar AA (2009) In vivo studies to elucidate the role of extracellular polymeric substances from *Azotobacter* in immobilization of heavy metals. Environ Sci Technol 43:5884–5889

Juwarkar AA, Nair A, Dubey KV, Singh SK, Devotta S (2007) Biosurfactant technology for remediation of cadmium and lead contaminated soils. Chemosphere 10:1996–2002

Kai M, Effmert U, Berg G, Piechulla B (2007) Volatiles of bacterial antagonists inhibits mycelia growth of the plant pathogen *Rhizoctonia solani*. Arch Microbiol 187:351–360

Kamnev AA, Lelie VD (2000) Chemical and biological parameters as tools to evaluate and improve heavy metal phytoremediation. Biosci Rep 20:239–258

Kamran MA, Syed JH, Eqani SAMAS, Munis MFH, Chaudhary HJ (2015) Effect of plant growth-promoting rhizobacteria inoculation on cadmium (Cd) uptake by *Eruca sativa*. Environ Sci Pollut Res 22:9275–9283

Kaya Z (2006) Pollution. In: Lal R (ed) Encyclopedia of soil science. CRC Press, Boca Raton, FL, pp 1343–1346

Khan Z, Doty S (2011) Endophyte-assisted phytoremediation. Curr Top Plant Biol 12:97e105

Khan AL, Lee IJ (2013) Endophytic *Penicillium funiculosum* LHL06 secretes gibberellins that reprogram *Glycine max* L. growth during copper stress. BMC Plant Biol 13:1–14

Khan MA, Ahmad I, Rahman IU (2007) Effect of environmental pollution on heavy metal content of *Withania somnifera*. J Chin ChemSoc 54:339–343

Khan S, Hesham AEL, Qiao M, Rehman S, He JZ (2010) Effects of Cd and Pb on soil microbial community structure and activities. Environ Sci Pollut Res 17:288–296

Khan AL, Waqas M, Hussain J, Al-Harrasi A, Hamayun M, Lee IJ (2015) Phytohormones enabled endophytic fungal symbiosis improve aluminum phytoextraction in tolerant *Solanum lycopersicum*: an examples of *Penicillium janthinellum* LK5 and comparison with exogenous GA3. J Hazard Mater 295:70–78

Khan AR, Ullah I, Waqas M, Park GS, Khan AL, Hong SJ et al (2017a) Host plant growth promotion and cadmium detoxification in *Solanum nigrum*, mediated by endophytic fungi. Ecotoxicol Environ Saf 136:180–188

Khan AR, Waqas M, Ullah I, Khan AL, Khan MA, Lee IJ et al (2017b) Culturable endophytic fungal diversity in the cadmium hyperaccumulator *Solanum nigrum* L. and their role in enhancing phytoremediation. Environ Exp Bot 135:126–135

Kim YC, Jung H, Kim KY, Park SK (2008) An effective biocontrol bioformulation against Phytophthora blight of pepper using growth mixtures of combined chitinolytic bacteria under different field conditions. Eur J Plant Pathol 120:373–382

Kloepper JW, Schroth MN (1978) Plant growth promoting rhizobacteria on radish. In: Station de pathologie végétale et phyto-bacteriologie (ed) Proceedings of the 4th Conference plant pathogenic bacteria, Angers, INRA, pp 879–882

Kraomer U (2005) Phytoremediation: novel approaches to cleaning up polluted soils. Curr Opin Biotechnol 16:133–141

Kuffner M, Maria SD, Puschenreiter M, Fallmann K, Wieshammer G, Gorfer M, Strauss J, Rivelli AR, Sessitsch A (2010) Culturable bacteria from Zn- and Cd-accumulating *Salix caprea* with differential effects on plant growth and heavy metal availability. J Appl Microbiol 108:1471–1484

Kuklinsky-Sobral J, Araujo WL, Mendes R, Geraldi IO, Pizzirani-Kleine AA, Azevedo JL (2004) Isolation and characterization of soybean-associated bacteria and their potential for plant growth promotion. Environ Microbiol 6:1244–1251

Lasat MM (2000) Phytoextraction of metals from contaminated soil: a review of plant/soil/metal interaction and assessment of pertinent agronomic issues. J Hazard Subst Res 5:1–25

Lewis AC (2006) Assessment and comparison of two phytoremediation systems treating slow-moving groundwater plumes of TCE. Master thesis, Ohio University pp 158

Li H, Wei D, He C, Shen M, Zhou Z, Mei T, Xu H-M (2012) Diversity and heavy metal tolerance of endophytic fungi from six dominant plant species in a Pb-Zn mine wasteland in China. Fungal Ecol 5:309e315

Liang Y, Zhang X, Dai D, Li G (2009) Porous biocarrier-enhanced biodegradation of crude oil contaminated soil. Int Biodeterior Biodegradation 63:80–87

Lin L, Ning B, Liao MA, Ren Y, Wang Z et al (2015) *Youngia erythrocarpa*, a newly discovered cadmium hyperaccumulator plant. Environ Monit Assess 187:4205

Louwies T, Nawrot T, Cox B, Dons E, Penders J, Provost E, Panis LI, deBoever P (2015) Blood pressure changes in association with black carbon exposure in a panel of healthy adults are independent of retinal microcirculation. Environ Int 75:81–86

Lugtenberg B, Kamilova F (2009) Plant growth promoting rhizobacteria. Annu Rev Microbiol 63:541–556

Luo ZB, Wu C, Zhang C, Li H, Lipka U, Polle A (2014) The role of ectomycorrhizas in heavy metal stress tolerance of host plants. Environ Exp Bot 108:47–62

Macek T, Mackova M, Kas J (2000) Exploitation of plants for the removal of organics in environmental remediation. Biotechnol Adv 18:23–34

Mann US, Dhingra A, Singh J (2014) Water pollution: causes, effects and remedies. IJATES 2:70–74

Marulanda A, Azcon R, Ruiz-Lozano JM (2003) Contribution of six arbuscular mycorrhizal fungal isolates to water uptake by *Lactuca sativa* plants under drought stress. Physiol Plant 119:526–533

Masiol M, Hofer A, Squizzato S, Piazza R, Rampazzo G, Pavoni B (2012) Carcinogenic and mutagenic risk associated to airborne particle-phase polycyclic aromatic hydrocarbons: a source apportionment. Atmos Environ 60:375–382

Mastretta C, Taghavi S, van der Lelie D, Mengoni A, Galardi F, Gonnelli C, Barac T, Boulet J, Weyens N, Vangronsveld J (2009) Endophytic bacteria from seeds of *Nicotiana tabacum* can reduce cadmium phytotoxicity. Int J Phytoremediation 11:251–267

Mathesius U, Schlaman HRM, Spaink HP, Sautter C, Rolfe BG, Djordjevic MA (1998) Auxin transport inhibition precedes root nodule formation in white clover roots and is regulated by flavonoids and derivatives of chitin oligosaccharides. Plant J 14:23–34

Mazurier S, Corberand T, Lemanceau P, Raaijmakers JM (2009) Phenazine antibiotics produced by fluorescent pseudomonads contribute to natural soil suppressiveness to Fusarium wilt. ISME J 3:977–991

Mcgrath SP, Zhao FJ, Lombi E (2001) Plant and rhizosphere processes involved in phytoremediation of metal-contaminated soils. Plant Soil 232:207–214

Mc-Guinness M, Dowling D (2009) Plant-associated bacterial degradation of toxic organic compounds in soil. Int J Environ Res Public Health 6:2226–2247

Meena K, Sorty A, Bitla U, Choudhary K, Gupta P, Pareek A et al (2017) Abiotic stress responses and microbe-mediated mitigation in plants: the omics strategies. Front Plant Sci 8:172–172

Memon AR, Aktoprakligil D, Ozdemir A, Vertii A (2001) Heavy metal accumulation and detoxification mechanisms in plants. Turk J Bot 25:111–121

Mesjasz-Przybylowicz J, Nakonieczny M, Migula P, Augustyniak M, Tarnawska M et al (2004) Uptake of cadmium, lead, nickel and zinc from soil and water solutions by the nickel hyperaccumulator *Berkheyacoddii*. Acta Biol Cracov Ser Bot 46:75–85

Monnet F, Vaillant N, Hitmi A, Coudret A, Sallanon H (2001) Endophytic *Neotyphodium lolii* induced tolerance to Zn stress in *Lolium perenne*. Physiol Plant 113:557–563

Moore FP, Barac T, Borremans B, Oeyen L, Vangronsveld J et al (2006) Endophytic bacterial diversity in poplar trees growing on a BTEX-contaminated site: the characterisation of isolates with potential to enhance phytoremediation. Syst Appl Microbiol 29:539–556

Moreira IT, Oliveira OM, Triguis JA, dos Santos AM, Queiroz AF, Martins CM et al (2011) Phytoremediation using *Rhizophora mangle* L. in mangrove sediments contaminated by persistent total petroleum hydrocarbons (TPH's). Microchem J 99:376–382

Morikawa H, Erkin OC (2003) Basic processes in phytoremediation and some applications to air pollution control. Chemosphere 52:1553–1558

Muthukumarasamy R, Revathi G, Seshadri S, Lakshminarasimhan C (2002) Gluconacetobacter *diazotrophicus* (syn. *Acetobacter diazotrophicus*), a promising diazotrophic endophyte in tropics. Curr Sci 83:137–145

Nadeem SM, Zahir ZA, Naveed M, Asghar HN, Arshad M (2010) Rhizobacteria capable of producing ACC-deaminase may mitigate salt stress in wheat. Soil Sci Soc Am J 74:533–542

Ndimele PE (2010) A review on the phytoremediation of petroleum hydrocarbon. Pak J Biol Sci 13:715–722

Neilands JB (1981) Iron absorption and transport in microorganisms. Annu Rev Nutr 1:27–46

Nowak DJ, Crane DE, Stevens JC (2006) Air pollution removal by urban trees and shrubs in the United States. Urban For Urban Green 4:115–123

Oelmüller R, Sherameti I, Tripathi S, Varma A (2009) *Piriformospora indica*, a cultivable root endophyte with multiple biotechnological applications. Symbiosis 49:1–17

Oleksyn J, Innes JL (2000) Air pollution and forests in heavily industrialized regions: an introduction. In: Innes JL, Oleksyn J (eds) Forest dynamics in heavily polluted regions, 1st edn. CABI Publishing, Oxon, pp 1–7

Ordentlich A, Elad Y, Chet I (1988) The role of chitinase of *Serratia marcescens* in biocontrol of Sclerotium rolfsii. Phytopathology 78:84–88

Ortega-Larrocea MP, Xoconostle-Cazares B, Maldonado-Mendoza IE, Carrillo-Gonzalez R, Hernandez-Hernandez J, Garduno MD et al (2010) Plant and fungal biodiversity from metal mine wastes under remediation at Zimapan, Hidalgo, Mexico. Environ Pollut 158:1922e1931

Padmavathiamma PK, Li LY (2007) Phytoremediation technology: hyper-accumulation metals in plants. Water Air Soil Pollut 184:105–126

Pandey S, Ghosh PK, Ghosh S, De TK, Maiti TK (2013) Role of heavy metal resistant *Ochrobactrum* sp. and *Bacillus* spp. strains in bioremediation of a rice cultivar and their PGPR like activities. J Microbiol 51:11–17

Panfili F, Schneider A, Vives A, Perrot F, Hubert P, Pellerin S (2009) Cadmium uptake by durum wheat in presence of citrate. Plant Soil 316:299–309

Paria S (2008) Surfactant-enhanced remediation of organic contaminated soil and water. Adv Colloid Interf Sci 138:24–58

Pastuszka JS (2007) Effect of particulate aerosols on air quality—methods of identification and assessment (Wplywaerozoliziarnistychnajakoscpowietrza—Metodyidentyfikacji I oceny. *Ekoprofit.FinanseNauka Technol. Prawo*). Ekoprofit Financ Sci Technol Law 2:7–15

Pereira SIA, Barbosa L, Castro PML (2015) Rhizobacteria isolated from a metal-polluted area enhance plant growth in zinc and cadmium-contaminated soil. Int J Environ Sci Technol 12:2127–2142

Pieterse CM, vanWees SC, vanPelt JA, Knoester M, Laan R, Gerrits H et al (1998) A novel signalling pathway controlling ISR in *Arabidopsis*. Plant Cell 10:1571–1580

Pieterse CMJ, Leon-Reyes A, van der Ent S, van Wees SCM (2009) Networking by small-molecule hormones in plant immunity. Nat Chem Biol 5:308–316

Pivetz BE (2001) Ground water issue: phytoremediation of contaminated soil and ground water at hazardous waste site, EPA/540/S-01/400. United States Environmental Protection Agency, Office of Research and Development Office of Solid Waste and Emergency Response, Washington DC, pp 1–36

Popek R, Gawronska H, Wrochna M, Gawronski SW, Saebo A (2012) Paticulate matter on foliage of 13 woody species: deposition on surfaces and phytostabilization in waxes—a 3-year study. Int J Phytoremediation 15:245–256

Pozo MJ, Azcon-Aguilar C (2007) Unraveling mycorrhiza–induced resistance. Curr Opin Plant Biol 10:393–398

Pozo MJ, VanDerEnt S, VanLoon LC, Pieterse CM (2008) Transcription factor MYC2 is involved in priming for enhanced defense during rhizobacteria ISR in *Arabidopsis thaliana*. New Phytol 180:511–523

Prasad MNV, Freitas H (2003) Metal hyperaccumulation in plants-biodiversity prospecting for phytoremediation technology. Electron J Biotechnol 6(3):285–321

Puente ME, Li CY, Bashan Y (2009) Endophytic bacteria in cacti seeds can improve the development of cactus seedlings. Environ Exp Bot 66:402–408

Pulford ID, Watson C (2003) Phytoremediation of heavy metal-contaminated land by trees—a review. Environ Int 29:529–540

Raaijmakers JM, Vlami M, de Souza JT (2002) Antibiotic production by bacterial biocontrol agents. Antonie Van Leeuwenhoek 81:537–547

Rabie GH (2005) Contribution of arbuscular mycorrhizal fungus to red kidney and wheat plants tolerance grown in heavy metal-polluted soil. Afr J Biotechnol 4:332–345

Rafati M, Khorasani N, Moattar F, Shirvany A, Moraghebi F et al (2011) Phytoremediation potential of *Populus alba* and *Morus alba* for cadmium, chromium and nickel absorption from polluted soil. Int J Environ Res 5:961–970

Raghukumar C (2008) Marine fungal biotechnology: an ecological perspective. Fungal Divers 31:5–19

Rahman MA, Hasegawa H (2011) Aquatic arsenic: phytoremediation using floating macrophytes. Chemosphere 83:633–646

Rajakaruna N, Tompkins KM, Pavicevic PG (2006) Phytoremediation: an affordable green technology for the clean-up of metal-contaminated sites in Sri Lanka. Ceylon J Sci (Biol Sci) 35:25–39

Rajkumar M, Nagendran R, Lee KJ, Lee WH, Kim SZ (2006) Influence of plant growth promoting bacteria and Cr6þ on the growth of Indian mustard. Chemosphere 62:741–748

Rajkumar M, Ae N, Freitas H (2009) Endophytic bacteria and their potential to enhance heavy metal phytoextraction. Chemosphere 77:153–160

Ramos JL, Molina L, Segura A (2009) Removal of organic toxic chemicals in the rhizosphere and phyllosphere of plants. Microb Biotechnol 2:144–146

Rascio N, Navari-Izzo F (2011) Heavy metal hyperaccumulating plants: how and why do they do it? And what makes them so interesting? Plant Sci 180:169–181

Raskin I, Ensley BD (2000) Recent developments for in situ treatment of metal contaminated soils. In: Phytoremediation of toxic metals: using plants to clean up the environment. Wiley, New York, NY, p 304

Redman RS, Sheehan KB, Stout RG, Rodriguez RJ, Henson JM (2002) Thermotolerance conferred to plant host and fungal endophyte during mutualistic symbiosis. Science 298:1581

Reeves RD, Baker AJM (2000) Metal-accumulating plants. In: Raskin I, Ensley BD (eds) Phytoremediation of toxic metals: using plants to clean up the environment. Wiley, New York, NY, pp 193–230

Ren AZ, Li CA, Gao YB (2011) Endophytic fungus improves growth and metal uptake of *Lolium arundinaceum* Darbyshire Ex. Schreb. Int J Phytoremediation 13:233–243

Richardson AE (2001) Prospects for using soil microorganisms to improve the acquisition of phosphorus by plants. Funct Plant Biol 28:897–906

Rivera-Becerril F, Calantzis C, Turnau K, Caussanel JP, Belimov AA, Gianinazzi S et al (2002) Cadmium accumulation and buffering of cadmium induced stress by arbuscular mycorrhiza in three *Pisum sativum* L genotypes. J Exp Bot 53:1177–1185

Robinson BH, Lombi E, Zhao FJ, McGrath SP (2003) Uptake and distribution of nickel and other metals in the hyperaccumulator *Berkheya coddii*. New Phytol 158:279–285

Rodríguez H, Fraga R (1999) Phosphate solubilizing bacteria and their role in plant growth promotion. Biotechnol Adv 17:319–339

Rodriguez RJ, White JF Jr, Arnold AE, Redman RS (2009) Fungal endophytes: diversity and functional roles. New Phytol 182:314–330

Rosa G, Peralta-Videa JR, Montes M, Parsons JG, Cano-Aguilera I, Gardea-Torresdey JL (2004) Cadmium uptake and translocation in tumbleweed (Salsola kali), a potential Cd-hyperaccumulator desert plant species: ICP/OES and XAS studies. Chemosphere 55:1159–1168

Ryu CM, Farag MA, Hu CH, Reddy MS, Wei HX, Paré PW, Kloepper JW (2003) Bacterial volatiles promote growth in *Arabidopsis*. Proc Natl Acad Sci U S A 100:4927–4932

Saebo A, Popek R, Nawrot B, Hanslin HM, Gawronska H, Gawronski SW (2012) Plant species differences in particulate matter accumulation on leaf surfaces. Sci Total Environ 427–428:347–354

Safronova VI, Stepanok VV, Engqvist GL, Alekseyev YV, Belimov AA (2006) Root-associated bacteria containing 1-aminocyclopropane-1-carboxylate deaminase improve growth and nutrient uptake by pea genotypes cultivated in cadmium supplemented soil. Biol Fertil Soils 42:267–272

Sakai Y, Ma Y, Xu C, Wu H, Zhu W et al (2012) Phytodesalination of a salt-affected soil with four halophytes in China. J Arid Land Stud 22:17–20

Salt DE, Smith RD, Raskin I (1998) Phytoremediation. Ann Rev Plant Physiol Plant MolBiol 49:643–668

Santos FS, Hernández-Allica J, Becerril JM, Amaral-Sobrinho N, Mazur N et al (2006) Chelate-induced phytoextraction of metal polluted soils with *Brachiaria decumbens*. Chemosphere 65:43–50

Sanusi SNA, Abdullah SRS, Idris M (2012) Preliminary test of phytoremediation of hydrocarbon contaminated soil using *Paspalum Vaginatum* Sw. Aust J Basic Appl Sci 6(1):39–42

Saravanan VS, Madhaiyan M, Thangaraju M (2007) Solubilization of zinc compounds by the diazotrophic, plant growth promoting bacterium *Gluconacetobacter diazotrophicus*. Chemosphere 66:1794–1798

Sawada A, Oyabu T, Chen LM, Li KZ, Hirai N, Yurimoto H, Orita I, Sakai Y, Kato N, Izui K (2007) Purification capability of tobacco transformed with enzymes from a methylotrophic bacterium for formaldehyde. Int J Phytoremediation 9:487–496

Schnoor JL, Licht LA, McCutcheon SC, Wolfe NL, Carreira LH (1995) Phytoremediation of organic and nutrient contaminants. Environ Sci Technol 29:318–323

Schoenborn L, Yates PS, Grinton BE, Hugenholtz P, Janssen PH (2004) Liquid serial dilution is inferior to solid media for isolation of cultures representative of the phylum-level diversity of soil bacteria. Appl Environ Microbiol 70:4363–4366

Schwitzguebel JP (2000) Potential of phytoremediation, an emerging green technology. Ecosyst Serv Sustain Watershed Manage Sci 9:210–220
Scullion J (2006) Remediating polluted soils. Naturwissenschaften 93:51–65
Sekara A, Poniedzialeek M, Ciura J, Jedrszczyk E (2005) Cadmium and lead accumulation and distribution in the organs of nine crops: implications for phytoremediation. Pol J Environ Stud 14:509–516
Sharma A, Johri BN, Sharma AK, Glick BR (2003) Plant growth-promoting bacterium *Pseudomonas* sp. strain GRP3 influences iron acquisition in mung bean (*Vigna radiata* L. Wilzeck). Soil Biol Biochem 35:887–894
Sheng XF, Xia JJ (2006) Improvement of rape (*Brassica napus*) plant growth and cadmium uptake by cadmium-resistant bacteria. Chemosphere 64:1036–1042
Sheng X, Chen X, He L (2008a) Characteristics of an endophytic pyrene-degrading bacterium of *Enterobacter* sp. 12J1 from *Allium macrostemon* Bunge. Int Biodeterior Biodegrad 62:88–95
Sheng X, Xia J, Jiang C, He L, Qian M (2008b) Characterization of heavy metal-resistant endophytic bacteria from rape (*Brassica napus*) roots and their potential in promoting the growth and lead accumulation of rape. Environ Pollut 156:1164–1170
Sheoran V, Sheoran AS, Poonia P (2009) Phytomining: a review. Miner Eng 22(12):1007–1019
Shi Y, Xie H, Cao L, Zhang R, Xu Z, Wang Z, Deng Z (2017) Effects of Cd- and Pb-resistant endophytic fungi on growth and phytoextraction of *Brassica napus* in metal-contaminated soils. Environ Sci Pollut Res Int 24:417–426
Siciliano SD, Germida JJ, Banks K, Greer CW (2003) Changes in microbial community composition and function during a polyaromatic hydrocarbon phytoremediation field trial. Appl Environ Microbiol 69:483–489
Singh PK (2012) Role of glomalin related soil protein produced by arbuscular mycorrhizal fungi: a review. Agric Sci Res J 2:119–125
Singh OV, Jain RK (2003) Phytoremediation of toxic aromatic pollutants from soil. Appl Microbiol Biotechnol 63:128–135
Singh PP, Shin YC, Park CS, Chung YR (1999) Biological control of Fusarium wilt of cucumber by chitinolytic bacteria. Phytopathology 89:92–99
Smith SE, Smith FA (2011) Roles of arbuscular mycorrhizas in plant nutrition and growth: new paradigms from cellular to ecosystem scales. Annu Rev Plant Biol 62:227–250
Soleimani M, Afyuni M, Hajabbasi MA, Nourbakhsh F, Sabzalian MR, Christensen JH (2010) Phytoremediation of an aged petroleum contaminated soil using endophyte infected and non-infected grasses. Chemosphere 81:1084e1090
Suresh B, Ravishankar GA (2004) Phytoremediation – a novel and promising approach for environmental clean-up. Crit Rev Biotechnol 24:97–124
Susarla S, Medina VF, McCutcheon SC (2002) Phytoremediation: an ecological solution to organic chemical contamination. Ecol Eng 18:647–658
Sziderics AH, Rasche F, Trognitz F, Sessitsch A, Wilhelm E (2007) Bacterial endophytes contribute to abiotic stress adaptation in pepper plants (*Capsicum annuum* L.). Can J Microbiol 53:195–1202
Taller BJ, Wong TY (1989) Cytokinins in *Azotobacter vinelandii* culture medium. Appl Environ Microbiol 55:266–267
Tank N, Saraf M (2009) Enhancement of plant growth and decontamination of nickel-spiked soil using PGPR. J Basic Microbiol 49:195–204
Tedersoo L, May TW, Smith ME (2010) Ectomycorrhizal lifestyle in fungi: global diversity, distribution, and evolution of phylogenetic lineages. Mycorrhiza 20:217–263
Tien T, Gaskin M, Hubbel D (1979) Plant growth substances produced by *Azospirillum* brasilense and their effect on the growth of pearl millet (*Pennisetum americanum* L.). Appl Environ Microbiol 37:1016–1024
Timmusk S, Nicander B, Granhall U, Tillberg E (1999) Cytokinin production by *Paenibacillus polymyxa*. Soil Biol Biochem 31:1847–1852

Tripathi M, Munot HP, Shouche Y, Meyer JM, Goel R (2005) Isolation and functional characterization of siderophore-producing lead- and cadmium-resistant *Pseudomonas putida* KNP9. Curr Microbiol 50:233–237

Trivedi P, Pandey A, Sa T (2007) Chromate reducing and plant growth-promoting activities of psychrotrophic *Rhodococcus erythropolis* MTCC 7905. J Basic Microbiol 47:513–517

Trotta A, Falaschi P, Cornara L, Minganti V, Fusconi A, Drava G, Berta G (2006) Arbuscular mycorrhizae increase the arsenic translocation factor in the As hyperaccumulating fern *Pteris vittata* L. Chemosphere 65:74–81

Ugrekhelidze D, Korte F, Kvesitadze G (1997) Uptake and transformation of benzene and toluene by plant leaves. Ecotoxicol Environ Saf 37:24–29

Upadhyaya H, Panda SK, Bhattacharjee MK, Dutta S (2010) Role of arbuscular mycorrhiza in heavy metal tolerance in plants: prospects for phytoremediation. J Phytology 2:16–27

USEPA (United States Environmental Protection Agency) (2000a) Introduction to phytoremediation, EPA 600/R-99/107. U.S. Environmental Protection Agency, Office of Research and Development, Cincinnati, OH

USEPA (United States Environmental Protection Agency) (2000b) Electrokinetic and phytoremediation in situ treatment of metal-contaminated soil: state of-the-practice. Draft for final review, EPA/542/R-00/XXX. US Environmental Protection Agency,Office of Solid Waste and Emergency Response Technology Innovation Office, Washington, DC

Van der Ent S, Van Wees SC, Pieterse CM (2009) Jasmonate signalling in plant interactions with resistance–inducing beneficial microbes. Phytochemistry 70:1581–1588

Van Loon LC, Bakker PA, Pieterse CM (1998) Systemic resistance induced by rhizosphere bacteria. Annu Rev Phytopathol 36:453–483

van Scholl L, Kuyper TW, Smits MM, Landeweert R, Holffland E, van Breemen N (2008) Rock-eating mycorrhizas: their role in plant nutrition and biogeochemical cycles. Plant Soil 303:35e47

Vangronsveld J, Herzig R, Weyens N, Boulet J, Adriaensen K et al (2009) Phytoremediation of contaminated soils and groundwater: lessons from the field. Environ Sci Pollut Res 16:765–794

Vassilev A, Berova M, Stoeva N, Zlatev Z (2005) Phytotechnologies for sustainable use management of metal contaminated soils: short review. Manag Sustain Dev 3(4):90–96. (In Bulgarian)

Venkatesh NM, Vedaraman N (2012) Remediation of soil contaminated with copper using rhamnolipids produced from Pseudomonas aeruginosa MTCC 2297 using waste frying rice bran oil. Ann Microbiol 62:85–91

Verhagen BWM, Glazebrook J, Zhu T, Chang HS, van Loon LC, Pieterse CMJ (2004) The transcriptome of rhizobacteria- induced systemic resistance in *Arabidopsis*. Mol Plant-Microbe Interact 17:895–908

Verma SCC, Ladha JKK, Tripathi AKK (2001) Evaluation of plant growth promoting and colonization ability of endophytic diazotrophs from deep water rice. J Biotechnol 91:127–141

Vidali M (2001) Bioremediation. An overview. Pure Appl Chem 73:1163–1172

Vijayarengan P (2005) Nitrogen and potassium status of greengram (*Vigna radiata*) cultivars under nickel stress. Nat Environ Pollut Technol 4:65–69

Vivas A, Bir O, Ruíz-Lozano B, Barea JM, Azcón R (2006) Two bacterial strains isolated from a Zn-polluted soil enhance plant growth and mycorrhizal efficiency under Zn-toxicity. Chemosphere 62:1523–1533

Wang J, Liu X, Zhang X, Liang X, Zhang W (2011) Growth response and phytoremediation ability of reed for diesel contaminant. Procedia Environ Sci 8:68–74

Wang W, Deng Z, Tan H, Cao L (2013) Effects of Cd, Pb, Zn, Cu-resistant endophytic *Enterobacter* sp. CBSB1 and *Rhodotorula* sp. CBSB79 on the growth and phytoextraction of *Brassica* plants in multimetal contaminated soils. Int J Phytoremediation 15:488e497

Wang L, Gong H, Liao W, Wang Z (2015) Accumulation of particles on the surface of leaves during leaf expansion. Sci Total Environ 532:420–434

Weyens N, van der Lelie D, Taghavi S, Newman L, Vangronsveld J (2009a) Exploiting plant–microbe partnerships for improving biomass production and remediation. Trends Biotechnol 27:591–598

Weyens N, van der Lelie D, Taghavi S, Vangronsveld J (2009b) Phytoremediation: plant-endophyte partnerships take the challenge. Curr Opin Biotechnol 20:248–254
Weyens N, Truyens S, Dupae J, Newman L, Taghavi S, Lelie D, Carleer R, Vangronsveld J (2010) Potential of the TCE degrading endophyte *Pseudomonas putida* W619-TCE to improve plant growth and reduce TCE phytotoxicity and evapotranspiration in poplar cuttings. Environ Pollut 158:2915–2919
Weyens N, Thijs S, Popek R, Witters N, Przybysz A, Espenshade J, Gawronska H, Vangronsveld J, Gawronski SW (2015) The role of plant–microbe interactions and their exploitation for phytoremediation of air pollutants. Int J Mol Sci 16:25576–25604
Wheeler CT, Hughes LT, Oldroyd J, Pulford ID (2001) Effect of nickel on *Frankia* and its symbiosis with *Alnus glutinosa* (L.) gaertn. Plant Soil 231:81–90
Whipps JM (1990) Carbon utilization. In: Lynch JM (ed) The rhizosphere. Wiley-Interscience, Chichester, pp 59–97
White PM, Wolf DC, Thoma GJ, Reynolds CM (2006) Phytoremediation of alkylated polycyclic aromatic hydrocarbons in a crude oil-contaminated soil. Water Air Soil Pollut 169:207–220
Williams PM, de Mallorca MS (1982) Abscisic acid and gibberellin-like substances in roots and root nodules of *Glycine max*. Plant Soil 65:19–26
Wilson D (1993) Fungal endophytes: out of sight but should not be out of mind. Oikos 68:379–384
World Health Organization (WHO) (2000) The right to healthy indoor air-report on a WHO meeting. World Health Organization (WHO): Bilthoven, The Netherlands
Wu SC, Cheung KC, Luo YM, Wong MH (2006) Effects of inoculation of plant growth-promoting rhizobacteria on metal uptake by *Brassica juncea*. Environ Pollut 140:124–135
Wu QS, He XH, Zou YN, Liu CY, Xiao J, Li Y (2012) Arbuscular mycorrhizas alter root system architecture of Citrus tangerine through regulating metabolism of endogenous polyamines. Plant Growth Regul 68:27–35
Wuana RA, Okieimen FE (2011) Heavy metals in contaminated soils: a review of sources, chemistry, risks and best available strategies for remediation. ISRN Ecol 2011:1–20
Yang X, Feng Y, He Z, Stoffella PJ (2005) Molecular mechanisms of heavy metal hyperaccumulation and phytoremediation. J Trace Elem Med Biol 18:339–353
Yeh TY, Pan CT (2012) Effect of chelating agents on copper, zinc, and lead uptake by sunflower, Chinese cabbage, cattail, and reed for different organic contents of soils. J Environ Anal Toxicol 2:2161–0525
Yoon J, Cao X, Zhou Q, Ma LQ (2006) Accumulation of Pb, Cu, and Zn in native plants growing on a contaminated Florida site. Sci Total Environ 368:456–464
Yousaf S, Andria V, Reichenauer TG, Smalla K, Sessitsch A (2010) Phylogenetic and functional diversity of alkane degrading bacteria associate with Italian ryegrass (*Lolium multiflorum*) and birds foot trefoil (*Lotus corniculatus*) in a petroleum oil-contaminated environment. J Hazard Mater 184:523–532
Zaidi S, Usmani S, Singh BR, Musarrat J (2006) Significance of *Bacillus subtilis* strain SJ-101 as a bioinoculant for concurrent plant growth promotion and nickel accumulation in *Brassica juncea*. Chemosphere 64:991–997
Zamioudis C, Pieterse CM (2012) Modulation of host immunity by beneficial microbes. Mol Plant-Microbe Interact 25:139–150
Zhang H, Zheng LC, Yi XY (2009) Remediation of soil co-contaminated with pyrene and cadmium by growing maize (*Zea mays* L.). Int J Environ Sci Technol 6:249–258
Zhang X, Xia H, Li ZA, Zhuang P, Gao B (2011a) Identification of a new potential Cd-hyperaccumulator *Solanum photeinocarpum* by soil seed bank-metal concentration gradient method. J Hazard Mater 189:414–419
Zhang Y, He L, Chen Z, Wang Q, Qian M, Sheng X (2011b) Characterization of ACC deaminase-producing endophytic bacteria isolated from copper-tolerant plants and their potential in promoting the growth and copper accumulation of *Brassica napus*. Chemosphere 83:57–62
Zhang X, Lin L, Chen M, Zhu Z, Yang W, Chen B, Yang X, An Q (2012) A nonpathogenic Fusarium *oxysporum* strain enhances phytoextraction of heavy metals by the hyperaccumulator *Sedum alfredii* Hance. J Hazard Mater 229:361–370

Microbiome: Effect on Plant System, Current Application and Future Aspect

7

Pragati Sahai and Vivek Kumar

Abstract

The Holobiont is the overall plant biochemistry that involves plant kingdom along with microbial world of bacteria and viruses interactive with plant world. The microbiome is the large Holobiont that gives plant health and stability as the microbes in its Holobiont provide resistance to abiotic and biotic stress and also to antibiotics and drugs apart from being part of plant nutrition and nutrient uptake. By understanding the plant and its microbiome, the application and modifications in the application can be studied.

Keywords

Holobiont · Microbiome · Plant nutrition · Antibiotic · Abiotic stress

The plant kingdom does not stand alone in this ecology, but it is surrounded inside and outside with microbes that are continuously interacting with them. These bacteria and viruses are part of plant biochemistry constituting a Holobiont. This Holobiont gives plant health and stability as the microbes in its Holobiont provide resistance to abiotic and biotic stress and also to antibiotics and drugs apart from being part of plant nutrition and nutrient uptake. But still it has the complexity that needs to be explored. The plant changes its microbiome with changing environment and so does the microbes in the microbiome who adapt to the environment by bringing structural changes in them. By understanding the plant and its microbiome,

P. Sahai (✉)
Amity Institute of Biotechnology, Amity University, Noida, India

V. Kumar
Himalayan School of Biosciences, Swami Rama Himalayan University, Jolly Grant, Dehradun, Uttarakhand, India

M. Kumar et al. (eds.), *In vitro Plant Breeding towards Novel Agronomic Traits*,
https://doi.org/10.1007/978-981-32-9824-8_7

the application and modifications in the application can be studied. These modifications lead to the future prospects of the plant and its microbiome, and a rationale of engineering the microbiome was possible that led to development of advanced agricultural practices for sustainable development and ecological stability.

7.1 Introduction

The world of microbes is spread all over the ecosphere. Their interactions with the environment have been subject to study for more than 100–200 years, and still there is scope of further study and research. The microbial interactions with the plant are known for plants strength, ecological diversity and nutritional stability to sustain in the changing environment. The interaction of host plant with its inhabitant microbes develops three nutritional zones also known as rhizosphere, phyllosphere and endosphere (Fig. 7.1). The three zones together constitute the *Holobiont* – the term

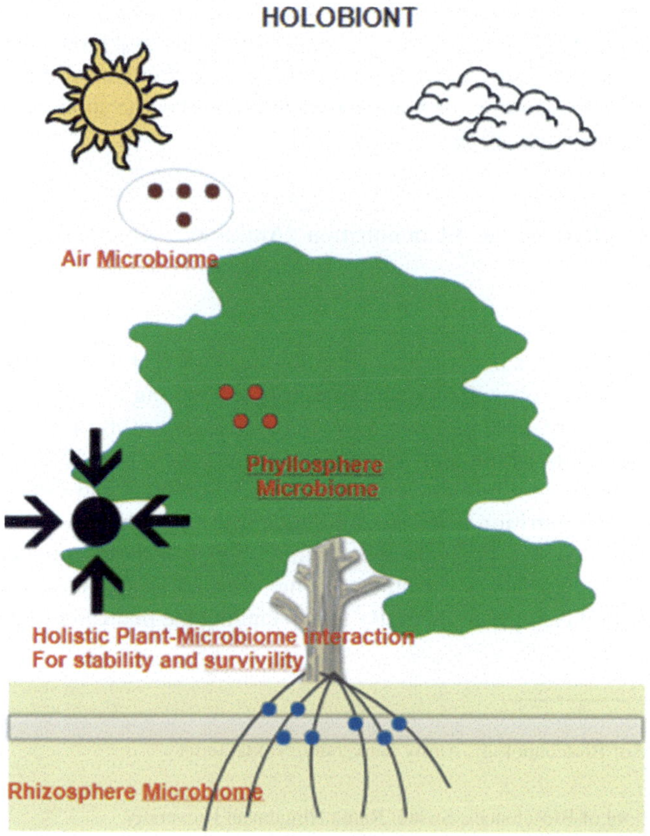

Fig. 7.1 The functionality of microbiome in the host plant

originally coined by Lynn Margulis (1991) in his book, *Symbiosis as a Source of Evolutionary Innovation* (Margulis and Fester 1991) (Fig. 7.1). But before this book, the intensive work was done on plant-microbe interaction by Lorenz Hiltner in 1901 that was then published in the research work of his co-worker later on (Hartmann et al. 2008).

The association of plants with microbes is very old and ancient. The presence of VAM fungi is the proof of mutualistic relationship between the host plant and the microbe that gave the nutritional strength and physical stability to the host plant that played a key role in formation of terrestrial habitat. Their presence also depicted the genetic diversity or changes in their genetic arrangements with the changes in the environment that brought about evolution and diversification in plants too (Heckman et al. 2001). Similarly the plant model of *Arabidopsis* demonstrated many plant microbial interactions by different research groups to further explore the biochemistry of Holobiont (Bulgarelli et al. 2013a; Lundberg et al. 2012).

The extensive long research in this field beside physical parameters also entered to molecular and proteomics and genomics level of exploration to study the changes in genetic arrangements in the host-associated microbes that brought about diversification in the host plant (Prestat et al. 2014). The suppression of soilborne pathogens and resistivity in plants to such pathogens when studied was found to be because of plants own rhizosphere (Cook et al. 1995; Mendes et al. 2013). Similarly, on studying the trans kingdom, different species and genus interactions as a subject of interest, it was found that even if the interaction was trans kingdom (Plant+Microbe) still the rhizosphere specificity could be seen (Hartmann et al. 2008; Berg and Smalla 2009). The study at genomic and molecular level of genome of endophytic microorganism or studying the signalling behaviour led to the formation of molecular markers that helped further in transcriptome analysis of any system (Müller et al. 2015; Fu et al. 2015).

Thus the molecular applications based on genomic sequencing and protein analysis have greatly enhanced the knowledge behind the diverse interaction of plant microbiome (Kiers et al. 2011; Werner et al. 2014); however this knowledge is yet insufficient to explain the functionality of plant microbiome with respect to vast ecological diversification. So it will be better to accept that despite the vast research in this field, still the study is primitive in context with Earth as large biome that opens the future prospects of great research that can be beneficial for improvement of crops and enhancing its production (Sánchez-Cañizares et al. 2017).

7.2 Effect of Microbiome on Plant System

The plant system largely depends on its microbiome for growth and for its survival. It was reported that in absence of fungi and bacteria, the plant transplants were not successful (Hardoim et al. 2008), and so the concepts of Holobiont came into picture stating a single dynamic entity that defines the complete function and interaction between the host plant and its associated microbiome. The word Holobiont came from the word Holobiome that means when the genome of the host was intercepted, the intervention of its associated microbial genome could be seen stating to be a single dynamic entity (Guerrero et al. 2013). So it was found that whenever there was change in environment, it brought about the change in Holobiont in whole, and it had an individualistic effect on plant and microbiome too. The stresses faced and the survival done in changing environment largely depend on the strength of the Holobiont being the unit of selection and adaptation (Zilber-Rosenberg and Rosenberg 2008). These adaptations bring heterogeneity and defence system that can be transitional if the relationship is symbiont or can be vertical transmission from seed to plant if there is genomic changes in the Holobiont that are functionally stable (Cankar et al. 2005; Berendsen et al. 2012).

7.3 Factors That Affect the Holobiont (Fig. 7.2)

7.3.1 Rhizosphere and Rhizoplane

The different planes decide the microbial consortia of that plane. Like in rhizosphere, the root-associated bacteria and soil microorganisms together decide the microbiome of that plane. The resistance to soilborne pathogens and the defence system of the plant largely depend on this microbiome. There can be a systemic resistance because of temporary association of microbiome with the host plant, or it can be a lateral transfer of resistivity from soil to seed and to plant further where the physical environment like soil properties, its pH, its humidity and temperature of the soil acts as growth chambers for the microbiome (Bouffaud et al. 2014). It also brings to the light the fact that instead of plant genome, it is the inhabitant microbial community of the plane that decides the microbiome of that plane largely, with host genome playing some role in selection of the microbiome due to its evolutionary significance (Reinhold-Hurek et al. 2015).

Another factor that affects the microbiome is segregation of microbes in different spheres in different planes, that is, the microbes in rhizosphere are difficult to be segregated from the soil microbes inhabiting the rhizoplane. The 16 SR sequencing of these soil inhabitants showed very little difference in their genome, so selection

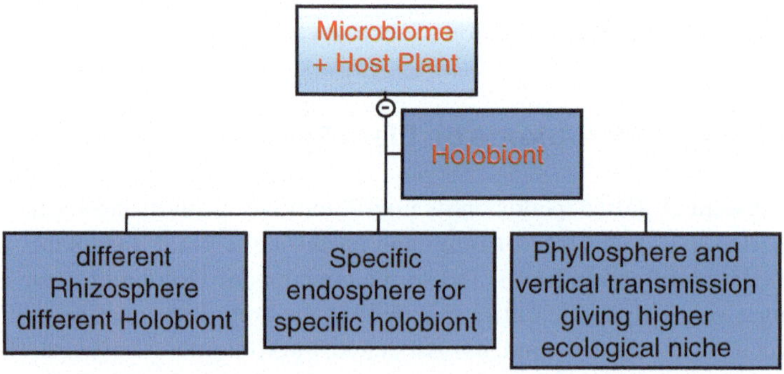

Fig. 7.2 Factors affecting Holobiont

largely depended on their interaction with their host plant, their competitors, and soil conditions of pH, temperature, and humidity (Bulgarelli et al. 2013b). The reports of CARD-FISH (catalyzed reporter deposition-fluorescent in situ hybridization) in rice plant also showed the colonization of bacteria in rhizoplane of rice, thus stating that rhizoplane plays an important role in deciding the microbiome of rhizosphere (Schmidt and Eickhorst 2013).

7.3.2 Endosphere and Host Specificity

After the microbe enters the tissues from rhizosphere or from air, they become part of endosphere. The order of endosphere microbiome is arbuscular mycorrhizal fungi that penetrate deeply followed by bacteria and least by archaea. The endospere microbiome unlike rhizosphere microbiome is highly specific and chiefly selected to host interaction. There is less diversity seen in the endosphere than in rhizosphere. When taxonomic units were studied of the endosphere microbiome and rhizosphere microbiome, it was seen that the microbial diversity decreased as one moved from rhizosphere to endosphere (Schlaeppi et al. 2014). The studies on field-grown popular trees on structural variability and niche differentiation in rhizosphere and endosphere bacterial microbiome stated that the rhizosphere microbiome is less structurally variable and holds lower ecological niche than well-structured endosphere microbiome. Thus the endosphere microbiome helps in better adaptability of the host plant and increased plant strength (Beckers et al. 2017).

7.3.3 Phyllosphere and Vertical Transmission

The endosphere microbiome can also penetrate leafy tissues and even reproductive parts probably for vertical transfer or lateral diffusion into next generation. These microbial denizens are residing either outside or inside the aerial plant tissues (Peñuelas and Terradas 2014). There have been few reports of finding microbiome in reproductive organs like anthers and carpels giving an evidence of vertical transmission from seed to new plant (Lopez-Velasco et al. 2013). The presence of large amount of microbiome on surface of host leaf and in the leaf internal tissues shows their growth in different environment, thus sharing a very different microbial population in the endosphere and ectosphere of plant tissues above the ground and below the ground (Romero et al. 2014). It was also reported that the microbiome on leaf surface or phyllosphere can comprise of archaea like methylotrophs as they are in much harsher conditions than rhizosphere because of desiccation, UV radiations and oligotrophy; therefore the microbiome is structurally more defined and has better ecological niche. They are capable of developing special proteins like microbial rhodopsins that help in absorbing green lights and act as proton pumps, facilitating photosynthesis (Atamna-Ismaeel et al. 2011; Atamna-Ismaeel et al. 2012). These methylotrophs and photoharvesters cause the competitive exclusion of plant pathogens and create a better survival for them in the space or have higher niche as compared to other microbiome. Thus the phyllosphere can also affect the Holobiont.

As stated earlier that the microbiome influences the strength of the plant, thus it participates in the host plants function. It was also reported that these microbiomes provide defence system to the plant increasing its immunity. It was also reported that because of this microbiome, the plants become more structurally advanced and have better ecological niche than the plants without their microbiome.

7.3.4 Core Microbiome and Ecological Niche

This microbiome provides functional plasticity in plants. That means, the microbiome that is part of Holobiont is after screening and competitive elimination of microbes that were weak and structurally poor. Such interactions with the environment lead to continuously elimination of weaker population and development of better microbial flora on the phyllosphere causing biodiversity. Because of this biodiversity, even the same species of the plant have different ecological niche resulting in different strains of the plant. The different strains of the same plant species have different Holobiont that facilitates different processes as survival in abiotic stresses, heterogeneity, mutualism sharing, resource accessibility and creating novel habitat (McIntire and Fajardo 2014; Wagner et al. 2014). Here it is important to mention about the term *core microbiome* that is generated out of a plant community, for example, if Leguminosae has a population of bacteria as its microbiome, then all the plants belonging to this family will have at least that set of population apart from some different ones. Again the different analysis at genomic and transcriptome level will help to understand the functionality of the core microbiome. That is the microbes of different Holobiont of the same family with similar genomic, and transcriptome analysis will constitute a core microbiome. It is this core microbiome that will place that particular family in the ecological map, so it becomes very important for a plant community (Finkel et al. 2017; Lemanceau et al. 2017).

7.3.5 Developing the Immunity in Plants

The plants have two types of innate immunity: *horizontal resistance* and *vertical resistance.*

Horizontal resistance is one in which the immunity is triggered due to pathogenic molecular patterns, and so it is also called *pathogen-associated molecular pattern triggered immunity or PTI*. Similarly vertical resistance or R gene-based immunity is triggered by effector cells also called *effector-triggered immunity or ETI (Jones and Dangl* 2006). The mechanism of defence in the former case depends on two types of molecular patterns: one that is generated by microbe and easily identified by plants' microbiome and second type of pattern is generated by cellular damage or any kind of biotic stress to the cell that is identified by host plant. So in this mechanism, the microbiome sends signal to its host, and the entire Holobiont prepares its defence system against its microbial enemy. The defence system and hypersensitive reaction are also triggered by effector cells that are produced by microbiome of the host plant. This helps in signalling to cell apoptosis or local necrosis in case of danger (Boller and Felix 2009). Thus both types of immunity because of microbiome help the plant to gain strength and survival. Many fungi have been reported to provide fitness and strength to tall trees, and similarly many bacteria have provided resistance against many soilborne bacteria (Hacquard et al. 2017).

7.4 Current Application of Microbiome and Their Future Aspects (Fig. 7.3)

The recent insight into application of microbiome revolves around the three habitats in the Holobiont, viz. rhizosphere, phyllosphere and endosphere, that harbours different microbes (Wagner et al. 2015; Vorholt 2012; Philippot et al. 2013; Hardoim et al. 2015). The rhizosphere apart from providing nutritional benefits to host plant also provides effective biocontrol to plant against phytopathogens. This biocontrol activity of microbiome is being extensively used to replace the existing pesticidal activities that are harmful to the environment. The specificity of microbiome to its host helps in defending the host from soilborne pathogens (Zachow et al. 2016; Cole et al. 2015). These associations form a Holobiont that can be considered as evolved microbial species and host assembly with better survival in the environment. The microbiome helps in preventing diseases because of its rich diversity acting as biomarkers in plant protection strategies, thus by identifying the biocontrol agents in rhizosphere, targeted and predictive protection strategies can be designed. These microbiome-based solutions are key to targeted and predictive biocontrol approaches that are required for effective plant breeding and ecosystem health (Berg et al. 2017). However in this symbiotic behaviour, there are some cheaters too, that is,

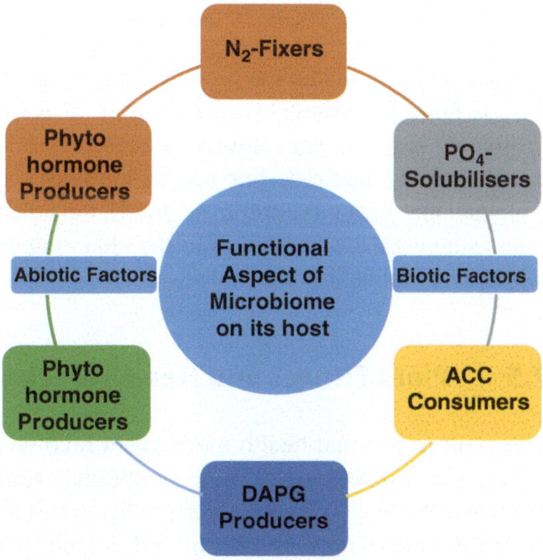

Fig. 7.3 Functional aspect of microbiome on the host plant

these are those microbes that take the gain from host plant at a rate that is so rapid that it makes the mutualism unstable. There are many hypotheses that support this fact. First hypothesis is the luxury resource exchange that allows both the host and symbiont to enjoy the surplus nutrients in the environment; however if this utilization shift tends more towards symbiont, then it will be stronger than the host, destroying the mutualism (Kiers and Van Der Heijden 2006). But this hypothesis does not hold true as the nutritional availability in the environment is very limited, so having a surplus is out of question. Second hypothesis states the partner choice, according to which the partner with better strength for better survival will be selected by the symbiont leading to disruption in mutualism with weaker partner. Such processes occur on the basis of molecular signalling that sends message to the symbiont to continue with the interaction or start another (Sachs et al. 2004; Sachs and Simms 2006). Again by studying these molecular signalling, a better plant symbiont relationship can be researched and experimented, and these molecular signalling is called as plant-microbe communication. This communication is largely seen in the rhizosphere of the Holobiont and is extensively studied (Van Overbeek and Van Elsas 2008; Venturi and Keel 2016). The third hypothesis is of forced cooperative behaviour that sounds very interesting in a way that the strong symbiont does the resource allocation and provides lesser resource to weaker symbiont, thereby forcing the weaker one to be always dependent on the host and the stronger symbiont. Prominently such behaviour is seen in legumes where N_2-fixing organism penalize the rhizobia that do not participate in fixation (Kiers et al. 2003) and similarly is seen in AM fungi where the stronger ones reallocate carbon and phosphorous uptake of the cheater AM fungi and transfer the phosphorous and carbon to the host plants forcing them to be a part of the symbiont (Kiers et al. 2011). Apart from the fact that the cheaters are in symbiosist, there are reports of them being free riders such as rhizobia in legumes that after sometime does not benefit its host but starts reallocating its lipid contents by producing rhizobitoxine, an ethylene inhibitor (Ratcliff and Denison 2009) and this symbiont has recently been reported to be selective for its partner (Porter and Simms 2014). The recent advances in cropping have largely eliminated the growth of these cheaters and free riders as the soil these days is nutrient enriched so reallocation of resource or selecting stronger host for symbiosis is no longer required; hence the cheater population is high in conventional farming methods but not in modern ways of agriculture (Johnson 1993; Duhamel and Vandenkoornhuyse 2013); thus these plant microbiomes acting as free riders suggest their adaptation to the changing environmental need, thus contributing to the plant health mechanism.

7.5 Plant Fitness and Health Mechanism

The plant fitness and health mechanism revolve around its endosphere and phyllosphere after rhizosphere as these symbionts even pass from soil into the plant tissue or inhabit the leaf surface from external environment. This collaboration is basically the collaboration of good like detoxifying enzymes, nutrients by and from host and its microbe and gene uptakes and expressions (Kemen 2014). According to Black Queen hypothesis, this symbiosis is automatic and for selfish needs. The hypothesis was supported by two facts that the microbial interactions involve two types of population, helper (host and strong microbe) and beneficiary (other microbe or host), where helper are responsible for major productions and beneficiaries are dependent on these helpers that result in loss of genes and functions in the latter case because of non-usability, and this shows the selfish trajectory that makes powerful symbiosis and accounts for plant health (Morris et al. 2012). The plants are highly adaptable and can survive with changing environmental conditions by bringing about the change in their microbiome. They can select their ectospheric and endospheric microbiome by mutual cooperation for better survival in the environment as stated earlier. This mutualism helps the Holobiont as a whole to improve its fitness and promote evolution in ecology.

The vertical transmission is also the cause of plant fitness. Not all microbes of microbiome can be part of vertical transmission process that involves the integration of microbial genome with host genome so that the trait can be transferred to the next generation, which otherwise is temporary and only in presence of that symbiont. Once that symbiont is removed that trait is also lost; however in case of vertical transmission, the trait enters the progeny and makes the plant better adapted to the environment. Various fungi like *Epichloe festucae* in *Festuca rubra* because of vertical transmission affect the plant's nutrient balance (Vázquez-de-Aldana et al. 2013). This also allows the partners to have fidelity check on each other (Frederickson 2013) and also to have an automatic fitness feedback (Sachs et al. 2011). Thus on the broader scale, the partner feedback is largely based on cooperation, so it completely collaborates with their personal feedbacks too, stating the relationship to be positive (Friesen 2012). The recent approaches to improve plant fitness thus largely revolves on these feedbacks that tells that plant works as a Holobiont, and these community dynamics help in modifying the intensity and nature of plant competition to see the survival rate of the plant (Pendergast et al. 2013), and this is the future of effective biocontrol and production of next-generation bio products as a part of sustainable agriculture that visualizes the plant as a metaorganism because of its ectosphere and endosphere mutualistic interactions that are even transferred to its seed and the traits because this spermosphere is transferred to the next generation. Hence the possibility of designing better agricultural strategies with the help of growing the plant in axenic environment by experimental botany tools helps in knowing the microbiome of the plant that is beneficial to it, and the rate of mutualism, the nature of collaboration, the different molecular signalling and developed defence system can be studied (Berg et al. 2013; Rybakova et al. 2016a, b).

7.6 Plant-Microbe Communication

The plant-microbe communication is another aspect that is of greater interest for designing the effective agricultural practices. As from the experiments on plant in axenic environment, the most common molecular signal that was identified was

IAA (indole acetic acid). The phylogenetic evidence suggests that IAA biosynthesis has an evolutionary significance of natural selection. Though the biosynthesis of IAA in bacteria to plant is independent of each other, the hypothesis of natural selection makes it a widespread physiological code in the microbiome and its participation in their interactions (Fu et al. 2015). But the recent research in plant-microbe communication revolves around secondary metabolites like volatile organic compounds (VOC) and its involvement in communication between symbiont and the host. It was reported that microbes use widespread of VOC to communicate among themselves and with the host (Bitas et al. 2013; Peñuelas et al. 2014). Besides VOC's the structural components of microbes like LPS, siderophores, flagella, and quorum-sensing molecules in Gram-negative molecule are found to activate the signalling pathway for plant's induced systemic resistance of defence system. Thus it's the entire microbiome whose continuous interaction and communication with each other and with the host if studied deeply will help in developing the effective defence system for plant that will lead to better crop productivity in the future (Ryu et al. 2003; Lee et al. 2012; Lugtenberg and Kamilova 2009; Hartmann and Schikora 2012; Schenk and Schikora 2015).

The Earth Microbiome Project helps in microbiome profiling of the soil that has enhanced the identification of microbiome in high and low susceptibility region (Gilbert et al. 2010). Such investigation assists in development of microbial probiotics and developing targeted approaches for habitat reclamation that is heavily invaded because of land pollution and wrong agricultural practices and focuses on efforts towards sustainable development (Berlec 2012).

7.7 Plant and Soil Microbiome Engineering

After the advancements in genome sequencing and meta-omic tools, the microbiome analysis without culturing has become feasible and thus further accelerates the microbiome research influencing the ecosystem (Mueller and Sachs 2015). After knowing that the microbiome composition directly influences the ecosystem, the recent advancement in this field is *microbiome engineering* that helps in shaping the microbiome to alter the ecosystem of interest.

When *plant microbiome engineering* is concerned, then it is basically engineering of root- associated microbes that include endosphere and rhizosphere microbiome since it has been proven that they play a crucial role in plants well-being and fitness (Agler et al. 2016). Thus by altering their composition and signalling pathways, improved phenotypes can be generated, and when these phenotypes become the part of the plant microbiome, they benefit the plant in its growth, fitness and health. The microbiome is engineered by simple molecular techniques of microbiome transfer (Zolla et al. 2013), formation of synthetic microbiome (Santhanam et al. 2015; Glick 2012) and host-mediated artificial selection of developing evolved microbiome (Swenson et al. 2000; Edwards et al. 2015). However even the best molecular technique cannot guarantee the successful microbiome modification due to lack of functional native microbiome that can be cultured easily. So the concept of *soil microbiome engineering* came into picture. The soil microbiome refers to the entire population outside rhizosphere that is mainly influenced by agricultural management practices than by the host plant. So the engineering of soil microbiome is basically using healthier green practices of agriculture like organic farming that bans the use of chemical fertilizers and pesticides (Hartmann et al. 2015) and other agricultural practices like cropping system (Xiong et al. 2015), tillage (Souza et al. 2016), rotational cropping (Souza et al. 2016), etc.

This balances the soil composition and the residents of soil bringing back the soil health and thereby plant fitness. The microbiome engineering was further facilitated by use of phages for microbiome transfer; however the utilization of phageome is very limited because of their limited host specificity. Still phages are part of agricultural practices as antibacterial agents and for antibacterial purposes (Dangl et al. 2013).

7.8 Conclusion (Fig. 7.4)

The conclusion of the study is to implicit the idea of considering plant as a metaorganism that consists of different types of microbial community and plant cells that is influenced by biotic and abiotic stress and follows the rule of natural selection. Thus a plant is a Holobiont where its fitness and health is interrelated to its microbiome, and they share a mutualistic relationship. Even the microbiome in the Holobiont is connected to each other and together helps in plant growth functioning. The above studies have clearly stated the importance and contribution of microbiome on the host physiology and ecological characteristics. The recent meta-omic tools and transcriptome analysis have shown recent advances in microbiome

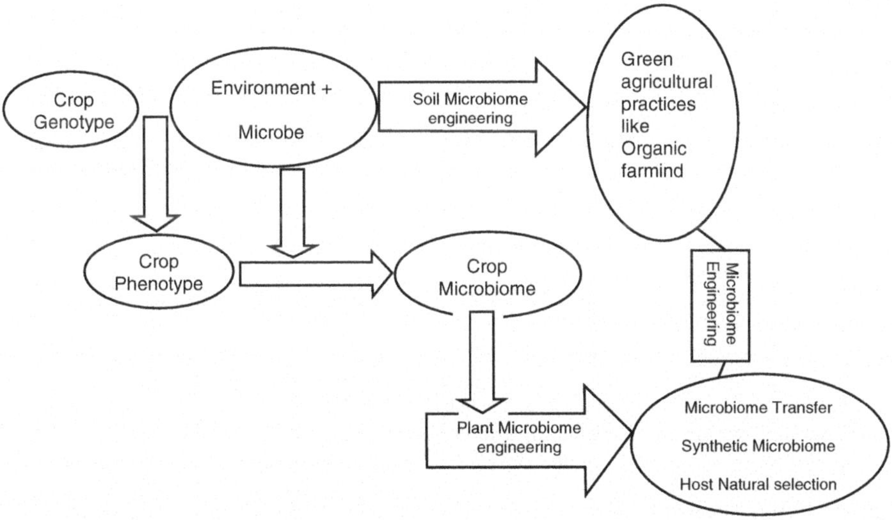

Fig. 7.4 Microbiome engineering: future prospective to sustainable agriculture

research that has helped in controlling pathogen burden. It was reported that the developed phenotype of plant (agriculture probiotic) had resistant genes for the plant pathogen (innate immune system) that led to chemical-free biocontrol of plant as a part of sustainable agriculture (Dangl et al. 2013). Another concept that came into picture was growing plants under axenic conditions or gnotobiotic conditions also called *neodomestication* of plant that helped in understanding the microbe and plant communication and their collaboration which was feasible in developing innovative and ecologically intensive agriculture (Duhamel and Vandenkoornhuyse 2013). Further the implementation of agricultural practices has recently seen lot of advancements that has enabled rational engineering of the soil (soil microbiome engineering) for crop improvement. Similarly capitalizing on the new found knowledge on microbiome led the start of plant microbiome engineering that has helped in development of plant protection strategies and reclamation of habitat. The microbiome engineering has largely been done on *A. thaliana* as host plant; however it raises a possibility of using other crops also as host plants. Therefore still the microbiome engineering is in its primitive stage, and yet lot of research is required on different plants with different microbiome so that stronger and drought resistant and wilt and rust resistant crops can be produced using green approaches. Even if the synthetic microbes are engineered, they will also contribute to sustainable agriculture as they possess novel functionalities.

So to conclude, with continuous efforts in gaining microbiome knowledge and its manipulation, the microbiome engineering will soon become an important tool in agriculture for sustainable development.

References

Agler MT, Ruhe J, Kroll S, Morhenn C, Kim ST, Weigel D, Kemen EM (2016) Microbial hub taxa link host and abiotic factors to plant microbiome variation. PLoS Biol 14(1):e1002352

Atamna-Ismaeel N, Finkel OM, Glaser F, Sharon I, Schneider R, Post AF, Spudich JL, von Mering C, Vorholt JA, Iluz D, Béjà O (2011) Microbial rhodopsins on leaf surfaces of terrestrial plantsemi_2554. Environ Microbiol 14:140–146

Atamna-Ismaeel N, Finkel OM, Glaser F, Sharon I, Schneider R, Post AF, Spudich JL, von Mering C, Vorholt JA, Iluz D, Béjà O (2012) Microbial rhodopsins on leaf surfaces of terrestrial plants. Environ Microbiol 14(1):140–146

Beckers B, De Beeck MO, Weyens N, Boerjan W, Vangronsveld J (2017) Structural variability and niche differentiation in the rhizosphere and endosphere bacterial microbiome of field-grown poplar trees. Microbiome 5(1):25

Berendsen RL, Pieterse CM, Bakker PA (2012) The rhizosphere microbiome and plant health. Trends Plant Sci 17(8):478–486

Berg G, Smalla K (2009) Plant species and soil type cooperatively shape the structure and function of microbial communities in the rhizosphere. FEMS Microbiol Ecol 68(1):1–13

Berg G, Zachow C, Müller H, Philipps J, Tilcher R (2013) Next-generation bio-products sowing the seeds of success for sustainable agriculture. Agronomy 3:648

Berg G, Köberl M, Rybakova D, Müller H, Grosch R, Smalla K (2017) Plant microbial diversity is suggested as the key to future biocontrol and health trends. FEMS Microbiol Ecol 93(5)

Berlec A (2012) Novel techniques and findings in the study of plant microbiota: search for plant probiotics. Plant Sci 193:96–102

Bitas V, Kim HS, Bennett JW, Kang S (2013) Sniffing on microbes: diverse roles of microbial volatile organic compounds in plant health. Mol Plant-Microbe Interact 26(8):835–843

Boller T, Felix G (2009) A renaissance of elicitors: perception of microbe-associated molecular patterns and danger signals by pattern-recognition receptors. Annu Rev Plant Biol 60:379–406

Bouffaud ML, Poirier MA, Muller D, Moënne-Loccoz Y (2014) Root microbiome relates to plant host evolution in maize and other Poaceae. Environ Microbiol 16(9):2804–2814

Bulgarelli D, Rott M, Schlaeppi K, van Themaat EVL, Ahmadinejad N, Assenza F, Rauf P, Huettel B, Reinhardt R, Schmelzer E, Peplies J (2013a) Revealing structure and assembly cues for Arabidopsis root-inhabiting bacterial microbiota. Nature 501(7468):S25

Bulgarelli D, Schlaeppi K, Spaepen S, van Themaat EVL, Schulze-Lefert P (2013b) Structure and functions of the bacterial microbiota of plants. Annu Rev Plant Biol 64:807–838

Cankar K, Kraigher H, Ravnikar M, Rupnik M (2005) Bacterial endophytes from seeds of Norway spruce (Picea abies L. Karst). FEMS Microbiol Lett 244(2):341–345

Cole BJ, Fletcher M, Waters J, Wetmore K, Blow MJ, Deutschbauer AM, Dangl JL, Visel A (2015) Genetic control of plant root colonization by the biocontrol agent, Pseudomonas fluorescens (No. LBNL-178476). Lawrence Berkeley National Laboratory (LBNL), Berkeley

Cook RJ, Thomashow LS, Weller DM, Fujimoto D, Mazzola M, Bangera G, Kim DS (1995) Molecular mechanisms of defense by rhizobacteria against root disease. Proc Natl Acad Sci 92(10):4197–4201

Dangl JL, Horvath DM, Staskawicz BJ (2013) Pivoting the plant immune system from dissection to deployment. Science 341(6147):746–751

Duhamel M, Vandenkoornhuyse P (2013) Sustainable agriculture: possible trajectories from mutualistic symbiosis and plant neodomestication. Trends Plant Sci 18(11):597–600

Edwards J, Johnson C, Santos-Medellín C, Lurie E, Podishetty NK, Bhatnagar S, Eisen JA, Sundaresan V (2015) Structure, variation, and assembly of the root-associated microbiomes of rice. Proc Natl Acad Sci 112(8):E911–E920

Finkel OM, Castrillo G, Paredes SH, González IS, Dangl JL (2017) Understanding and exploiting plant beneficial microbes. Curr Opin Plant Biol 38:155–163

Frederickson ME (2013) Rethinking mutualism stability: cheaters and the evolution of sanctions. Q Rev Biol 88(4):269–295

Friesen ML (2012) Widespread fitness alignment in the legume–rhizobium symbiosis. New Phytol 194(4):1096–1111

Fu SF, Wei JY, Chen HW, Liu YY, Lu HY, Chou JY (2015) Indole-3-acetic acid: A widespread physiological code in interactions of fungi with other organisms. Plant Signal Behav 10(8):e1048052

Gilbert JA, Meyer F, Jansson J, Gordon J, Pace N, Tiedje J, Ley R, Fierer N, Field D, Kyrpides N, Glöckner FO (2010) The earth microbiome project: meeting report of the "1 st EMP meeting on sample selection and acquisition" at Argonne National Laboratory October 6 th 2010. Stand Genomic Sci 3(3):249

Glick BR (2012) Plant growth-promoting bacteria: mechanisms and applications. Scientifica 2012:1

Guerrero R, Margulis L, Berlanga M (2013) Symbiogenesis: the holobiont as a unit of evolution. Int Microbiol 16(3):133–143

Hacquard S, Spaepen S, Garrido-Oter R, Schulze-Lefert P (2017) Interplay between innate immunity and the plant microbiota. Annu Rev Phytopathol 55(1):565

Hardoim PR, van Overbeek LS, van Elsas JD (2008) Properties of bacterial endophytes and their proposed role in plant growth. Trends Microbiol 16(10):463–471

Hardoim PR, Van Overbeek LS, Berg G, Pirttilä AM, Compant S, Campisano A, Döring M, Sessitsch A (2015) The hidden world within plants: ecological and evolutionary considerations for defining functioning of microbial endophytes. Microbiol Mol Biol Rev 79(3):293–320

Hartmann A, Schikora A (2012) Quorum sensing of bacteria and trans-kingdom interactions of N-acyl homoserine lactones with eukaryotes. J Chem Ecol 38(6):704–713

Hartmann A, Rothballer M, Schmid M (2008) Lorenz Hiltner, a pioneer in rhizosphere microbial ecology and soil bacteriology research. Plant Soil 312(1–2):7–14

Hartmann M, Frey B, Mayer J, Mäder P, Widmer F (2015) Distinct soil microbial diversity under long-term organic and conventional farming. ISME J 9(5):1177

Heckman DS, Geiser DM, Eidell BR, Stauffer RL, Kardos NL, Hedges SB (2001) Molecular evidence for the early colonization of land by fungi and plants. Science 293(5532):1129–1133

Johnson BJ (1993) Response of tall fescue to plant growth regulators and mowing frequencies. J Environ Hortic 11(4):163–167

Jones JD, Dangl JL (2006) The plant immune system. Nature 444(7117):323

Kemen E (2014) Microbe–microbe interactions determine oomycete and fungal host colonization. Curr Opin Plant Biol 20:75–81

Kiers ET, Van Der Heijden MG (2006) Mutualistic stability in the arbuscular mycorrhizal symbiosis: exploring hypotheses of evolutionary cooperation. Ecology 87(7):1627–1636

Kiers ET, Rousseau RA, West SA, Denison RF (2003) Host sanctions and the legume-rhizobium mutualism. Nature 425(6953):78

Kiers ET, Duhamel M, Beesetty Y, Mensah JA, Franken O, Verbruggen E, Fellbaum CR, Kowalchuk GA, Hart MM, Bago A, Palmer TM (2011) Reciprocal rewards stabilize cooperation in the mycorrhizal symbiosis. Science 333(6044):880–882

Lee B, Farag MA, Park HB, Kloepper JW, Lee SH, Ryu CM (2012) Induced resistance by a long-chain bacterial volatile: elicitation of plant systemic defense by a C13 volatile produced by Paenibacillus polymyxa. PloS One 7(11):e48744

Lemanceau P, Blouin M, Muller D, Moënne-Loccoz Y (2017) Let the core microbiota be functional. Trends Plant Sci 22:583

Lopez-Velasco G, Carder PA, Welbaum GE, Ponder MA (2013) Diversity of the spinach (Spinacia oleracea) spermosphere and phyllosphere bacterial communities. FEMS Microbiol Lett 346(2):146–154

Lugtenberg B, Kamilova F (2009) Plant-growth-promoting rhizobacteria. Annu Rev Microbiol 63:541–556

Lundberg DS, Lebeis SL, Paredes SH, Yourstone S, Gehring J, Malfatti S, Tremblay J, Engelbrektson A, Kunin V, Del Rio TG, Edgar RC (2012) Defining the core Arabidopsis thaliana root microbiome. Nature 488(7409):86

Margulis L, Fester R (eds) (1991) Symbiosis as a source of evolutionary innovation: speciation and morphogenesis. MIT Press, Cambridge, MA/London

McIntire EJ, Fajardo A (2014) Facilitation as a ubiquitous driver of biodiversity. New Phytol 201(2):403–416

Mendes R, Garbeva P, Raaijmakers JM (2013) The rhizosphere microbiome: significance of plant beneficial, plant pathogenic, and human pathogenic microorganisms. FEMS Microbiol Rev 37(5):634–663

Morris JJ, Lenski RE, Zinser ER (2012) The Black Queen Hypothesis: evolution of dependencies through adaptive gene loss. MBio 3(2):e00036–e00012

Mueller UG, Sachs JL (2015) Engineering microbiomes to improve plant and animal health. Trends Microbiol 23(10):606–617

Müller H, Berg C, Landa BB, Auerbach A, Moissl-Eichinger C, Berg G (2015) Plant genotype-specific archaeal and bacterial endophytes but similar Bacillus antagonists colonize Mediterranean olive trees. Front Microbiol 6

Pendergast TH, Burke DJ, Carson WP (2013) Belowground biotic complexity drives aboveground dynamics: a test of the soil community feedback model. New Phytol 197(4):1300–1310

Peñuelas J, Terradas J (2014) The foliar microbiome. Trends Plant Sci 19(5):278–280

Peñuelas J, Asensio D, Tholl D, Wenke K, Rosenkranz M, Piechulla B, Schnitzler JP (2014) Biogenic volatile emissions from the soil. Plant Cell Environ 37(8):1866–1891

Philippot L, Raaijmakers JM, Lemanceau P, Van Der Putten WH (2013) Going back to the roots: the microbial ecology of the rhizosphere. Nat Rev Microbiol 11(11):789–799

Porter SS, Simms EL (2014) Selection for cheating across disparate environments in the legume-rhizobium mutualism. Ecol Lett 17(9):1121–1129

Prestat E, David MM, Hultman J, Taş N, Lamendella R, Dvornik J, Mackelprang R, Myrold DD, Jumpponen A, Tringe SG, Holman E (2014) FOAM (functional ontology assignments for metagenomes): a hidden Markov model (HMM) database with environmental focus. Nucleic Acids Res 42(19):e145–e145

Ratcliff WC, Denison RF (2009) Rhizobitoxine producers gain more poly-3-hydroxybutyrate in symbiosis than do competing rhizobia, but reduce plant growth. ISME J 3(7):870

Reinhold-Hurek B, Bünger W, Burbano CS, Sabale M, Hurek T (2015) Roots shaping their microbiome: global hotspots for microbial activity. Annu Rev Phytopathol 53:403–424

Romero FM, Marina M, Pieckenstain FL (2014) The communities of tomato (Solanum lycopersicum L.) leaf endophytic bacteria, analyzed by 16S-ribosomal RNA gene pyrosequencing. FEMS Microbiol Lett 351(2):187–194

Rybakova D, Cernava T, Köberl M, Liebminger S, Etemadi M, Berg G (2016a) Endophytes-assisted biocontrol: novel insights in ecology and the mode of action of Paenibacillus. Plant Soil 405(1–2):125–140

Rybakova D, Schmuck M, Wetzlinger U, Varo-Suarez A, Murgu O, Müller H, Berg G (2016b) Kill or cure? The interaction between endophytic Paenibacillus and Serratia strains and the host plant is shaped by plant growth conditions. Plant Soil 405(1–2):65–79

Ryu CM, Farag MA, Hu CH, Reddy MS, Wei HX, Paré PW, Kloepper JW (2003) Bacterial volatiles promote growth in Arabidopsis. Proc Natl Acad Sci 100(8):4927–4932

Sachs JL, Simms EL (2006) Pathways to mutualism breakdown. Trends Ecol Evol 21(10):585–592

Sachs JL, Mueller UG, Wilcox TP, Bull JJ (2004) The evolution of cooperation. Q Rev Biol 79(2):135–160

Sachs JL, Skophammer RG, Regus JU (2011) Evolutionary transitions in bacterial symbiosis. Proc Nat Acad Sci 108(supplement 2):10800–10807

Sánchez-Cañizares C, Jorrín B, Poole PS, Tkacz A (2017) Understanding the holobiont: the interdependence of plants and their microbiome. Curr Opin Microbiol 38:188–196

Santhanam R, Weinhold A, Goldberg J, Oh Y, Baldwin IT (2015) Native root-associated bacteria rescue a plant from a sudden-wilt disease that emerged during continuous cropping. Proc Natl Acad Sci 112(36):E5013–E5020

Schenk ST, Schikora A (2015) AHL-priming functions via oxylipin and salicylic acid. Front Plant Sci 5:784

Schlaeppi K, Dombrowski N, Oter RG, van Themaat EVL, Schulze-Lefert P (2014) Quantitative divergence of the bacterial root microbiota in Arabidopsis thaliana relatives. Proc Natl Acad Sci 111(2):585–592

Schmidt H, Eickhorst T (2013) Spatio-temporal variability of microbial abundance and community structure in the puddled layer of a paddy soil cultivated with wetland rice (Oryza sativa L.). Appl Soil Ecol 72:93–102

Souza RC, Mendes IC, Reis-Junior FB, Carvalho FM, Nogueira MA, Vasconcelos ATR, Vicente VA, Hungria M (2016) Shifts in taxonomic and functional microbial diversity with agriculture: how fragile is the Brazilian Cerrado? BMC Microbiol 16(1):42

Swenson W, Wilson DS, Elias R (2000) Artificial ecosystem selection. Proc Natl Acad Sci 97(16):9110–9114

Van Overbeek L, Van Elsas JD (2008) Effects of plant genotype and growth stage on the structure of bacterial communities associated with potato (Solanum tuberosum L.). FEMS Microbiol Ecol 64(2):283–296

Vázquez-de-Aldana BR, García-Criado B, Vicente-Tavera S, Zabalgogeazcoa I (2013) Fungal endophyte (Epichloë festucae) alters the nutrient content of Festuca rubra regardless of water availability. PloS One 8(12):e84539

Venturi V, Keel C (2016) Signaling in the rhizosphere. Trends Plant Sci 21(3):187–198

Vorholt JA (2012) Microbial life in the phyllosphere. Nat Rev Microbiol 10(12):828–840

Wagner MR, Lundberg DS, Coleman-Derr D, Tringe SG, Dangl JL, Mitchell-Olds T (2014) Natural soil microbes alter flowering phenology and the intensity of selection on flowering time in a wild Arabidopsis relative. Ecol Lett 17(6):717–726

Wagner MR, Lundberg DS, Coleman-Derr D, Tringe SG, Dangl JL, Mitchell-Olds T (2015) Corrigendum to Wagner et al.: natural soil microbes alter flowering phenology and the intensity of selection on flowering time in a wild Arabidopsis relative. Ecol Lett 18(2):218–220

Werner GD, Strassmann JE, Ivens AB, Engelmoer DJ, Verbruggen E, Queller DC, Noë R, Johnson NC, Hammerstein P, Kiers ET (2014) Evolution of microbial markets. Proc Natl Acad Sci 111(4):1237–1244

Xiong W, Li Z, Liu H, Xue C, Zhang R, Wu H, Li R, Shen Q (2015) The effect of long-term continuous cropping of black pepper on soil bacterial communities as determined by 454 pyrosequencing. PloS One 10(8):e0136946

Zachow C, Berg C, Müller H, Monk J, Berg G (2016) Endemic plants harbour specific Trichoderma communities with an exceptional potential for biocontrol of phytopathogens. J Biotechnol 235:162–170

Zilber-Rosenberg I, Rosenberg E (2008) Role of microorganisms in the evolution of animals and plants: the hologenome theory of evolution. FEMS Microbiol Rev 32(5):723–735

Zolla G, Badri DV, Bakker MG, Manter DK, Vivanco JM (2013) Soil microbiomes vary in their ability to confer drought tolerance to Arabidopsis. Appl Soil Ecol 68:1–9

Health Implications of a Plant Beneficial and Probiotic *Lactobacillus casei* in Foods Containing the Isoflavone-Daidzein

8

Moushumi Ghosh, Krishnaswamy Balamurugan, and Abhijit Ganguli

Abstract

This study reports the behavior of *Lactobacillus casei*, a probiotic strain upon exposure to isoflavones commonly consumed in Indian diet. In vitro studies with daidzein, formononetin, genistein, coumesterol, and biochanin A indicated the strain to be tolerant to these with little degradation, but daidzein was preferred as a carbon source with 99.23% degradation. Exposure to daidzein did not alter essential probiotic and other technological attributes of the strain. The ability of the strain to release isoflavones by fermentation of soil plant material may facilitate rhizospheric interaction and is beneficial for plant growth. Besides, the *Lactobacillus casei* offers additional advantages as an established probiotic, for instance, its importance in alleviation of malathion toxicity (Kamaladevi et al. 2016) which is of relevance in food plants. The results of our study indicate the potential importance of applicability of this strain both for human and plant benefits.

Keywords

probiotic · plant health · isoflavones · Lactobacillus casei · human health

M. Ghosh
Department of Biotechnology, Thapar Institute of engineering and Technology, Patiala, India

K. Balamurugan
Department of Biotechnology, Alagappa University, Karaikudi, Tamil Nadu, India

A. Ganguli (✉)
SSD Project Conglomerate, PHFI, Gurgaon, India

M. Kumar et al. (eds.), *In vitro Plant Breeding towards Novel Agronomic Traits*,
https://doi.org/10.1007/978-981-32-9824-8_8

8.1 Introduction

Isoflavones are natural flavonoids, estrogen-related compounds present in relatively large amounts in many plant food products. Some of the health benefits of isoflavones are extensive easing of menopause symptoms, reduction of heart disease risk, protection against prostate problems, improvement of bone health, and reduction of cancer risk. The bioavailability of these isoflavones compounds depends upon the gut bacteria like lactic acid bacteria. Also in soil, plants use isoflavones and their derivatives as phytoalexin compounds to ward off disease-causing pathogenic fungi and other microbes. In addition, soybean uses isoflavones to stimulate soil-microbe rhizobium to form nitrogen-fixing root nodules. Several microorganisms which are ubiquitously distributed facilitate rapid conversion of complex flavonoid-containing sources to isoflavones, in which lactic acid bacteria occupy a prominent position. Intestinal microflora especially probiotic lactic acid bacteria play a key role in the metabolism and bioavailability of isoflavones as they hydrolyze the glycoside components using their indigenous β-glucosidase and β-galactosidases in the jejunum, releasing the bioactive aglycone isoflavone form (Setchell 2000). Aglycone forms have also been found to absorb faster and in higher amounts in human than their respective glycoside forms (Izumi et al. 2000). Besides, intestinal bacteria are known to convert daidzein to its metabolite, equol, which is a more potent estrogenic compound than its precursor (Setchell et al. 1998). The Asian diet containing highly processed foods made from legumes, such as tofu, retains most of their isoflavone content, with the exception of fermented miso, which has increased levels. Asians consume soybeans in many forms, including miso soup, tofu, natto (fermented soybean), soy sauce, soya cheese, soya nuts, soya flour, barley miso, ice beans, and tempeh. The Indian diet contains isoflavones including chick pea (biochanin A), peanut (genistein), soybean (Daidzein and Genistein), spinach (coumestrol), and green beans (formononetin). It is worthwhile to note that the Asian populations with their high intake (50–70 mg/day) of soy-derived isoflavones are known to have the lowest incidence of osteoporosis, menopausal symptoms, cardiovascular disease, and cancer (Adlercreutz et al. 1992; Izumi et al. 2000).

The chemical form in which isoflavones occur is an important consideration because it may influence the biological activity, the bioavailability, and therefore the physiological effects of these dietary constituents (Kurzer and Xu 1997; Setchell and Cassidy 1999). However recent reports have indicated that higher amount of intake of isoflavones can be detrimental to health, especially during the more susceptible stages like pregnancy. Thus the use of probiotic bacteria to improve the biological activity of high isoflavone-based products during processing is an important study. Few studies have, however, evaluated the behavior/metabolic response of probiotic lactic acid bacteria to isoflavones originating from traditional diets. It may be envisaged that a combination of isoflavone-containing diet containing lactic acid bacteria may play an important role in bioavailability of isoflavones, thus enhancing the nutritional benefits. In the present study, we investigated the growth and expression of probiotic properties of *L. casei*. The strain of *L. casei* has been extensively evaluated for probiotic properties and its role in *E. sakazakii* (foodborne pathogen) as well as in malathion toxicity in our earlier studies (Sivamaruthi et al. 2011; Kamaladevi et al. 2016). The behavior of *L. casei* was demonstrated through in vitro studies with pure isoflavones and later its fate in soya samples determined.

8.2 Materials and Method

8.2.1 Reference Standards and Materials

Reference isoflavones were purchased from Sigma-Aldrich (MO, USA) and were of HPLC grade. All other solvents and chemicals were of highest analytical grade. Soya bean samples were purchased from the local market.

8.2.2 Bacterial Strain and Culture Media

A strain of *Lactobacillus casei (L. casei)* isolated from vegetables and fermented beverages (fermented rice and *Madhuca longifolia* flowers) and characterized as potential new probiotic strain (Kumar et al. 2010) was used in this study. The strain was stored as stock solutions in 20% (v/v) glycerol at −70 °C. Prior to all experiments, the culture was propagated in MRS medium at least thrice; purity of the strain was confirmed microscopically and by gram staining and catalase activity periodically.

8.2.3 Tolerance and Growth of *Lactobacillus casei* in the Presence of Pure Daidzein

Overnight grown culture of *L. casei* on MRS (HiMedia, Mumbai, India) broth was challenged with different concentrations (30, 20, 15, 10, 5 μg/ml) of daidzein dissolved in methanol. A high-throughput screening (Bioscreen, Thermo Labsystems, Helsinki) was performed by growing the *L. casei* in MRS in microtiter plates, shaking the plates after 30 min and measuring the OD at 600 nm. Viability was recorded as log CFU/ml on MRS agar media by plate count method. The growth of *L. casei* in basal minimal medium was adopted from the protocol described by Cerning et al. (1994). The ability of *L. casei* to grow in basal minimal media with glucose (control) and isoflavones as carbon source was examined. The minimal media contained KH_2PO_4 454 mg/l, $Na_2HPO_4.12H_2O$ 1.19 g/l, NaCl 0.2 g/l, $MgSO_4.7H_20$ 0.1 g/l, $CaCl_2.2H_2O$ 0.002 g/l, $MnSO_4$ 0.004 g/l, $FeSO_4$ 0.005 g/ml, and peptone 0.002 g/l. The controls comprised of glucose as carbon source and daidzein, genistein, and formononetin as sole carbon sources.

8.2.4 Analysis of Daidzein by High-Performance Liquid Chromatography (HPLC)

All HPLC analyses were performed using a reverse phase column (C-18) octadecylsilane (4 mm I.D × 30 cm). Detector used was UV-VIS variable wavelength detector. Column temperature was maintained at 30 °C. Detection of phytoestrogens was carried out at 254 nm (optical bandwidth 8 nm). Metahnol and water used were HPLC grade. Separation was achieved by using a linear methanol-water gradient system at a flow rate of 1.0 ml/min. Methanol and water reservoirs each contained 1% glacial acetic acid (v/v) and 0.01 M ammonium acetate HPLC grade (Casteele et al. 1982; Wong 1962). The gradient was programmed to increase from 53% to 58% reservoir B (methanol) over 30 min total analysis time per sample, including column equilibration which was 60 min. Preliminary peak identification was based on a comparison of retention times of phytoestrogen standards and unknown peaks in the sample extracts.

8.2.5 Fate of Daidzein in Soya Samples Fermented with *L. casei*

One gram of powdered soya sample containing daidzein was dissolved in freshly prepared phosphate buffer solution. Overnight grown culture of *L. casei* was dissolved in the sample containing PBS and incubated at 37 °C. The 0 and 3 h soya sample was removed and extracted with 3 × 20 ml petroleum ether (PE) till it settled down. Then the upper layer of petroleum ether was removed. Finally extraction was completed with 4 × 20 ml diethyl ether and the aqueous layer discarded. The sample was then dissolved in methanol and kept in at 5 °C until analysis. Preliminary analysis of daidzein and formononetin was performed by thin layer chromatography (Beck 1964).

8.2.6 Tolerance of Artificial Gastric Juice

The ability of the *L. casei* with daidzein to survive under gastric conditions was examined according to the method of Casey et al. (2004). Overnight cultures were washed with phosphate-buffered saline (PBS), resuspended in synthetic gastric juice (pH 1.85 adjusted using HCl), and incubated at 37 °C. The artificial gastric juice consisted of 3.5 gl-1 D-glucose, 2.05 g l-1 NaCl, 0.6 g l-1 KH2PO4, 0.11 g l-1 CaC12, 0.37 g l-1, KCl, 0.05 g l-1 Ox bile, 0.1 g l-1 lysozyme, and 13.3 mg l-1 pepsin. Samples containing control and daidzein were withdrawn at regular intervals, serially diluted in PBS, and enumerated on MRS agar to enumerate the viable cells.

8.2.7 Glutamate Decarboxylase Activity

A rapid assay (Chou and Weimer 1998) was used to determine qualitatively the glutamate decarboxylase (GAD) activity in the isolate with isoflavones (daidzein and formononetin). One milliliter volumes of overnight cultures in MRS were washed in quarter strength of Ringer's solution and resuspended in 0.5 ml of test reagent adjusted to pH 3.0 with 1 M HCl. The reagent consisted of 90 g l-1 NaCl, 1 g L-glutamic acid, 0.3 g l-1 Triton X-100, and 0.05 g l-1 bromocresol green, made up in distilled water. The development of a blue coloration after 4 h of incubation at 37 °C indicated a positive result.

8.2.8 Microbial Adhesion to Solvents (MATS)

Microbial adhesion to solvents was measured according to the method of Rosenberg et al. (1980) with some modifications (Fontaine et al. 1996). *L. casei* treated with isoflavones were harvested in the stationary phase by centrifugation at 5000 g for 15 min, washed twice, and resuspended in 0.1 mol-1 KNO3 (pH 6.2) to approximately 108 CFU ml-1. The absorbance of the cell suspension was measured at 600 nm (A0). One milliliter of solvent was added to 3 ml of cell suspension. After a 10 min preincubation at room temperature, the two-phase system was mixed by vortexing for 2 min. The aqueous phase was removed after 20 min of incubation at room temperature, and its absorbance at 600 nm (A1) was measured. The percentage of bacterial adhesion to solvent was calculated as (1 – A1/A0)∗100. Three different isoflavones were tested in this study. The adhesion of *L. casei* to xylene reflects cell surface hydrophobicity.

8.2.9 β-Glucosidase Activity in the Presence of Isoflavones

The determination of β-glucosidase activity was by the method of Otieno and Shah (2007) with some modifications. The *L. casei* was activated in rehydrated de Man, Rogosa, and Sharpe (MRS) broth at 37 ° C for 20 h twice followed by a third activation in soya hydrolysate using an inoculum level of 5% (v/v). The enzyme was determined in soya extract with isoflavones by measuring the rate of hydrolysis of p-nitrophenyl-β-d-glucopyranoside (pNPG). Of the 20-ml aliquot, 1 ml was used for the determination of β-glucosidase activity according to the method by Otieno and Shah (2007). Briefly, 1 ml of 5 mmol/l of pNPG prepared in 100 mmol/l of sodium phosphate buffer (pH 7.0) was added to 1 ml of aliquot and incubated at 37 °C for 15 min. Exactly 0.5 ml of 1 mol/l cold sodium carbonate was added to stop the reaction. The aliquots were then placed in 1.8-ml Eppendorf centrifuge tubes followed by centrifugation (14,000 g for 30 min). The amount of p-nitrophenol released was measured using a spectrophotometer at 420 nm. One unit of the enzyme activity was defined as the amount of β-glucosidase that released 1 mol of p-nitrophenol from the substrate pNPG per milliliter per minute under assay conditions.

8.3 Result and Discussion

In recent years, the probiotic effects of lactic acid bacteria have gained overwhelming interest in terms of their functional aspect and have been a subject of continuous research and concurrent development of numerous health beneficial formulations as well as technologies aimed in substantially improving human nutrition and health.

One of the most important criteria for selection of probiotic organisms is their ability to survive in the acidic environment of the product and in the stomach, where the pH can reach as low as 1.5. Similarly, the organisms must be able to survive in the bile concentrations encountered in the intestine. The concentration of bile in the human gastrointestinal system is variable and is difficult to predict at any given moment (Chou and Weimer 1998). After the bacteria pass through the stomach, they enter the upper intestinal tract where bile is secreted into the gut. Following its journey through this harsh environment, the organism colonizes the epithelium of the lower intestinal tract. Thus, strains selected for use as probiotic bacteria should be able to tolerate acid for at least 90 min, tolerate bile acids, attach to the epithelium, and grow in the lower intestinal tract before they can start providing any health benefits. They should also have the ability should also have the ability to resist the digestion process in the stomach intestinal tract. Kumar et al. (2010) reported the time from entrance to release from the stomach to be 90 min. However, further digestive processes have longer residence times; hence, there is a need for the bacteria to be resistant to the stressful conditions of the stomach and upper intestine, which contain bile.

Many previous studies on the physicochemistry of microbial cell surfaces have demonstrated that the presence of glycoproteinaceous material at the cell surface results in higher hydrophobicity, whereas hydrophilic surfaces are associated with the presence of polysaccharides (Pelletier et al. 1997). It is known that only pronase- and pepsin-sensitive surface molecules are responsible for cell surface hydrophobicity in bacteria. In the previous studies, the autoaggregation ability of *L. acidophilus* M92 was also reduced by proteolytic treatment, while metaperiodate did not affect hydrophobicity. As bacterial cells subjected to proteolytic attack weaken their autoaggregation ability, it is reasonable to consider proteins as mediators in the aggregation process. In the present study, *L. casei* was found to tolerate isoflavones tested with significant degradation of daidzein. The daidzein-exposed cells indicated the ability of the culture to exhibit probiotic properties described in Fig. 8.1.

8.3.1 Growth Kinetics of *L. casei* in Presence of Daidzein

The growth kinetics of probiotic strain *Lactobacillus casei* in MRS supplemented with varying concentrations of daidzein was evaluated over a period of 24 h. Each experiment was performed in duplicate, and the result represents the average of two readings. The highest cell number of test organism achieved was 6.68 CFU/ml. The specific growth rate achieved was 0.54/h. The *L. casei* tolerated different concentrations of daidzein provided in the MRS media (Fig. 8.2).

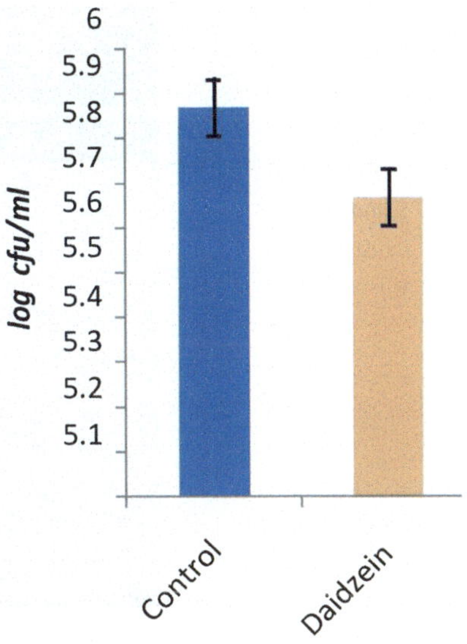

Fig. 8.1 Tolerance of *L. casei* to the isoflavone-Daidzein

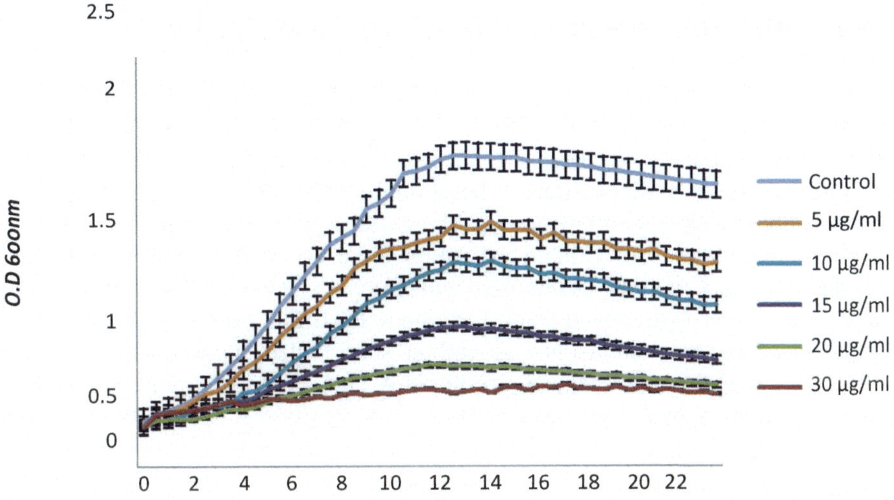

Fig. 8.2 Growth kinetics of *L. casei* with daidzein

8.3.2 Fate of Daidzein

The fate of daidzein consumed by the *L. casei* in MRS media was analyzed by thin layer chromatography as described by Beck (1964). The isoflavones degraded were extracted from *L. casei* on MRS media and compared with standard isoflavones on silica gel plates. The spot of degraded daidzein was visualized under UV (260 nm) (Fig. 8.3).

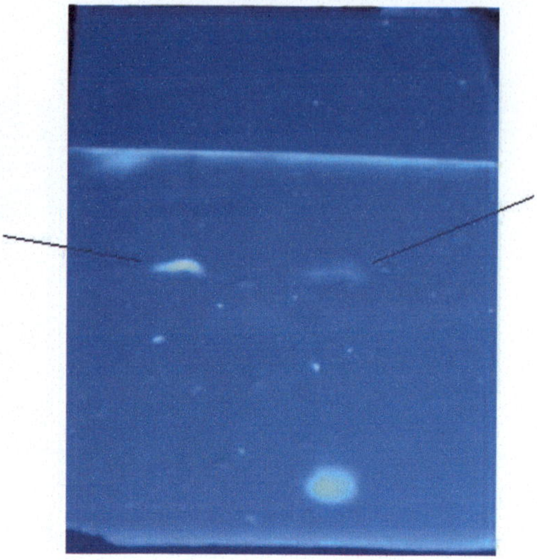

Fig. 8.3 Degradation of Daidzein by *L. casei*

8.3.3 Utilization of Isoflavones by *L. casei* in Minimal Media

Minimal media allowed the growth of *L. casei* with glucose (control) and isoflavones (daidzein, genistein, formononetin) during the time period of 24 and 48 h. MRS media contain all the nutrients for growth, whereas basal minimal media contain only minimum nutrients; thus in basal minimal media, isoflavones acted as a carbon source for the growth of *L. casei*, these observations being similar to that of Cerning et al. (1994). A final viable population of the test organism ranging between 6.04 and 5.61 log CFU/ml was observed with glucose and isoflavones as the carbon source. Specific growth ranged from 0.15/min to 0.11/min. These results indicated that *L. casei* utilized isoflavones as carbon source also suggesting that under nutrient-depleted conditions which are possibly simulated in the gut, isoflavones can be metabolized by *L. casei* (Fig. 8.4).

8.3.4 Real-Time Analysis of Soya Sample with *L. casei*

Soya sample was used as model food plant; samples were treated with *L. casei* dissolved in the sterile PBS and inoculated with overnight cultures of *L. casei* and incubated at 37 °C for fermentation. Viable counts of *L. casei*, β-glucosidase, and β-galactosidase activities and residual daidzein were monitored hourly for up to 3 h by TLC. Figure 8.5 depicts the daidzein levels indicating complete utilization at the end of incubation period.

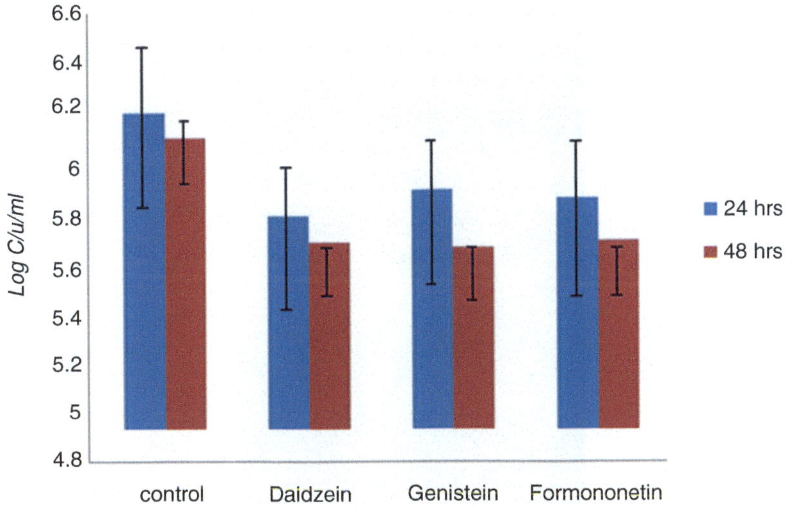

Fig. 8.4 Utilization of isoflavones by *L. casei* in minimal media

8.3.5 Quantitative Analysis of Daidzein

It is important to ascertain whether *L. casei* grown in daidzein retained the probiotic attributes with isoflavones upon being subjected to gastric juice. The small intestine and colons of humans and animals contain relatively high concentrations of bile acid, which can inhibit growth of many bacteria. Final viable population of the test organism ranging between 5.8 and 3.4 log CFU/ml was observed. A decrease in 2 log CFU/ml was observed (Figs. 8.6a, 8.6b, and 8.6c). The results were in concurrence to those reported by Casey et al. (2004) that *L. casei* survived the gastric juice treatment upon growth in isoflavones.

8.3.6 Microbial Adhesion to Solvents

MATS method was used to evaluate the hydrophobic/hydrophilic cell surface properties of *L. casei* and compare them with the cell surface properties of *L. casei* grown with isoflavones. The hydrophobicity observed was in the order control >dai dzein>genistein>formononetin. The percentage adhesion with xylene was found to be control 57%, daidzein 45.6%, genistein 35.12%, and formononetin 37.9%, respectively. The fact that a high percentage of *L. casei* cells adhered to xylene, a polar solvent, demonstrated hydrophobic cell surface of this strain. The results indicated that *L. casei* was less hydrophobic with some isoflavones; however that with daidzein was comparable to the control.

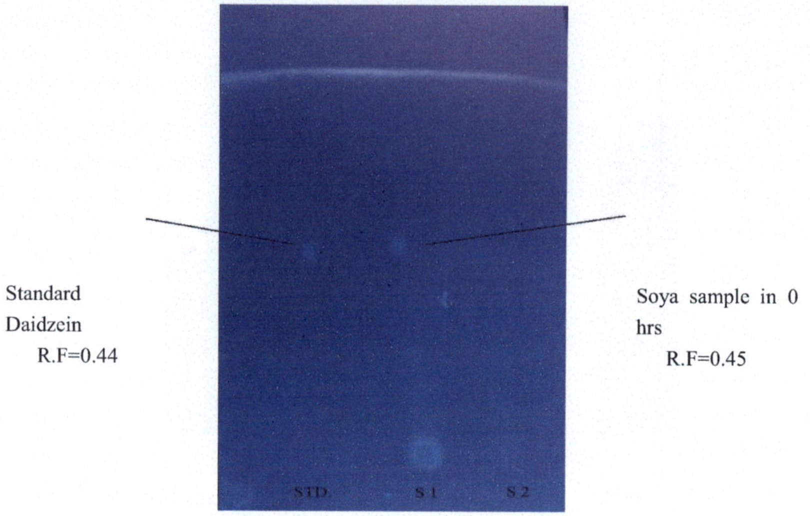

Fig. 8.5 Thin layer chromatographic analysis of soya sample inoculated with *L. casei*. Std, standard daidzein; RF = 0.44; S1, soya sample in 0 h; RF = 0.45; and S2 soya sample after 3 h

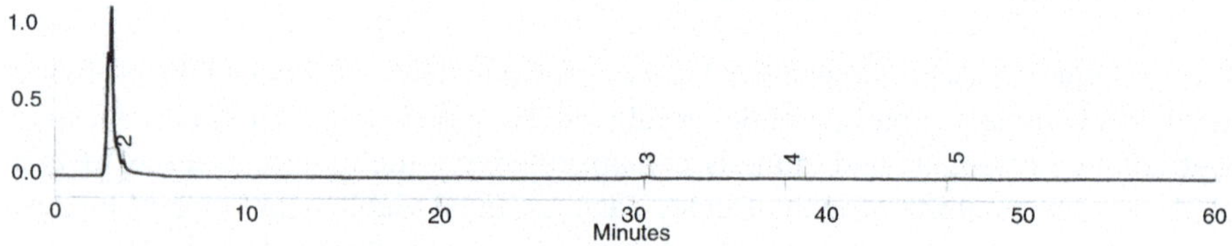

Fig. 8.6a RP- HPLC analysis of standard daidzein

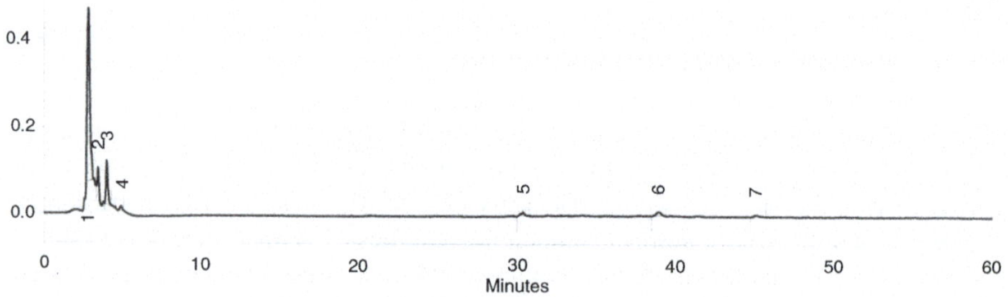

Fig. 8.6b Quantitative analysis of soya sample through RP-HPLC analysis. The amount of daidzein present in the soya sample was calculated as 15 mg/gm of extract

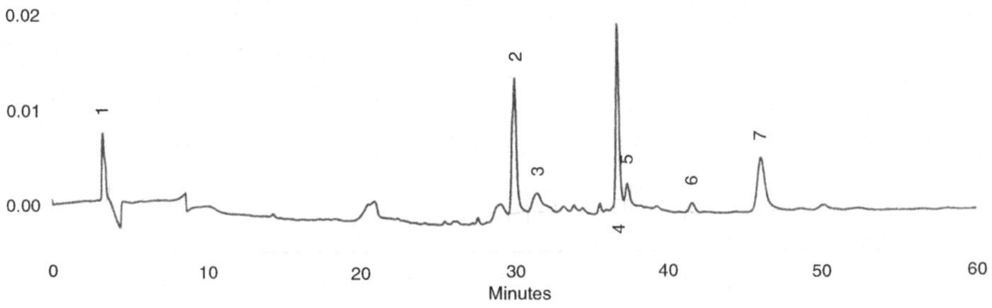

Fig. 8.6c The RP- HPLC analysis of degraded daidzein by *L. casei*. The amount degraded by *L. casei* was 99.23%

8.3.7 Glutamate Decarboxylase (GAD) Activity

Among gram-positive bacteria, GAD acid resistance system is the only amino acid decarboxylation system that has been associated with acid response. The GAD system as an acid defense mechanism has been described for *L. lactis* among lactic acid bacteria, although GAD activity was also detected in *L. casei* with isoflavone—daidzein. A strong GAD activity by *L.* casei with isoflavones indicated the presence of this acid defense mechanism in this strain. Our results agreed with those reported by Lee and No (1997, Matsura et al. 1995) who reported that *L. lactis* retain GAD activity upon growth in isoflavones (Fig. 8.7).

β-glucosidases hydrolyze the β-glycosides into isoflavone aglycones. β-glucosidase activity was over 15-fold higher than β-galactosidase activity (Otieno et al. 2006). β-glucosidase is the chief enzyme responsible for bioconversion of isoflavones glycosidase to aglycones (Donkor and Shah 2008; Matsura and Obat 1993). Peak activities occurred after 15 h for *L. casei*; notable increase in the enzyme activity occurred after 1 h of fermentation. The culminating peak activity was observed after 15 h.

Overall, the present study indicated that the probiotic strain *L. casei* was able to tolerate the isoflavones commonly occurring in food plants, but daidzein was degraded effectively, presumably by glucosidases. Importantly, the probiotic characteristics of the strain remained unaltered upon exposure to the isoflavone, daidzein. Our findings have multiple implications for the use of *L. casei* as a potential probiotic strain for human use; besides, it is able to carry out fermentation of complex isoflavone-containing plant materials in soil. The robust fermentation ability of plant materials of this strain under ambient conditions leads to the release of isoflavones in soil rhizosphere. In addition to stimulation of the rhizospheric bacteria, isoflavones are taken up by plants for producing phytoalexins which are important molecules for plant defense.

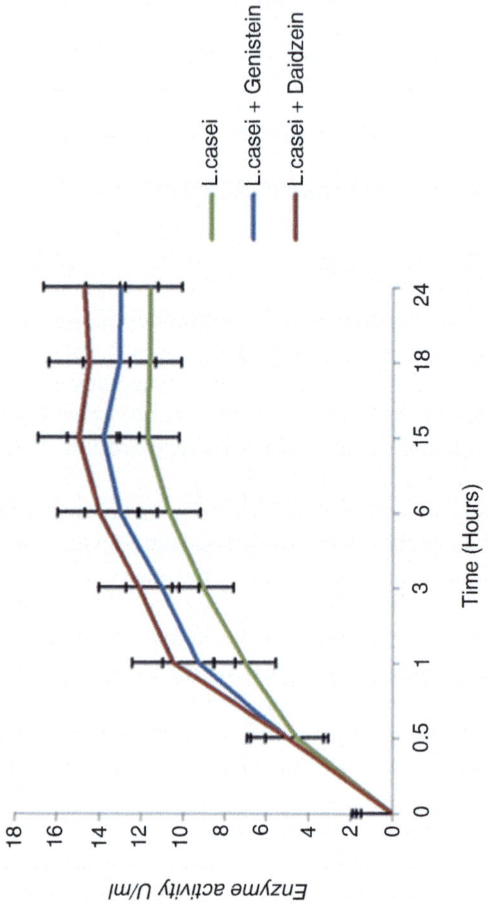

Fig. 8.7 β-glucosidase activity in *L. casei* in the presence of isoflavones

8.4 Conclusion

Our study focused on interaction of the probiotic strain of *L. casei* with predominant isoflavones commonly present in the Indian diet. Though tolerance to all isoflavones was observed, daidzein could be utilized preferentially by *L. casei*, its probiotic attributes remained unaltered upon exposure/growth of *L. casei*, and high β-glucosidase production by the strain could be detected substantiating its role in degradation of daidzein. By virtue of its strong fermentative ability of plant materials, *L. casei* strain can lead to release of isoflavones from plant material existent in soil, enabling stimulation of soil-microbe rhizobium interaction to form nitrogen-fixing root nodules. The overarching beneficial implications for both human and plants suggest potential practical importance of *L. casei*.

Acknowledgment The authors are grateful to Dr. Sutapa B Neogi, Additional Professor, Indian Institute of Public Health, Delhi, PHFI, for her expert suggestions and comments.

References

Adlercreutz A, Mousavi Y, Clark J, Hockerstedt K, Hamalainen E, Wahala K, Makela T, Hase T (1992) Dietary phytoestrogens and cancer: in vitro and in vivo studies. J Steroid Biochem Mol Biol 41:331–337

Beck AB (1964) The oestrogenic isoflavones of Subterranean clover. Aust J Agric Res 15:223–230

Casey PG, Casey GD, Gardiner GE, Tangney M, Stanson C, Ross RP, Hill C, Fitzgerald GF (2004) Isolation and characterization of anti-*Salmonella* lactic acid bacteria from the porcine gastrointestinal tract. Lett Appl Microbiol 39:431–438

Casteele KV, Geiger H, Sumere CFV (1982) Separation of flavonoids by reverse-phase high-performance liquid chromatography. J Chromatogr 240:81–87

Cerning J, Renard CMGC, Thibualt JF, Bouillanne C, Landon M, Desmazeaud M, Topisirovic L (1994) Carbon source requirements for exopolysaccharide production by *Lactobacillus casei* CG11 and partial structure analysis of the polymer. Appl Environ Microbiol 60(11):3914–3919

Chou L, Weimer B (1998) Isolation and characterization of acid and bile tolerant isolates from strains of *Lactobacillus acidophilus*. J Dairy Sci 82:23–31

Donkor ON, Shah NP (2008) Production of β-glucosidase and hydrolysis of isoflavone phytoestrogens by *Lactobacillus acidophilus*, *Bifidobacterium lactis*, and *Lactobacillus casei* in soymilk. J Food Sci 73(1):15–20

Fontaine MNB, Rault J, Vanoss CJ (1996) Microbial adhesion to solvents: a novel method to determine the electron-donor/ electron-acceptor or Lewis acid-base properties of microbial cells. Colloids Surf 7:47–53

Izumi T, Piskula MK, Osawa S, Obata A, Tobe K, Saito M, Kataoka S, Kubota Y, Kikuchi M (2000) Soy isoflavone aglycones are absorbed faster and in higher amounts than their glucosides in humans. J Nutr 130:1695–1699

Kamaladevi A, Ganguli A, Balamurugan K (2016) *Lactobacillus casei* stimulates phase-II detoxification system and rescues malathion-induced physiological impairments in *Caenorhabditis elegans*. Comp Biochem Physiol C 179:19–28

Kumar M, Ghosh M, Ganguli A (2010) Mitogenic response and probiotic characteristics of lactic acid bacteria isolated from indigenously pickled vegetables and fermented beverages. World J Microbiol Biotechnol 28:703–711

Kurzer MS, Xu X (1997) Dietary phytoestrogen. Annu Rev Nutr 17:353–381

Lee SH, No MJ (1997) Viability in artificial gastric and bile juice and antimicrobial activity of some lactic acid bacteria isolated from Kimchi. Korean J Appl Microbiol Biotechnol 28:279–284

Matsura M, Obata A (1993) β-Glucosidases from soybeans hydrolyze daidzin and genistin. J Food Sci 58:144–147

Matsura M, Sasaki J, Murao S (1995) Studies on β-glucosidases from soybeans that hydrolyze daidzein and genistin: isolation and characterization of an isozyme. Biosci Biotechnol Biochem 59:1623–1627

Otieno DO, Ashton JF, Shah NP (2006) Evaluation of enzymic potential for biotransformation of isoflavone phytoestrogen in soymilk by *Bifidobacterium animalis*, *Lactobacillus acidophilus* and *Lactobacillus casei*. Food Res Int 39:394–407

Otieno DO, Shah NP (2007) Endogenous β-glucosidase and β-galactosidase activities from selected probiotic micro-organisms and their role in isoflavone biotransformation in soymilk. J Appl Microbiol 103(4):910–917

Pelletier C, Bouley C, Cayuela C, Bouttier S, Bourlioux P, Bellon-Fontaine MN (1997) Cell surface characteristics of *Lactobacillus casei* subsp. *casei*, *Lactobacillus paracasei* subsp. *paracasei*, and *Lactobacillus rhamnosus* strains. Appl Environ Microbiol 63:1725–1731

Rosenberg M, Gutnick D, Rosenberg E (1980) Adherence of bacteria to hydrocarbons: a simple method for measuring cell-surface hydrophobicity. FEMS Microbiol Lett 9:29–33

Setchell KDR (2000) Absorption and metabolism of soy isoflavones-from food to dietary supplements and adults to infants. Am Soc Nutr Sci J Nutr 130:654S–655S

Sivamaruthi BS, Ganguli A, Kumar M, Bhaviya S, Pandian SKand Balamurugan K (2011) *Caenorhabditis elegans* as a model for studying *Cronobacter sakazakii* ATCC BAA-894 pathogenesis. J Basic Microbiol 51:540–549

Setchell KDR, Cassidy A (1999) Dietary isoflavones: biological effects and relevance to human health. J Nutr 129(3):758S–767S

Setchell KDR, Zimmer-Nechemias L, Cai J, Heubi JE (1998) Isoflavone content of infant formulas and the metabolic fate of these phytoestrogens in early life. Am J Clin Nutr 68:1453S–1461S

Wong E (1962) Detection and estimation of oestrogenic constituents in red clover. J Sci Food Agric 13:231–253

Microbial Diversity and Their Role in Plant and Soil Health Under Stress Conditions

9

Zubair A. Dar, Rouf Ahmad Bhat, Javeed I. A. Bhat, Shafat A. Mir, Azra Amin, Asmat Rashid, Bhat Rifat, and Rafiq Lone

Abstract

Microbial diversity is an important component of ecosystem. Intellectual strategy and careful selection of microbes are the key steps in developing innovative technologies for effective exploitation of microorganisms to solve sustainably environmental and economic issues and improvement in the quality of life without compromising yields. Appreciativeness of these factors might be one way to change the public perception towards microorganisms by showing that the sustainable use of microbial diversity has positive ecological and economic value. This review encapsulates the several microbial applications such as biofertilizers, biopesticides, and bioherbicides to solve the major agricultural issues in the form of crop productivity and soil and plant health. Furthermore, the review emphasizes on pivotal microbial activities that lead to neutralize the harmful impacts of toxic and hazardous chemical substances in eco-friendly manner. Microbial technology therefore not only could be an efficient and vital tool for treating waste water and industrial wastes but can also play an imperative and significant role in cycling of dead organic matter and thereby can form useful products in the form of compost which has potential significance with regard to sustainable environment.

Keywords

Microbial diversity · Sustainable environmental · Biofertilizers · Biomonitoring; nutrient cycling

Z. A. Dar · R. A. Bhat · J. I. A. Bhat · S. A. Mir · A. Amin · A. Rashid
Division of Environmental Sciences, Sher-e-Kashmir University of Agricultural Sciences and Technology, Shalimar, Kashmir, India

B. Rifat
Department of Biotechnology, University of Kashmir, Srinagar, India

R. Lone (✉)
Department of Botany, SBBS University, Khiala/Jalandhar, Punjab, India

M. Kumar et al. (eds.), *In vitro Plant Breeding towards Novel Agronomic Traits*,
https://doi.org/10.1007/978-981-32-9824-8_9

9.1 Introduction

Microorganisms are defined by their size: any organism that is so small to be observed by the unaided human eye. Microbial diversity is an important constituent of the overall global biological diversity. Recent technological advances in exploring microbial diversity have revealed that a large proportion of microorganisms are still undiscovered and their ecological roles are largely unknown. Careful selection of microbes and intelligent design of test assays are the key steps in developing new technologies for effective utilization of microorganisms for sustainable agriculture, environmental protection, and human and animal health (Ahmad et al. 2011). Several microbial applications are broadly known in resolving major agricultural (i.e., crop productivity, plant health protection, and soil health maintenance) and environmental issues (i.e., bioremediation of soil and water from organic and inorganic pollutants). Furthermore, they have many positive features which can be used in different fields of environment. They require fewer amounts of nutrients for their optimum growth available throughout year and can survive under extreme conditions of temperature, pressure, etc. (Rothschild and Mancinelli 2001). In environmental microbiology bacteria have attracted the attention because they are easy to cultivate and study and have a great role to play for betterment of our environment. As most of the early bacteriological studies focus only at medical bacteriology to know the causes of common diseases affecting people, now, the focus has shifted towards improving our degraded environment. Various biological agents (biopesticides, biofertilizers, biocontrol, and PGPRs) are now becoming vital for the environmental management (Gerhardson 2002). As progress is being made in controlling disease-producing bacteria, environmental microbiologists have shifted their focus from pathogenic bacteria to helpful bacteria, such as those used in the treatment of municipal wastewater and industrial wastewater in order to protect and preserve the polluted environment. Apart from this efforts are being made to use microbes in agricultural fields to improve the agriculture productivity. Some bacteria fix nitrogen in symbiotic associations with plants, while some nitrogen-fixing bacteria are free-living in soil and aquatic habitats (Gray and Smith 2005; Bhatti et al. 2017). Also, soil microbiologists and microbial ecologists have differentiated soil microorganisms as constructive or destructive according to their functions and how they affect soil quality, plant growth and health, and yield (Zolla et al. 2013). Constructive microorganisms are those that can fix atmospheric nitrogen, detoxify pesticides, decompose organic wastes and residues, suppress plant and soilborne pathogens and diseases, improve nutrient cycling, and produce bioactive chemicals such as hormones, vitamins, and enzymes that stimulate plant growth (Lugtenberg 2015; Bhatti et al. 2017). Even though bacteria are related to fungi, the problem that lies with fungi is their complex nature when compared with bacteria. The discovery of antibiotics and their application in the control of numerous diseases facilitated to increase interest in the biochemistry of fungi. They also play an important role in the treatment of some industrial wastewaters and in composting of solid wastes so as to protect the environment, for example, yeast cells are being used in some industrial wastewater treatment systems (Daiss et al. 2008). In current era, various environmental microbiologists are currently looking over the possible use of algae in various fields. The study has revealed that algae play an important role in controlling environmental pollution (Stevenson and Bahls 1999). They have high potential to secure nitrogen and phosphorus in their protoplasm. Algae when added to soil can improve soil characteristics and increase fertility for higher plants. Growth of algae in ponds has proven beneficial in stimulating fish farming. The present era demands the better management and utilization of these bio-resources, so as to take best advantage and benefits of their best characteristics.

9.2 Conservation of Microbial Diversity

The problem of biodiversity conservation is usually a conflict between the humankind and living organisms inhabiting different environments (Kaushik 2015). In modern era, the conservation of microbial diversity is of great concern, as they play an important role in sustainability of better environment. Both ex situ and in situ techniques can be employed to preserve the biodiversity (Brutting et al. 2013).

9.3 Ex Situ Preservation

Ex situ conservation involves the preservation and protection of the habitat from destruction or degradation (Kjaer et al. 2001). It is the most effective mechanism for conserving the biodiversity of a habitat (Veteläinen et al. 2006). Because of the various qualms linked with in situ conservation of microorganisms, ex situ preservation includes gene banks, culture collections, and microbial resource centers, thereby establishing the storehouse for microbial isolates (Borokini et al. 2010). The convection on biodiversity diversity (CBD) inspires to adopt the measures of ex situ for biodiversity conservation, mostly in the country of origin (Antofie 2011).

9.4 In Situ Preservation

This type of preservation involves on-site conservation of the microbial flora by conserving their natural habitats and ecosystems. It also includes the recovery and maintenance of viable populations of species in their natural habitat and in surroundings where they have developed their characteristic properties. This type of preservation can be achieved by avoiding deforestation and planting trees (afforestation) that will protect the surface soil to be washed out by torrential rains, which comprises diverse microflora (Melfi 2012). Further, preventing water bodies such as oceans, lakes, or rivers from the pollution will preserve phytoplanktons and zooplanktons (microalgae, diatoms, dinoflagellates, etc.).

9.5 Utility of Microbial Diversity

Microbial diversity present in natural ecosystems has the following major applications.

9.6 Microbes as Biofertilizers

The excessive use of chemical fertilizers/inorganic fertilizers in agriculture has resulted in numerous environmental problems like eutrophication, poor soil health, decline in natural microflora, and ozone depletion and has harmful impacts on human health (Chowdhury 2004). In order to overcome these problems, use of biofertilizers has been found effective and efficient. As most of the microorganisms have been found to be useful in eradicating problems associated with the use of chemical fertilizers and pesticides (Bhatti et al. 2017; Khanday et al. 2016), they are now widely applied in nature farming and organic agriculture (Nogueira et al. 2001). Certain species of microorganism are widely used which have unique properties to provide natural products and serve as a good substitute of chemical fertilizers (Rania et al. 2015). Biofertilizers are supposed to be safe alternative to chemical fertilizers to minimize or reduce the ecological disturbance (Vaghela et al. 2014). They are cost-effective and eco-friendly and, when required in bulk (Dutta et al. 2003), can be produced at the farm itself. They help in the breakdown of organic matter and secretion of growth-promoting substances (Akande et al. 2005). They are cheap source of nutrients and are suppliers of essential microelements and micronutrients (Youssef and Eissa 2005). They help to increase the crop yield by 10–40%, fix nitrogen up to 40–50%, and improve soil texture, pH, and other properties of soil (Youssef and Eissa 2005; Bhatti et al. 2017). They also control soilborne diseases and maintain the soil structure in cultivable fields (Gaur 1999). The main sources of biofertilizers are bacteria, fungi, and cyanobacteria (Pidwirny 2002). Biofertilizers enable plants to use all of the food available in the soil and air, thus allowing farmers to reduce the amount of chemical fertilizers. This helps in preserving the environment for the next generations to come. Phosphate and nitrogen compounds occur naturally in the environment and are essential for the growth of plant (Akande et al. 2003; Ginting et al. (2003). Phosphate plays a significant role in crop stress tolerance and maturity and in nitrogen fixation directly or indirectly (He et al. 2002). A fungi *Penicillium bilaiae* releases an organic acid which dissolves the unlocked phosphate in the soil so that the roots can use it (Bashan et al. 2013). Biofertilizer made from this organism is applied by either seed coating with the fungus as inoculation or putting it directly into the ground (Shen et al. 2011). Rhizospheric microorganisms facilitate soil processes such as decomposition, nitrogen fixation and denitrification, nutrient mobilization and mineralization, and storage release of nutrients and water (Khan et al. 2007).

9.7 Nitrogen-Fixing Bacteria

9.7.1 Rhizobia

Legume plants having root nodules fix atmospheric nitrogen in association with bacteria belonging to genera *Rhizobium*, *Bradyrhizobium*, *Sinorhizobium*, *Azorhizobium*, and *Mesorhizobium* collectively called as rhizobia. Rhizobium is helpful in fixing 15–20 kg N/ha which can increase the crop yield by 20 folds reported by (Dubey et al. 2006). Species of the genus *Rhizobium*, such as *Rhizobium trifolii*, *Rhizobium giardinii*, *Rhizobium leguminosarum*, and *Rhizobium*, have been reported to nodulate and establish a symbiotic association with different common bean varieties in tropical areas (Gilbert et al. 2017).

9.7.2 Azorhizobium

It is a stem nodule-forming bacteria and fixes nitrogen symbiont soft stem nodule. It produces large amount of indole acetic acid (IAA) that promotes plant growth (Wani et al. 2005). It has the high affinity to form a nitrogen-fixing symbiotic relationship with the host plant *Sesbania rostrata* which is a small semiaquatic leguminous tree (Capoen et al. 2010).

9.7.3 Azotobacteracae (*Azotobacter*)

These free-living aerobic bacteria occurred in the roots of *Paspalum notatum* (tropical grasses). They secrete acetic acid, gibberellins, and naphthalene and secrete substances that inhibit root pathogens and enhance root growth (El-Haddad et al. 2011). It can be applied in cereals, millets, vegetables, and flowers through seed, seedling, and soil treatments.

9.7.4 Bradyrhizobium

They are good nitrogen fixers; *Bradyrhizobium* strain with mucuna seeds increases total organic carbon and increases plant growth and consequently plant biomass, nitrogen, phosphorus, and potassium in the soil (Saidi et al. 2011). It also plays an important role in reduction of weed population and improves soil microbial population (Pagano 2008). Bacteria of the genus *Bradyrhizobium* are reported as tolerant to low pH and fertility of tropical soils (Menna and Hungria 2011).

9.7.5 *Azolla-Anabaena* Symbiosis

It is a free-floating, aquatic fern having a cyanobacterial symbiont. *Azolla-Anabaena* live and reproduce in cavities of leaves, but the association remains extracellular (Shridhar 2012). The main role of this symbiont is to fix atmospheric nitrogen and excrete organic nitrogen in water during its growth mostly in rice fields. *Azolla* increases 10–20% yield of paddy crops and suppresses the weed growth (Carrapiço et al. 2000). It absorbs traces of potassium from irrigation water and thus can be used as a green manure before rice planting (Reyes et al. 2002).

9.7.6 Spirillaceae (*Azospirillum*)

These are aerobic gram-negative, free-living, associative symbiotic bacteria and occur in the roots of dicots and monocot plants, i.e., corn, sorghum, wheat, etc. *Azospirillum* is very effective in increasing yield of cereal crops and produces gibberellins and cytokinins which enhance the root development and uptake of plant nutrients (Carrillo et al. 2002; Gonzalez et al. 2005).

9.7.7 Cyanobacteria (Blue-Green Algae)

Cyanobacteria play an important role in preservation and maintenance of soil fertility, increasing rice growth and yield, and improve soil porosity and production and excretion of growth-promoting substances such as hormones (auxin, gibberellin), vitamins, and amino acids. Cyanobacteria fix 20–30 Kg/N/ha and increase 10–15% crop yield (Rodriguez et al. 2006). Cyanobacteria play an important role in maintenance and build-up of soil fertility, consequently increasing rice growth and yield as a natural biofertilizer (Song et al. 2005). Beneficial effects of cyanobacterial inoculation were also reported on a number of other crops such as barley, oats, tomato, radish, cotton, sugarcane, maize, chili, and lettuce (Thajuddin and Subramanian 2005).

9.7.8 *Acetobacter diazotrophicus*

It occurs in roots, stem, and leaves of sugarcane and sugar beet crops. It also produces growth promoters, e.g., indole acetic acid, and helps in seed germination and nutrient uptake (Muthukumarasamy et al. 2006). This bacterium enhanced up to 0.5–1% crop yield (Gahukar 2005).

9.8 Phosphate-Solubilizing Bacteria

There are numerous bacteria species which have phosphate-solubilizing ability, e.g., *Pseudomonas fluorescens*, *Aerobacter aerogenes*, *Escherichia freundii*, *Nitrobacter* spp., *Paenibacillus polymyxa*, *Serratia* spp., and *Pseudomonas striata* (Whitelaw 2002). They increased about 10–20% crop yield (Khan et al. 2007). They have the ability to produce plant growth-promoting hormones which play an important role in phosphate-solubilizing activity of soil (Igual et al. 2001). Treatment with *Bacillus megaterium* resulted in 40% reduction in nematode penetration and gall formation compared with non-treated rice plants (Padgham and Sikora 2007). Nematode disease can be controlled or managed by using phosphate-solubilizing microorganisms in place of inorganic fertilizers (Khan et al. 2010).

9.9 Phosphate-Solubilizing Fungi

Several strains of fungal species have been described and investigated in detail for their phosphate-solubilizing capabilities (He et al. 1997). These include, e.g., *Aspergillus niger*, *Aspergillus awamori*, *Penicillium digitatum*, etc. (Wakelin et al. 2004). The viable microbial preparations possessing P-solubilizing activity are generally termed as microphos (Zaidi et al. 2009). The phosphate-solubilizing microbes showing greater solubilization (both qualitatively and quantitatively) of insoluble P under in vitro conditions are selected for field trials prior to production in bulk for ultimate transmission as a biofertilizer (Collavino et al. 2010). Among filamentous fungi that solubilize phosphate, the genera *Aspergillus* and *Penicillium* (Fenice et al. 2000; Khan and Khan 2002; Reyes et al. 1999, 2002) are the most representative although strains of *Trichoderma* (Altomare et al. 1999) and *Rhizoctonia solani* (Jacobs et al. 2002) have also been reported as P solubilizers.

9.10 Microorganisms in the Conversion of Agricultural Wastes to Compost

Crop residues are produced in huge quantities and constitute an ample but underutilized source of renewable biomass in agriculture. Recycling agriculture wastes is essential for environment as well as economical saving (Abou Hussein and Sawan 2010). This recycling will not only increase agricultural production but also will improve its quality (Bernal et al. 2009). Half of its quantity is used as roofing material, animal feed, fuel, and packing material, while the other half is disposed of by burning in the field. Burning of agro-residues is considered as a cheap and easy means of disposal of excess residues (Hoornweg et al. 2010). This practice leads to air pollution, causes respiratory problems, increases fog incidences even in distant places, increases soil erosion, and decreases the efficacy of soil-applied herbicides (Walia et al. 1999). Incorporation of agro-residues like paddy straw increases the methane emission from field, especially in irrigated soils, which in turn leads to global warming (Conrad 2002). Microorganisms play an important role in the recycling of agricultural wastes (Margesin et al. 2006). Several fungi like *Trichoderma harzianum*, *Pleurotus ostreatus*, *Polyporus ostriformis*, and *Phanerochaete chrysosporium* are known to play important role in composting of lignocellulosic materials (Singh and Nain 2012). The composting of agricultural residues rich in lignocellulose like paddy straw generally takes 180 days to obtain good and mature compost. Hence, the capacity of microorganisms to assimilate organic matter depends on their ability to produce the enzymes needed for degradation of the substrate components, i.e., cellulose, hemicellulose, and lignin (Fauziah and Agamuthu 2009). The compost generated by bioconversion of agro-residues offers several benefits such as enhanced soil fertility and soil health which can lead to increased agricultural productivity, improved soil biodiversity, reduced ecological risks, and a healthier environment. These virtues make composting an ideal option for processing of the enormous quantities of agro-wastes that are generated in the world (El-Meniawy 2003). Recent reports also indicate multifarious uses of bio-augmented compost as carrier for bioinoculants and biocontrol agents for disease suppressions and bioremediation. The recycled material when applied to soil improves soil fertility and soil health which can lead to increased agricultural productivity, improved soil biodiversity, a healthier environment, and reduced ecological risks (Badr El-Din et al. 2000).

9.11 Bioremediation and Biodegradation of Xenobiotics

In order to eliminate the unwanted weeds, pests, or pathogenic microorganisms, humans are constantly and increasingly using pesticides such as DDT, 2,4-D, BHC, and 2,4,5-T (Labana et al. 2005). Eliminating these toxic chemicals from the environment can be done by ease and in an environment-friendly manner using biological methods that involve use of microbes and plants to degrade a xenobiotic compound and thus decontamination of the polluted site (Elekwachi et al. 2014). Bioremediation, the use of microorganisms to remove toxic pollutants from the environment, is the most promising technology that is eco-friendly and effective even if the pollutants are present at low concentrations (Schroll et al. 2004; Zafar et al. 2007; Lal et al. 2010). Bioremediation is cost-effective compared with conventional techniques such as incineration (Kuiper et al. 2004). Bioremediation methods are performed either in situ (bio-augmentation, bio-stimulation, and bioventing) or ex situ (land farming, bio-piles, and bioreactors) (Li et al. 2004). Bioremediation uses biological agents, mainly microorganisms, i.e., yeast, fungi, or bacteria, to clean up contaminated soil and water (Strong and Burgess 2008). In bioremediation processes, microorganisms use the contaminants as nutrient or energy sources (Tang et al. 2007). Many site-specific microorganisms are capable of carrying out bioremediation reactions, and many have already been used at sites previously contaminated with polycyclic aromatic hydrocarbons (PAHs), nitro-aromatic compounds, chlorinated organic, etc. (Carvalho et al. 2005). Recently, biological remediation process has also been devised to either precipitate effectively immobilize inorganic pollutants such as heavy metals (Bhatnagar and Kumari 2013). Stimulation of microorganisms is achieved by the addition of growth substances, nutrients, terminal electron acceptor/donors, or some combination, thereby resulting in an increase in organic pollutant degradation and biotransformation. The energy and carbon are obtained through the metabolism of organic compounds by the microbes involved in bioremediation processes (Fulekar 2009). Bioremediation process involves biotransformation and biodegradation by transforming contaminants to non-hazardous or less hazardous chemicals (Singh et al. 2014). Biotransformation is any modification of the molecule or structure of a compound by microorganisms. Biodegradation is the breaking down of organic or inorganic compounds into environmental friendly compounds (Gupta and Mahapatra 2003). Biodegradation of a compound is often a result of the activities of multiple organisms. When microorganisms are introduced to a contaminated site to augment degradation, the process is called as "bio-augmentation."

The various methods of bioremediation are mycoremediation, phytoremediation, and bacteria bioremediation.

9.12 Mycoremediation

In this method fungi are used to remove chemical contaminants from the soil (Khanday et al. 2016). The fungus naturally secretes certain enzymes and acids to decompose the hazardous chemicals into less or nontoxic compounds (Bhattacharya et al. 2012; Khanday et al. 2016).

9.13 Phytoremediation

This method mainly uses plants to control and eliminate pollutants from the soil, air, and water. By this method various organic and inorganic wastes such as salts, metals, sewage sludge, leachates, and xenobiotic contaminants can be removed (Eneh and Owo 2008).

9.14 Bacteria Bioremediation

Bacteria are used to clean up environmental pollutants such as oil spills, heavy metals, mine effluents, and even human waste through its natural metabolic process (Salem et al. 2000; Ahluwalia and Goyal 2007).

9.15 Desulfurization of Fossil Fuels by Microbial Oxidation

The elimination of inorganic sulfur from coal is assisted by microbial oxidation using *Thiobacillus* sp. In the oxidation of inorganic sulfur, *Thiobacillus* sp. is used for a membrane-bound reaction. A variety of pure and mixed cultures of microorganisms (heterotrophic bacteria) can be used to remove organic sulfur from coal and oil (Okpokwasili 2007).

Microbial biosensors have become one of the most beneficial means of monitoring environmental, food, and clinical samples and emphasis on new technologies and applications of microbial biosensors in different aspects of ecosystem.

9.16 Environmental Monitoring

Pollutants in the environment are great risks for the health of human beings. Several microbial biosensors for detection of organic and inorganic toxicities are shown in Table 9.1.

9.17 Food and Fermentation

Microorganisms, despite their small size, play an important role in the planning and maintenance of food quality and safety in food processing and manufacturing industries. Preservation of foods by fermentation is a widely practiced and ancient technology (Magnuson and Lasure 2004). Fermentation is a biochemical process which involves the conversion of simple sugar to acid, ethanol, and carbon dioxide through the metabolic pathways, e.g., the conversion of lactose to lactic acid by *Lactobacillus bulgaricus* and *Streptomyces thermophilus* during yogurt production. Fermentation is extensively used for the production of foodstuffs and drinks, which requires a cautiously performed fermentation system operation (Kim et al. 2009). Microbial biosensors have been used for profound determination of ethanol in order to monitor the fermentation process. Table 9.2 shows the recent developed microbial biosensors used in food and fermentation.

9.18 Microbes Used in Extraction of Metals from Electronic Wastes

Among major groups of bacteria, the most commonly used for metal extraction from electronic wastes are acidophilus and chemolithotrophic microbial consortia of *Acidithiobacillus ferrooxidans*, *Acidithiobacillus thiooxidans*, and *Leptospirillum*

Table 9.1 Target Based Microbial Biosensors

Target	Microorganism	Transducer	References
Hg	*E. coli*	Bioluminescence	Barrocas et al. (2010)
Hg	*Chlorella* sp.	*Chlorella* sp.	Singh and Mittal (2012)
Cr	Sulfur-oxidizing bacteria (SOB)	Conductimetric	Gurung et al. (2002)
Zinc	*Pseudomonas putida* X4	Fluorescence	Liu et al. (2012)
Methyl parathion	*Flavobacterium* sp.	Optical fiber	Kumar et al. (2006)

Table 9.2 Microbial biosensor for detection in food and fermentation industry

Target analyte	Microorganism	Transducer	References
Ethanol	*Candida tropicalis*	Amperometric	Akyilmaz and Dinckaya (2005)
Ethanol	*Gluconobacter oxydans*	Amperometric	Valach et al. (2009)
Ethanol	Eukaryote double-mediator (EDM)	Amperometric	Nakamura et al. (2009)
Ethanol	*Methylobacterium organophilum*	Oxygen electrode	Wen et al. (2012)
Volatile alcohols	*Pichia pastoris*	Amperometric	Hammerle et al. (2011)

Table 9.3 Examples of industrial waste treated with bacterial leaching

Waste	Leached metal	Used microorganisms
Lithium batteries	Li, Co	*Acidithiobacillus ferrooxidans*
Tannery	Cr	*Acidithiobacillus thiooxidans*
Sewage sludge	Cu, Ni, Zn, Cr	Iron-oxidizing bacteria
Electronic scrap	Cu, Ni, Al, Zn	*Acidithiobacillus ferrooxidans*

Willner and Fornalczyk (2013)

Table 9.4 Shows a summary of these bio-products and their application in oil recovery

Product	Microorganism	Application in oil recovery
Biomass	*Bacillus licheniformis*, *Leuconostoc mesenteroides*, *Xanthomonas campestris*	Selective plugging, viscosity reduction, oil degradation, wettability alteration
Biosurfactants	*Acinetobacter calcoaceticus*, *Arthrobacter paraffineus*, *Bacillus* sp., *Clostridium* sp., *Pseudomonas* sp.	Emulsification, interfacial tension reduction, viscosity reduction
Biopolymers	*Paenibacillus polymyxa*, *Brevibacterium viscogenes*, *Leuconostoc mesenteroides*, *Xanthomonas campestris*, *Enterobacter* sp.	Injectivity profile modification, mobility control, viscosity modification
Bio-acids	*Clostridium* sp.	Permeability increase, emulsification

Sulaimani et al. (2011)

ferrooxidans. In addition, fungi such as *Penicillium* sp. and *Aspergillus niger* are examples of some eucaryotic microorganisms used in bioleaching during metal recovery from industrial wastes (Olson et al. 2003).

The technology is cheaper and easier to conduct in comparison to conventional techniques to recover valuable metals from wastes. Bio-hydrometallurgical processing of solid waste is similar to natural biogeochemical metal cycles and reduces the demand of resources, such as ores, energy, and landfill space. This technology is environmentally friendly in comparison to chemical method and is considered a green technology because it generates less amount of waste (Mishra et al. 2008). This is why more and more scientists become interested in bio-hydrometallurgy (Tables 9.3 and 9.4).

9.19 Microbial Enhanced Oil Recovery (MEOR)

Microbial enhanced oil recovery (MEOR) is one of the technologies that can be potentially implemented with an exceptionally low operating cost. It has several advantages compared to conventional EOR processes where it does not consume large amounts of energy as do thermal processes, nor does it depend on the oil price as do many chemical processes (Yousef et al. 2007). Microorganisms can be genetically engineered in the petroleum industry to produce chemicals useful for enhanced oil recovery (Daini 2000). On this basis, microbiological methods based on the metabolic activities of bacteria seem to be attractive (Jinfeng et al. 2005). The various useful microorganisms used in MEOR include *Pseudomonas* sp., *Bacillus* sp., *Brevibacillus* sp., *Agrobacterium* sp., *Sphingomonas* sp., *Rhizobium* sp., *Coprothermobacter* sp., and *Thermolithobacter* sp. (Zhang 2010). Microorganisms produce a variety of metabolites that are potentially useful for oil recovery. There are six main bio-products or metabolites produced by microbes.

9.20 Role of Microbes for the Removal of Toxic Chemicals of Industries from the Environment

Bioremediation is the use of microorganisms or microbial processes to detoxify and reduce environmental contaminants. Microorganisms are being used for the routine treatment and transformation of waste products from several years. The fixed-film and activated sludge treatment systems depend on the metabolic activities of microorganisms which degrade the wastes to enter in the treatment facility.

We can summarize the most important activities of microorganisms in the environment as follows (BiróAtalaj 2005):

(i) Conversation of organic material, via mineralization and humification
(ii) Enhancement of the available plant nutrient with greater root surface and phosphorus mobilization
(iii) Hormone release of microorganisms helps plant growth (plant growth regulators (PGR))
(iv) Biological nitrogen fixation
(v) Improving drought tolerance by active water absorption of the greater root mass (mycorrhizal fungi and plant growth-promoting rhizobacteria (PGPR))
(vi) Biological plant protection, biostatic, biocidal ferro-chelate agents, antagonist organisms
(vii) Micropollutants, partial or full breakdown of xenobiotics

In addition, the more general activities of living organisms in the soil:

(i) Several groups of organism take part in the biological weathering processes of rock (algae, fungi, and lichens)
(ii) Development of soil structure (bacteria, fungi, and worms)
(iii) Ensure the cycle of nutrients and elements between the living and nonliving environment
(iv) The symbiosis is very vital among the organisms, mutualism, which is mutually beneficial for the individual organisms
(v) Control and influence the composition of the atmospheric air

9.21 Conclusion

This review highlights the role of microorganisms in sustaining the dynamic equilibrium and integrity of the biosphere. Presently, plants and microbes are linked through recycling of different materials. In environmental microbiology microbes have gained lot of attention because they are easy to cultivate and play crucial role for improvement of our environment. Microbial biotechnology is an important area that promotes for advances in food safety, food security, human nutrition and functional foods, plant and animal protection, and overall fundamental research in the environmental sciences. Microorganisms are widely applied test species in different bioassays because of the ease and low costs of their culturing as well as the lack of ethical issues often accompanying the use of higher organisms. Combining biology to engineering skills has enabled the development of biosensors and biomonitoring. The present era of information big bang thus stresses the better management and utilization of these bio-resources. Microorganisms are important gene pools and so must be viewed as cultural heritage and should be transferred to next generation in a normal and healthy condition.

References

Abou Hussein SD, Sawan OM (2010) The utilization of agricultural waste as one of the environmental issues in Egypt (a case study). J Appl Sci Res 6:1116–1124
Ahluwalia SS, Goyal D (2007) Microbial and plant derived biomass for removal of heavy metals from wastewater. Bioresour Technol 98:2243–2257
Ahmad IM, Khan AS, Singh M (2011) Microbial applications in agriculture and the environment: a broad perspective. Microbes Microbial Technol:1–27
Akande MO, Oluwatoyinbo FI, Adediran JA, Buari KW, Yusuf IO (2003) Soil amendments affect the release of P from rock phosphate and the development and yield of okra. J Veg Crop Prod 19:3–9
Akande MO, Adediran JA, Oluwatoyinbo FI (2005) Effects of rock phosphate amended with poultry manure on soil available P and yield of maize and cowpea. Afr J Biotechnol 4:444–448
Akyilmaz E, Dinckaya E (2005) An amperometric microbial biosensor development based Candida tropicalis yeast cells for sensitive determination of ethanol. Biosens Bioelectron 20:1263–1269

Altomare C, Norvell WA, Borjkman T, Harman GE (1999) Solubilization of phosphates and micronutrients by the plant growth promoting and biocontrol fungus *Trichoderma harzianum*. Appl Environ Microbiol 65:2926–2933

Antofie M (2011) Current political commitments challenges for ex situ conservation of plant genetic resources for food and agriculture. Analele Universităţii din Oradea Fascicula Biologie 18:157–163

Badr El-Din SMS, Attia M, Abo-Sedera SA (2000) Field assessment of composts produced by highly effective cellulolytic microorganisms. Biol Fertil Soils 32:35–40

Barrocas PRG, Landing WM, Hudson RJM (2010) Assessment of mercury (II) bioavailability using a bioluminescent bacterial biosensor: practical and theoretical challenges. J Environ Sci 22:1137–1143

Bashan Y, Kamnev AA, de Bashan LE (2013) A proposal for isolating and testing phosphate-solubilizing bacteria that enhance plant growth. Biol Fertil Soils 49:1–2

Bernal MP, Alburquerque JA, Moral R (2009) Composting of animal manures and chemical criteria for compost maturity assessment. Rev Bioresour Technol 100:5444–5453

Bhatnagar S, Kumari R (2013) Bioremediation: a sustainable tool for environmental management-a review. Annu Rev Res Biol 3:974–993

Bhattacharya S, Angayarkanni J, Das A, Palaniswamy M (2012) Mycoremediation of Benzo[a] pyrene by Pleurotus ostreatus isolated from Wayanad district in Kerala. India Int J Pharm Bio Sci 2:84–93

Bhatti AA, Haq S, Bhat RA (2017) Actinomycetes benefaction role in soil and plant health. Microb Pathog 111:458–467

BiróAtalaj B (2005) mint a mikroszervezetekélettere. In: Magyarországazezredfordulón. Stratégiaitanulmányok a Magyar Tudományos Akadémián. II. Azagráriumhelyzeteésjövője. A talajokjelentősége a 21.században. Budapest, Társadalomkutató Központ, pp 141–169

Borokini TI, Okere AU, Giwa AO, Daramola BO, Odofin WT (2010) Biodiversity and conservation of plant genetic resources in Field Gene-bank of the National Centre for Genetic Resources and Biotechnology, Ibadan, Nigeria. Int J Biodiver Conserv 2:37–50

Brutting C, Hensen I, Wesche K (2013) Ex situ cultivation affects genetic structure and diversity in arable plants. Plant Biol 15:505–513

Capoen W, Oldroyd G, Goormachtig S, Holsters M (2010) *Sesbania rostrata*: a case study of natural variation in legume nodulation. New Phytol 186:340–345

Carrapiço F, Teixeira G, Diniz MA (2000) Azolla as biofertiliser in Africa. A challenge for the future. Revista de Ciências Agrárias 23:120–138

Carrillo AE, Li CY, Bashan Y (2002) Increased acidification in the rhizosphere of cactus seedlings induced by *Azospirillum brasilense*. Naturwissenschaften 89:428–432

Carvalho MF, Ferreira JR, Pacheco CC, De Marco P, Castro PML (2005) Isolation and properties of a pure bacterial strain capable of fluorobenzene degradation as sole carbon and energy source. Environ Microbiol 7:294–298

Chowdhury R (2004) Effects of chemical fertilizers on the surrounding environment and the alternative to the chemical fertilizers. IES-ENVIS Newsl 7:4–5

Collavino MM, Sansberro PA, Mroginski LA, Aguilar OM (2010) Comparison of in vitro solubilization activity of diverse phosphate-solubilizing bacteria native to acid soil and their ability to promote *Phaseolus vulgaris* growth. Biol Fertil Soils 46:727–738

Conrad R (2002) Control of microbial methane production in wetland rice fields. Nutr Cycl Agroecosyst 64:59–69

Daini OA (2000) Fundamentals of genetic engineering. Samrol Ventures and Printing Co., Ijebu-Igbo

Daiss N, Lobo MG, Socorro AR, Bruckner U, Heller J, Gonzales M (2008) The effect of three organic pre-harvest treatments on Swiss Chard (*Beta vulgaris*. L var. cycla) quality. Eur Food Res Technol 226:345–353

Dubey RC (2006) A textbook of biotechnology, 4th edn. S. Chand & Co. Ltd, New Delhi, p 732. isbn:81-219-2608-4

Dutta S, Pal R, Chakeraborty A, Chakrabarti K (2003) Influence of integrated plant nutrient phosphorus and sugarcane and sugar yields. Field Crop Res 77:43–49
Elekwachi CO, Andresen J, Hodgman TC (2014) Global use of bioremediation technologies for decontamination of ecosystems. J Bioremed Biodegr 5:1–9
El-Haddad ME, Mustafa I, Selim SM, El-Tayeb TS, Mahgoob AE, Abdel-Aziz NH (2011) The nematicidal effect of some bacterial biofertilizers on Meloidogyne incognita in sandy soil. Brazil J Microbiol 42:105–113
El-Meniawy MA (2003) Microbiological studies on the production of non- traditional animal feed from agricultural wastes. MSc, thesis, Faculty of Agriculture, Al-Azhar University, Cairo, Egypt
Eneh OC, Owo NJ (2008) Sustainable development: a review. Int J Dev Stud 3:100–103
Fauziah SH, Agamuthu P (2009) Sustainable household organic waste management via vermicomposting. Malays J Sci 28:135–142
Fenice M, Seblman L, Federici F, Vassilev N (2000) Application of encapsulated *Penicillium variabile* P16 in solubilization of rock phosphate. Bioresour Technol 73:157–162
Fulekar MH (2009) Bioremediation of fenvalerate by *Pseudomonas aeruginosa* in a scale up bioreactor. Rom Biotechnol Lett 14:4900–4905
Gahukar RT (2005) Potential and use of bio-fertilizers in India. Evermans Sci XL(5):354–361
Gaur AC (1999) Microbial technology for composting of agricultural residues by improved method. Indian council of Agricultural Research, New Delhi, pp 43–53
Gerhardson B (2002) Biological substitutes for pesticides. Trends Biotechnol 20:338–343
Gilbert K, Mburu SW, Njeru EM, Kimiti JM, Ombori O, Maingi JM (2017) Potential of native rhizobia in enhancing nitrogen fixation and yields of climbing beans (Phaseolus vulgaris L.) in contrasting environments of eastern Kenya. Front Plant Sci 8:443
Ginting D, Kessavalou A, Eghball B, Doran JW (2003) Greenhouse gas emissions and soil indicators four years after manure and compost applications. J Environ Qual 32:23–32
Gonzalez LJ, Rodelas B, Pozo C, Salmeron V, Martnez MV, Salmeron V (2005) Liberation of amino acids by heterotrophic nitrogen fixing bacteria. Amino Acids 28:363–367
Gray EJ, Smith DL (2005) Intracellular and extracellular PGPR: commonalities and distinctions in the plant-bacterium signaling processes. Soil Biol Biochem 37:395–412
Gupta R, Mahapatra H (2003) Microbial biomass: an economical alternative for removal of heavy metals from waste water. Indian J Exp Biol 41:945–966
Gurung A, Oh S, Kim KD, Shin B (2002) Semi-continuous detection of toxic hexavalent chromium using a sulfur-oxidizing bacteria biosensor. J Environ Manag 106:110–112
Hammerle M, Hilgert K, Marcus AH, Ralf M (2011) Analysis of volatile alcohols in apple juices by an electrochemical biosensor measuring in the headspace above the liquid. Sensor Actuators B Chem 158:313–318
He ZL, Wu J, O'Donnell AG, Syers JK (1997) Seasonal responses in microbial biomass carbon, phosphorus and sulphur in soils under pasture. Biol Fertil Soils 24:421–428
He ZL, Bian W, Zhu J (2002) Screening and identification of microorganisms capable of utilizing phosphate adsorbed by goethite. Commun Soil Sci Plant Anal 33:647–663
Hoornweg DL, Thomas L, Otten L (2010) Composting and its applicability in developing countries. The International Bank for Reconstruction and Development/The World Bank, Washington, DC
Igual JM, Valverde A, Cervantes E, Velazquez E (2001) Phosphate solubilizing bacteria as an inoculants for agriculture: use of updated molecular techniques in their study. Agronomie 21:561–568
Jacobs H, Boswell GP, Ritz K, Davidson FA, Gadd GM (2002) Solubilization of calcium phosphate as a consequence of carbon translocation by *Rhizoctonia solani*. FEMS Microbiol Ecol 40:65–71
Jinfeng L, Lijun M, Bozhong M, Rulin L, Fangtian N, Jiaxi Z (2005) The field pilot of microbial enhanced oil recovery in a high temperature petroleum reservoir. J Pet Sci Eng 48:265–271
Kaushik G (2015) Applied environmental biotechnology: present scenario and future trends. Springer, New Delhi

Khan MR, Khan SM (2002) Effects of root-dip treatment with certain phosphate solubilizing microorganisms on the fusarial wilt of tomato. Bioresour Technol 85:213–215

Khan MS, Zaidi A, Wani P (2007) Role of phosphate solubilizing microorganisms in sustainable agriculture-a review. Agron Sustain Dev 27:29–43

Khan MS, Zaidi A, Ahemad M, Oves M, Wani PA (2010) Plant growth promotion by phosphate solubilizing fungi-current perspective. Arch Agron Soil Sci 56:73–98

Khanday M, Bhat RA, Haq S, Dervash MA, Bhatti AA, Nissa M, Mir MR (2016) Arbuscular mycorrhizal Fungi boon for plant nutrition and soil health. In: Hakeem KR et al (eds) Soil science: agricultural and environmental Prospectives. Springer, Cham, pp 317–332

Kim Y, Park J, Jung H (2009) An impedimetric biosensor for real-time monitoring of bacterial growth in a microbial fermentor. Sensor Actuators B Chem 138:270–277

Kjaer ED, Graudal L, Nathan I (2001) Ex situ conservation of commercial tropical trees: strategies, options and constraints. Danida Forest Seed Centre, Humlebaek

Kuiper I, Lagendijk EL, Bloemberg GV, Lugtenberg BJJ (2004) Rhizoremediation: a beneficial plant-microbe interaction. Mol Plant-Microbe Interact 17:6–15

Kumar J, Jhaadn SK, Souza SFD (2006) Optical microbial biosensor for detection of methyl parathion pesticide using *Flavobacterium sp.* whole cells adsorbed on glass Fiber filters as disposable biocomponent. Biosens Bioelectron 21:2100–2105

Labana S, Pandey G, Paul D, Sharma NK, Basu A, Jain RK (2005) Pot and field studies on bioremediation of *p*-nitrophenol contaminated soil using *Arthrobacter protophormiae* RKJ100. Environ Sci Technol 39:3330–3337

Lal R, Pandey G, Sharma P, Kumari K, Malhotra S, Pandey R, Raina V, Kohler HPE, Holliger C, Jackson C, Oakeshott JG (2010) Biochemistry of microbial degradation of hexachlorocyclohexane and prospects for bioremediation. Microbiol Mol Biol Rev 74:58–80

Li L, Cunningham C, Pas V, Philp J, Barry D (2004) Field trial of a new aeration system for enhancing biodegradation in a biopile. Waste Manag 24:127–137

Liu P, Huang Q, Chen W (2012) Construction and application of a zinc-specific biosensor for assessing the immobilization and bioavailability of zinc in different soils. Environ Pollut 164:66–72

Lugtenberg B (2015) Life of microbes in the rhizosphere. In: Lugtenberg B (ed) Principles of plant-microbe interactions. Springer International Publishing Switzerland, Heidelberg, pp 7–15

Magnuson JK, Lasure LL (2004) Organic acid production by filamentous fungi. In: Tkacz J, Lange L (eds) Advances in fungal biotechnology for industry, agriculture and medicine. Klewer Academic/Plenum, New York, pp 307–340

Margesin R, Cimadon J, Schinner F (2006) Biological activity during composting of sewage sludge at low temperatures. Int Biodeterior Biodegrad 57:88–92

Melfi VA (2012) Ex situ gibbon conservation: status, management and birth sex ratios. Int Zoo Yearb 46:241–251

Menna P, Hungria M (2011) Phylogeny of nodulation and nitrogen-fixation genes in *Bradyrhizobium*: supporting evidence for the theory of monophyletic origin, and spread and maintenance by both horizontal and vertical transfer. Int J Syst Evol Microbiol 61:3052–3306

Mishra D, Kim DJ, Ralph DE, Jong-hwan A, Young HR (2008) Bioleaching of spent hydroprocessing catalyst using acidophilic bacteria and its kinetics aspect. J Hazard Mater 152:1082

Muthukumarasamy R, Revathi G, Vadivelu M (2006) In: Recent advances in bio-fertilizer technology (Yadav AK, Motsara MR, Ray Chaudhuri S, eds), Society for Promotion and Utilization of Resources and Technology, New Delhi, pp 126–153

Nakamura H, Tanaka R, Suzuki K, Yataka M, Mogi Y (2009) A direct determination method for ethanol concentrations in alcoholic beverages employing a eukaryote double-mediator system. Food Chem 117:509–513

Nogueira EM, Vinagre F, Masuda HP, Vargas C, Muniz de Pádua VL, Da Silva FR, Dos Santos RV, Baldani JI, Ferreira PCG, Hemerly AS (2001) Expression of sugarcane genes induced by inoculation with Gluconacetobacter diazotrophicus and Herbaspirillum rubrisubalbicans. Genet Mol Biol 24:199–206

Okpokwasili GC (2007) Biotechnology and clean environment. Proceedings of the 20th annual conference of the Biotechnology Society of Nigeria (BSN), at the Ebonyi State University, Abakaliki, Nigeria
Olson GJ, Brierley JA, Brierley CL (2003) Bioleaching review. Part B. Progress in bioleaching: applications of microbial processes by the minerals industries. Appl Microbiol Biotechnol 63:249
Padgham JL, Sikora RA (2007) Biological control potential and modes of action of *Bacillus megaterium* and *Meloidogyne graminicola* on rice. Crop Prot 26:971–977
Pagano MC (2008) Rhizobia associated with Neotropical tree *Centrolobium tomentosum* used in riparian restoration. Plant Soil Environ 54:498–508
Pidwirny MJ (2002) Fundamentals of physical geography. Introduction to 117 biogeography and ecology, the nitrogen cycle, 2nd edn. British Columbia Canada
Rania MAN, Mohamed SB, Dalia AS (2015) Effect of mineral and bio-fertilizers on vegetative growth, mineral status, seed yield, tropane alkaloids and leaf anatomy of thornapple plant (*Datura stramonium* L.). Middle East J Agric Res 4:754–768
Reyes I, Bernier L, Simard RR, Antoun H (1999) Effect of nitrogen source on the solubilization of different inorganic phosphates by an isolate of *Penicillium rugulosum* and two UV-induced mutants. FEMS Microbiol Ecol 28:281–290
Reyes I, Bernier L, Antoun H (2002) Rock phosphate solubilization and colonization of maize rizosphere by wild and genetically modified strains of *Penicillium rugulosum*. Microb Ecol 44:39–48
Rodriguez AA, Stella AA, Storni MM, Zulpa G, Zaccaro MC (2006) Effects of cyanobacterial extracelular products and gibberellic acid on salinity tolerance in *Oryza sativa* L. Saline Syst 2:7
Rothschild LJ, Mancinelli RL (2001) Life in extreme environments. Nature 409:1092–1101
Saidi S, Mnasri B, Mhamdi R (2011) Diversity of nodule-endophytic agrobacteria-like strains associated with different grain legumes in Tunisia. Syst Appl Microbiol 34:524–530
Salem HM, Eweida EA, Farag A (2000) Heavy metals in drinking water and their environmental impact on human health. In: ICEHM 2000; Cairo University, Giza, pp 542–556
Schroll R, Brahushi R, Dorfler U, Kuhn S, Fekete J, Munch JC (2004) Biomineralisation of 1,2,4-trichlorobenzene in soils by an adapted microbial population. Environ Pollut 127:395–401
Shen J, Yuan L, Zhang J, Li H, Bai Z, Chen X, Zhang W, Zhang F (2011) Phosphorus dynamics: from soil to plant. Plant Physiol 156:997–1005
Shridhar BS (2012) Review: nitrogen fixing microorganisms. Int J Microbiol Res 3:46–52
Singh J, Mittal SK (2012) *Chlorella sp.* based biosensor for selective determination of mercury in presence of silver ions. Sensors Actuators B Chem 165:48–52
Singh S, Nain L (2012) Microorganisms in the conversion of agricultural wastes to compost. Proc Indian Nat Sci Acad 80:473–481
Singh R, Singh P, Sharma R (2014) Microorganism as a tool of bioremediation technology for cleaning environment: a review. Int Acad Ecol Environ Sci 4:1–6
Song T, Martensson L, Eriksson T, Zheng W, Rasmussen U (2005) Biodiversity and seasonal variation of the cyanobacterial assemblage in a rice paddy field in Fujian, China. Fed Eur Mater Soc Microbiol Ecol 54:131–140
Stevenson RJ, Bahls LL (1999) Periphyton protocols. In: Rapid bioassessment protocols for use in wadeable streams and rivers: periphyton, benthic macroinvertebrates, and fish, 2nd edn. U.S. Environmental Protection Agency, Washington, DC, pp 6–22. EPA 841-B-99-002
Strong PJ, Burgess JE (2008) Treatment methods for wine-related and distillery wastewaters review. Biorem J 12:70–87
Sulaimani H, Joshi S, Al-Wahaibi Y, Al-Bahry S, Elshafie A, Al-Bemani A (2011) Microbial biotechnology for enhancing oil recovery: current developments and future prospects. Soc Appl Biol 1:147–158
Tang CY, Criddle QS, Fu CS, Leckie JO (2007) Effect of flux and technique. Biol Med 1:1–6
Thajuddin N, Subramanian G (2005) Cyanobacterial biodiversity and potential applications in biotechnology. Curr Sci 89:47–57

Vaghela PO, Sutariya DA, Prajapati DV, Parmar SK (2014) Biofertilizers: a enemy of chemical fertilizer. Rashtriya Krishi 9:15–16

Valach M, Katrlik J, Sturdik E, Gemeiner P (2009) Ethanol Gluconobacter biosensor designed for flow injection analysis: application in ethanol fermentation off-line monitoring. Sensors Actuators B Chem 138:581–586

Veteläinen M, Helgadóttir A, Weibull J (eds) (2006) Climatic change and genetic resources in northern Europe. Report of a Workshop on Genetic resources in the northern parts of Europe: Rovaniemi, Finland, European Cooperative Programme for Plant Genetic Resources (ECPGR)

Wakelin SA, Warren RA, Harvey PR, Ryder MH (2004) Phosphate solubilization by *Penicillium* sp. closely associated with wheat roots. Biol Fertil Soils 40:36–43

Walia U, Kumar N, Brar H (1999) Indian. J Weed Sci 31:225–228

Wani PA, Zaidi A, Khan AA, Khan MS (2005) Effect of phorate on phosphate solubilization and indole acetic acid (IAA) releasing potentials of rhizospheric microorganisms. Ann Plant Prot Sci 13:139–144

Wen GM, Shuang SM, Dong C, Choi MF (2012) An ethanol biosensor based on a bacterial cell-immobilized eggshell membrane. Chin Chem Lett 23:481–483

Whitelaw MA (2002) Growth promotion of plants inoculated with phosphate solubilizing fungi. Adv Agron 69:99–151

Willner J, Fornalczyk A (2013) Extraction of metals from electronic waste by bacterial leaching. Environ Prot Eng 39:197–208

Yousef N, Simpson DR, Duncan KE (2007) In situ biosurfactant production by *Bacillus* strains injected into a limestone petroleum reservoir. Appl Environ Microbiol 73:1239–1247

Youssef MM, Eissa MFM (2005) Biofertilizers and their role in management of plant parasitic nematodes. J Biotechnol Pharm Res 5:1–6

Youssef MMA, Eissa MFM (2014) Biofertilizers and their role in management of plant parasitic nematodes: a review. J Biotechnol Pharm Res 5:1–6

Zafar S, Aqil F, Ahmad I (2007) Metal tolerance and biosorption potential of filamentous fungi isolated from metal contaminated agricultural soil. Bioresour Technol 98:2257–2261

Zaidi A, Khan MS, Ahemad M, Oves M, Wani PA et al (2009) Recent advances in plant growth promotion by phosphate-solubilizing microbes. In: Khan MS et al (eds) Microbial strategies for crop improvement. Springer-Verlag, Berlin/Heidelberg, pp 23–50

Zhang F (2010) Response of microbial community structure to microbial plugging in mesothermic petroleum in China. Appl Microbiol Biotechnol 88:1413–1422

Zolla G, Bakker MG, Badri DV, Chaparro JM, Sheflin AM, Manter DK, Vivanco J (2013) Understanding root-microbiome interactions. In: de Bruijn FJ (ed) Molecular microbial ecology of the rhizosphere, vol 2. Wiley Blackwell, Hoboken, pp 745–754

Role of miRNAs in Plant-Microbe Interaction

10

Ravi Rajwanshi, Karam Jayanandi Devi, Gopa Rani Sharma, and Beche Lal

Abstract

MicroRNAs are a class of small RNAs that play a pivotal role in post-transcriptional gene regulation. The role of miRNA in fine-tuning gene expression is essential for growth and development of the plant as well as to cope up with various abiotic and biotic stress conditions. Plant immunity is a well-regulated and complex system. Plants initiate a series of host immune responses upon infestation by microbes. miRNAs have been implicated to play a role in plant-microbe interaction not only in regulation of plant-pathogen interaction but also during symbiosis which is a beneficial interaction. The expression of resistance genes, transcription factors, hormone signaling, and nutrient homeostasis genes is fine-tuned by miRNAs to bring a balance between plant growth and defense. The present chapter will provide an insight into the role of miRNAs in response to bacterial, fungal, viral, aphid infection as well as during symbiosis.

Keywords

miRNAs · Plant-microbe interaction · Target gene · Plant defense

10.1 Introduction

MicroRNAs (miRNAs) comprise a family of endogenous, non-coding regulatory RNAs of 19–28 nt in length, generated from double-stranded RNAs (dsRNAs). miRNAs regulate the expression of gene in a sequence-specific manner and

R. Rajwanshi (✉) · K. J. Devi · G. R. Sharma
Department of Biotechnology, Assam University, Silchar, Assam, India

B. Lal
National Bureau of Plant Genetic Resources, Plant Quarantine Division, Indian Council of Agricultural Research (ICAR), New Delhi, India

M. Kumar et al. (eds.), *In vitro Plant Breeding towards Novel Agronomic Traits*, https://doi.org/10.1007/978-981-32-9824-8_10

ultimately cleave or inhibit translation of the target mRNA. Plant *MIR* genes are transcribed by RNA polymerase II to primary-miRNAs (pri-miRNA) that forms a fold back structure and is further processed into a stem-loop hairpin precursor-miRNA (pre-miRNA). DICER-like protein 1 (DCL1) is recruited by DAWDLE (DDL) and plays an important role in downstream processing of pri-miRNA and pre-miRNA. DCL1 along with HYPONASTIC LEAVES 1 (HYL1) and SERRATE (SE) generate duplex miRNA which is further methylated at its 3′ end by HUA ENHANCER-1 (HEN1). The methylated duplex miRNA is exported out of the nucleus by the exportin-like protein HASTY (HST). One strand of the mature miRNA is preferentially loaded into ARGONAUTE (AGO), the central regulators in RNA-induced silencing complex (RISC). AGO directs miRNA to either cleave or inhibit translation of the target mRNA by base pairing, a mechanism called post-transcriptional gene silencing (PTGS). A large number of miRNAs have been uncovered from different plants species through high-throughput sequencing technologies as well as computational approaches (Kumar et al. 2017; Luan et al. 2015; Devi et al. 2016, 2018; Chakraborty et al. 2016).

Plants being sessile and live in an unpredictable environment are often exposed to various types of abiotic and biotic stress. The key element to remain adaptive in such conditions is regulation of gene expression by RNA silencing. Plants are highly evolved organisms which are constantly exposed to various pathogenic microorganisms. They have adaptive layers of immunity against pathogenic microbes to counteract with them at various levels. Plants recognize these pathogens via pathogen-associated molecular patterns (PAMPs) or microbes-associated molecular patterns (MAMPs) also known as elicitors which is due to presence of receptors called pattern recognition receptor (PRR) on its cell surface. PAMPs in turn switch on the pathogen-triggered immunity (PTI), which enhances the response including callose deposition, reactive oxygen species (ROS) production, protein phosphorylation, and expression of pathogen-related (*PR*) genes (Nejat and Mantri 2017). In response, plants have adapted a secondary immunity called the effector-trigger immunity (ETI) through proteins encoding resistance (*R*) genes that recognize the pathogens-effector proteins and counteract them. R proteins generally trigger vigorous and specific response such as hypersensitive response (HR), which mediates the cell death to the site of infection to inhibit the pathogen growth. Pathogens have diversified their effectors to initiate another pathway of effector trigger susceptibility (ETS), while plants have evolved new resistance proteins encoded by *R* genes to neutralize the effectors. Moreover, phytohormones such as salicylic acid (SA), ethylene (ET), jasmonic acid (JA), abscisic acid (ABA) play significant role in resistance to pathogens and are well established in plants (Baldrich et al. 2015). The interactions between the plant and pathogens would provide a distinctive comprehension of pathogen effectors and resistance genes interplay. miRNAs play a pivotal role in various biological processes in plants, including development (Yang et al. 2007), stress response to abiotic and biotic factors (Rajwanshi et al. 2014), and metabolism (Singh et al. 2016; Boke et al. 2015). The broad impact of miRNAs in counteracting bacteria, virus, fungi, and herbivores has been reported in multiple studies (Table 10.1) (Navarro et al. 2006; Bazzini et al. 2009; Chen et al. 2004;

Table 10.1 List of miRNAs and their targets involved in plant-microbe interactions

miRNA	Pathogen	Plant	Target	Role of target	Tissue	References
miR393	*Pseudomonas syringae*	*Arabidopsis*	TIR1, AFB2 and AFB3	Auxin signaling	Leaf	Navarro et al. (2006)
	Xanthomonas axonopodis	Cassava				Pérez-Quintero et al. (2012)
	Candidatus Leberibcater asiaticus	Citrus				Zhao et al. (2013)
	Aphis gossypii	Melon				Sattar et al. (2016)
	Macrosiphoniella sanbourni (Gillette)	Chrysanthemum				Xia et al. (2015)
	Mungbean yellow mosaic India virus	Beans				Kundu et al. (2017)
	Soybean mosaic virus	Soybean				Yin et al. (2013)
	Piriformospora indica	*Oncidium*			Root	Ye et al. (2014)
miR156	Rice black-streaked dwarf virus	Rice	ARF transcription factor	Auxin signaling	Roots	Sun et al. (2015)
	Mungbean yellow mosaic India virus	Beans	Squamosa promoter binding protein		Leaf	Kundu et al. (2017)
	Alternaria alternaria f. sp. *mali* (ALT1)	Apple	WRKY transcription factors	Defense response		Zhang et al. (2017)
miR160	Soybean mosaic virus	Soybean	TIR1, ARF	Auxin signaling pathway	Leaf	Yin et al. (2013)
	Piriformospora indica	Orchid			Root	Ye et al. (2014)
miR164	Southern rice black-streaked dwarf virus	Rice	NAC1, NAM and SA	Plant development		Xu et al. (2014a, b)
	Rice black-streaked dwarf virus				Root	Sun et al. (2015)
	Phythium aphanidermatum	Turmeric			Rhizome	Chand et al. (2016)

(continued)

Table 10.1 (continued)

miRNA	Pathogen	Plant	Target	Role of target	Tissue	References
miR167	*Pseudomonas syringae*	*Arabidopsis*	ARF transcription factors	Auxin signaling	Leaf	Fahlgren et al. (2007)
	Xanthomonas axonopodis	Cassava				Pérez-Quintero et al. (2012)
	Aphis gossypii	Melon				Sattar et al. (2016)
miR158	*Pseudomonas syringae*	*Arabidopsis*	PPR domain-containing proteins	Defense response	Leaf	Zhang et al. (2011)
miR159	*Pseudomonas syringae*	*Arabidopsis*	MYB33, MYB65, and MYC10	ABA signaling	Leaf	Zhang et al. (2011)
	Macrosiphoniella sanbourni	Chrysanthemum	GAMYB	Gibberellin (GA) signaling pathway	Leaf	Xia et al. (2015)
miR171	Soybean mosaic virus	Soybean	Scarecrow-like transcription factors	Signaling pathway	Leaf	Chen et al. (2016)
	Rhizophagus irregularis	Barrelclover	LOM1	Root colonization	Root	Couzigou et al. (2017)
	Mesorhizobium loti	Birdsfoot trefoil	NSP2	Nodulation signaling pathway	Root	De Luis et al. (2012)
miR172	*Bradyrhizobium japonicum*	Soybean	ENOD 40	Nodulation	Root	Wang et al. (2009)
	Rhizobium etli	Bean	APE2	Nodule senescence	Root	Nova-Franco et al. (2015)
miR396	Southern rice black-streaked dwarf virus	Rice	AtGRF5 and AtGRF4	Leaf symptoms	Leaf	Xu et al. (2014a, b)
	Puccinia cucumerina	*Arabidopsis*	GRF	ROS detoxification		Soto-Suarez et al. (2017)
				Callose deposition		
	Mungbean yellow mosaic India virus	Beans	LOX	SA signaling pathway		Kundu et al. (2017)
miR397	*Heraspirillum seropedicae*	Maize	LAC	Copper homeostasis	Root	Thiebaut et al. (2014)
	Mesorhizobium loti	Birdsfoot trefoil				De Luis et al. (2012)

miR399	*Candidatus* Leberibcaters asiaticus	Citrus	PHO2	Nutrient homeostasis	Leaf	Zhao et al. (2013)s
miR400 (D)	*Pseudomonas syringae*	*Arabidopsis*	PPR1, PPR2	Post-transcriptional gene regulation	Leaf	Fahlgren et al. (2007)
						Park et al. (2014)
miR403	*Pseudomonas syringae*		AGO2/AGO3	RNA silencing		Zhang et al. (2011)
miR408	*Herbaspirillum seropedicae* *Azospirillum brasilense*	Maize	CSD; LAC	Facilitate colonization	Root	Theibaut et al. (2014)
miR482	Mungbean yellow mosaic India virus	Beans	NBS-LRR	Defense response	Leaf	Kundu et al. (2017)
	Xanthomonas axonopodis	Cassava				Pérez-Quintero et al. (2012)
	Verticillium dahliae	Potato	NBS-LRR			Yang et al. (2015)
miR398	*Pseudomonas syringae*	*Arabidopsis*	CSD1	ROS detoxification	Leaf	Li et al. (2010)
	Rice black-streaked dwarf virus	Rice				Sun et al. (2015)
	Xanthomonas axonopodis	Cassava				Pérez-Quintero et al. (2012)
	Herbaspirillum seropedicae	Maize	CSD; LAC	Copper homeostasis	Root	Thiebaut et al. (2014)
	Rhizobium tropici	Beans	CSD1; Nod 19	Facilitate colonization	Root	Naya et al. (2014)
miR863-3p	*Pst* (*avrRpt2*)	*Arabidopsis*	ARLPK1/ARLPK2 and SE	Immune signaling; miRNA biogenesis	Leaf	Niu et al. (2016)
miR2118	Rice stripe virus	Rice	NBS-LRR	Defense response	Leaf	Lian et al. (2016)
miR1507	*Bradyrhizobium japonicum*	Soybean	NB-LRR	Defense response	Root	Yan et al. (2016)

TIR1 transport inhibitor response, *AFB* auxin F-box 2, *ARF* auxin response factor, *NAC1* NAM, ATAF and CUC, *NAM* no apical meristem, *SA* salicylic acid, *PPR* peptatricopeptide repeat, *MYB* myeloblastosis, *MYC* myelocytomatosis, *GA* gibberellic acid, *ABA* abscisic acid, *LOM* lost meristem 1, *NSP2* nodulation signaling pathway 2, *ENOD40* early nodulin 40, *APE2* apetala 2, *GRF* growth-regulating factor, *LOX* lipoxgenases, *LAC* laccase, *PHO2* phosphate 2, *AGO* Argonaute, *CSD* copper-zinc superoxide dismutase, *NBS-LRR* nucleotide-binding site leucine-rich repeat, *Nod* nodulin, *SE* serrate, *ARLPK* atypical receptor-like protein kinase

Barah et al. 2013; Sattar et al. 2016). The studies have shown the involvement of miRNAs in plant signaling pathway during infection by microbes either pathogenic or beneficial and hence have enabled to provide an insight into role of miRNA during plant-pathogen interaction (Fig. 10.1). The miRNAs regulated in leaves and root during infestation by different microbes are depicted in Fig. 10.2. The current knowledge of miRNA as regulators of plant defense in response to pathogen infection has been summarized in the present chapter.

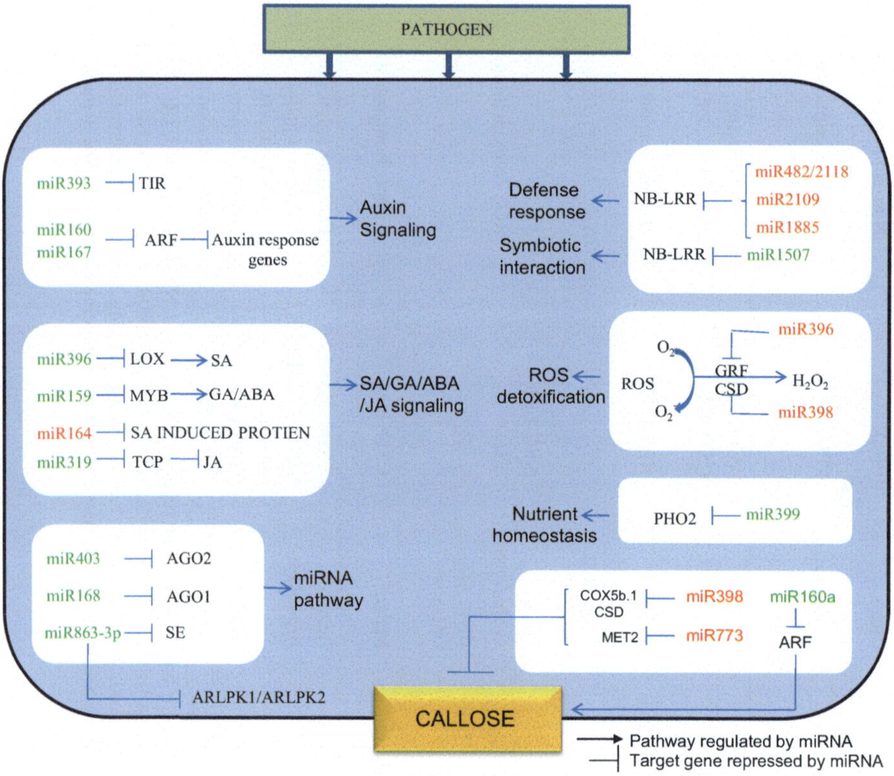

Fig. 10.1 Overview of role of miRNAs in relation to plant-microbe interaction
Infection of plant by bacteria, viruses, fungi, aphids, and even nonpathogenic rhizobium leads to alteration in the expression pattern of miRNAs. The role of miRNAs during plant-microbe interaction reaches from fine-tuning of hormone pathways, defense responses, symbiotic interaction to miRNA pathway regulation. Red font decrease in abundance of miRNA, green font increase in abundance of miRNA, *TIR* transport inhibitor response, *ARF* auxin response factor, *NB-LRR* nucleotide-binding leucine rich repeat, *LOX* lipoxygenases, *MYB* myoblastosis, *SA* salicylic acid, *GA* gibberellic acid, *ABA* abscisic acid, *ROS* reactive oxygen species, *GRF* growth-regulating factor, *CSD* copper-zinc superoxide dismutase, *AGO* argonaute, *SE* serrate, *ARLPK* atypical receptor-like protein like kinase, *PHO2* phosphate 2, *TCP* teosinte branched 1, cycloidea, PCF1, *JA* jasmonic acid, *COX5b.1* cytochrome c oxidase subunit 5B, *MET2* methyltransferase 2

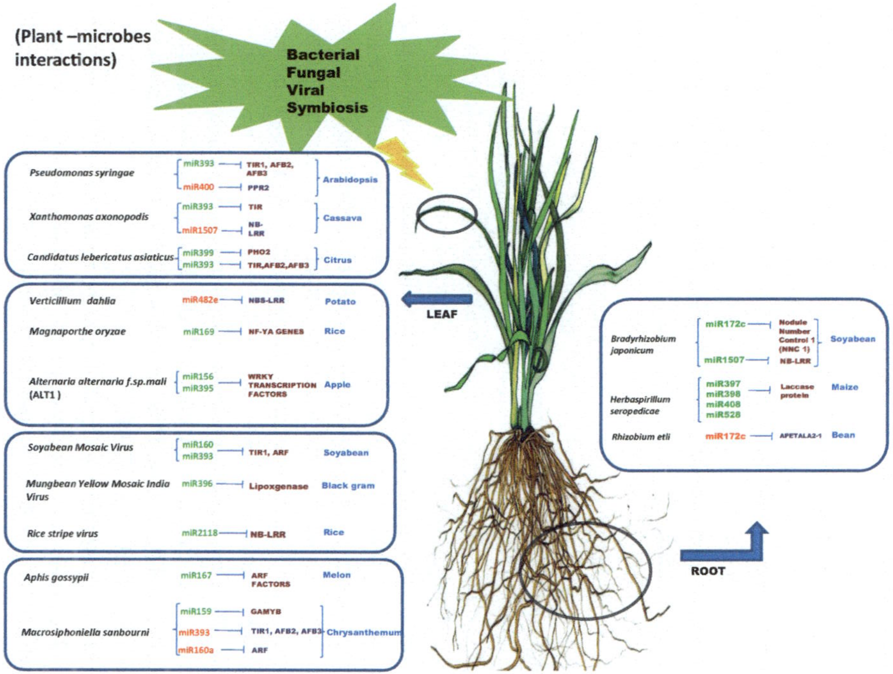

Fig. 10.2 miRNA regulation in leaf and roots of plant during pathogen induction
Various pathogens interact with plants resulting in alteration of miRNA level which are responsible for regulating the expression level of respective targets. The red and green font of the miRNA level reveals the low abundance and high abundance, respectively, in various crop plants (blue font) during microbe (black font) interaction. The target genes are either repressed (brown font) or increased (purple font) when regulated by their respective miRNAs. *TIR* transport inhibitor response, *ARF* auxin response factor, *NB-LRR* nucleotide-binding leucine-rich repeat, *PHO2*-phosphate 2, *NF-YA* nuclear transcription factor Y subunit alpha, *PPR* pentatricopeptide repeats

10.1.1 Bacteria

miR393 was the first miRNA to be discovered to play a role in plant-bacteria interaction. Flg22, a 22 amino acid peptide from the N-terminus of eubacterial flagellin triggers rapid downregulation of the auxin response genes, viz., transport inhibitor response 1 (TIR1) and auxin signaling F-Box proteins 2 and 3 (AFB2 and AFB3). In *Arabidopsis*, flg22 induces miR393 through transcriptional activation of *At-miR393a*, where TIR1, AFB2, and AFB3 are targets of miR393. However, AFB1 repression was miRNA-independent as it showed partial resistance to miR393-directed cleavage due to the presence of a single mismatch between miR393 complementary sites. Overexpression of miR393 represses auxin signaling which enhances bacterial resistance, indicating the downregulation of auxin signaling to be a part of plant-induced immune response (Navarro et al. 2006). The complementary strand miR393* has also been reported to play a role in antibacterial immunity by negatively regulating the expression of MEMB12 (SNARE), a protein involved in membrane fusion, thereby promoting the exocytosis of pathogenesis-related protein (PR1). Profiling of small RNAs (smRNAs) from *Arabidopsis* leaves infected by *Pseudomonas syringae* pv. Tomato (DC3000*hrcC*) predicted the increase in abundance of miR393 as well as two additional miRNA families, miR160 and miR167 which targets TIR1 and members of ARF family of transcription factors, respectively (Fahlgren et al. 2007). A miRNA, miR400, that targets pentatricopeptide repeat (PPR) mRNA was reported to be decreased in abundance upon infection by *Pst* DC3000hrcC mutant (Fahlgren et al. 2007) or by the virulent strain *Pst* DC3000 (Zhang et al. 2011). Overexpression of miR400 rendered the plants susceptible to *P. syringae* pv. tomato (*Pst*) DC3000. Transgenic plants overexpressing miR400 and ppr2 knockdown mutants increased the susceptibility to *Pst* DC3000 indicating that regulation by miR400 downregulates PPR2 hence negatively affecting the defense response of *Arabidopsis*. Therefore, the decrease in abundance of miR400 accompanied by increase in abundance of PPR1 and PPR2 is a defense response of *Arabidopsis* to bacteria. PPR proteins have been earlier reported to play various roles in post-transcriptional gene regulation, circadian clock, seed development, and seed germination, and the highly diverse PPR proteins are localized either to the mitochondria or chloroplast (Park et al. 2014).

A gram-negative bacteria, *Xanthomonas axonopodis* pv. manihotis (*Xam*), causes cassava bacterial blight disease and severely affects the production of cassava (*Manihot esculenta*) which is a staple crop and main source of calories in tropical regions (El-Sharkawy 2004). In response to *Xam* infection, the expressions of ten conserved miRNA (miR160, miR167a, miR167b, miR390, miR393, miR394, miR165, miR171, miR197b, and miRE) were highly increased in abundance, while seven miRNA families (miR535, miR395, miR482, miR397, miR399, miR408, miR1507) decreased in abundance. As reported in previous studies, miR167, miR160, miR393, and miR390 regulate auxin signaling, while miR394, miR165, and miR171 were shown to target F-box family protein, *Arabidopsis thaliana* homeobox (ATHB) transcription factor, and a scarecrow-like transcription factor. miR535, miR395, and miR482 were reported to target various NB-LRR and LRR disease resistance gene while miR397, miR398, and miR408 targeted laccases, copper superoxide dismutases, and plantacyanins, respectively, hence involved in the regulation of copper (Pérez-Quintero et al. 2012).

Jagadeeswaran et al. (2009) reported that *Pseudomonas syringae* infection in leaves accumulates the ROS which is detoxified by an increase of Cu/Zn superoxide dismutase (SOD). In *Arabidopsis*, it was reported that upon infection by avirulent strains of *Pst* DC3000 (avrRpm1) or *Pst* DC3000 (avrRpt2), miR398 decreased in abundance associated with an increase in the transcript level of its target copper-zinc superoxide dismutase (CSD1), but not CSD2 (Jagadeeswaran et al. 2009). In another study, it was reported that PAMP-induced callose deposition, defense gene expression, and seedling growth inhibition were positively regulated by AGO1 contributing to PAMP-induced disease resistance to *P. syringae*. PAMP-induced callose deposition is positively regulated by overexpression of miR160a, while flg22 suppressed the expression of miR398a, and miR773 showed reduced callose deposition and increased disease susceptibility to *P. syringae* DC3000 and DC3000 hrcC^{-} (Li et al. 2010). Zhang et al. (2011) reported 15, 27, and 30 miRNAs that were differentially expressed upon infection by *Pst* DC3000 *hrcC*, *Pst* DC3000 *EV*, and *Pst* DC3000 *avrRpt2*, respectively. miR158 was reported to be highly induced by all the three strains and target genes encoding PPR domain-containing proteins and glycosyltransferases, while miR403 which targets AGO2 and AGO3 decreased in abundance suggesting the possible role of AGO2 and AGO3 in disease resistance. Several miRNAs involved in plant hormone signaling and biosynthesis which includes miR160, miR167, miR390, and miR393 were differentially expressed upon bacterial infection. Bacterial infection downregulated miR159 and targets transcription factors MYB33, MYB65, and MYC10 which activate gibberellin (GA) signaling pathways as well as positive regulators of ABA signaling pathways in *Arabidopsis*. Stress-related miRNAs such as miR398 and miR408 were also decreased in abundance upon bacterial infection (Zhang et al. 2011). The miRNA expression in the study suggested that miRNA mediates specific regulation in nonpathogenic, compatible, and incompatible interactions between host and bacteria. The level of miRNA induction was higher by the avirulent strain *Pst* DC3000 (avrRpt2) than the nonpathogenic *Pst* DC3000 (hrcC) and virulent strain *Pst* DC3000 (EV).

Ralstonia solanacearum (Smith) induces bacterial wilt in ginger and is one of the most important factors that imposes production constraint in ginger-cultivated region of the world (Kumar and Sarma 2012). So far, resistance source to *Ralstonia*-induced bacterial wilt as well as to *Pythium* soft root and *Fusarium* yellow could be identified in *Zingiber* genus; however, *Curcuma amada* Roxb., the Indian mango ginger, displayed resistance to *R. solanacearum* and *Pythium* species, thus providing an opening for developing bacterial wilt resistance in ginger. A total of 192 potential miRNA targets were predicted which also include defense-related genes nucleotide-binding site-leucine-rich repeat (NBS-LRR), −1, 3 glucanase, cytochrome P450, and S-adenosyl methionine as well as transcription factors. The transcriptome data generated suggest the role of hormone signaling in the defense response (Prasath et al. 2014).

Huanglongbing (HLB) or citrus greening disease is the most destructive disease in all commercial citrus such as oranges, grapefruit, and tangerines caused by gram-negative bacteria and phloem-restricted *Candidatus* Liberibacter, vectored by the Asian citrus psyllid (*Diaphorina citri* Kuwayama) (Bové 2006). *Candidatus* Liberibacter asiaticus (*Ca*Las) is the most prevalent, while *Ca.* L. africanus (*Ca*Laf) and *Ca.* L. americanus (*Ca*Lam) are more geographically restricted. The expression of citrus sRNAs in response to HLB has been reported by Zhao et al. (2013) and identified a total of 76 conserved and 10 novel csi-miRNAs. Upon Las infection, csi-miRNAs were differentially expressed where csi-miR159, 399, and 393 increased in abundance while csi-miR396 decreased in abundance. The induction of csi-miR399 and csi-miR159 were reported to be specific to HLB as no apparent change in the expression level of csi-miR399 and csi-miR159 was observed after inoculation with *Spiroplasma citri*, a phloem-limited bacteria. Las-positive plants exhibited a significant induction of miR399 and a decreased level of phosphorous. Las infection possibly caused P deficiency and subsequently induces miR399 expression which leads to the silencing of its target gene *PHO2* sequentially causing an increase in expression of P transporters (PTs) hence indicating P deficiency to be a factor associated with HLB (Zhao et al. 2013). A nested network of five HLB-responsive miRNAs classes (csi-miR393, csi-miR166, csi-miR156, csi-miR167, and csi-miR172) were identified by Rawat et al. (2015). Plant nutrients which are cationic in nature such as calcium, potassium, magnesium, iron, copper, manganese, and zinc have been reported to play important roles in citrus-HLB interaction. The miRNA nested network also showed the linking of csi-miR167 to a hub probe set that codes for K transport, which is further linked to the nodes of sulfate, phosphate, metal, peptide, and metabolic transport-related probe sets. The study suggested that upon *Ca*Las infection, csi-miR396 regulates citrus proteases, while csi-miR167 regulates ion transport in susceptible plants implicating the role of csi-miR396 and miR167 in defense pathway and ion transport (Rawat et al. 2015).

miR863-3p was reported to be highly induced by infection by the avirulent bacterial strain *Pst* (avrRpt2), and through two different modes, it silences the negative and positive regulators of plant immunity. The study indicated miR863-3p to be true miRNAs as its accumulation was dependent on DCL1 and SE, but independent of RDR6. At initial stage, the negative regulators- atypical receptor-like pseudokinase1 (ARLPK1) and ARLPK2 are suppressed by miR863-3p through mRNA degradation to enhance defense response rapidly following infection. ARLPK1 and ARLPK2 are receptor-like kinases that belong to the LRR-RLK subfamily (Shiu et al. 2004). The expression of antimicrobial pathogenesis-related protein (PR1), a marker for salicylic acid-dependent disease responses, was higher in arlpk1–1 arlpk2–1 double mutant plants upon infection with *Pst* (EV) and *Pst* (avrRpt2), hence pointing out the negative regulation of plant immune response by ARLPK1 and ARLPK2. During subsequent stages of infection, defense signaling is attenuated by downregulation of SE through translational inhibition by miR863-3p; consequently the level of miR863-3p also reduces as the accumulation of miR863-3p is dependent on SE. SE, a C_2H_2-type zinc-finger protein, plays a role in the pivotal step of miRNA maturation and accumulation. Thus, the attenuation of immune response after successful defense in addition to the prompt upregulation of plant defense upon infection by pathogen is regulated and fine-tuned by miR863-3p (Niu et al. 2016).

10.2 Aphids

Aphids are major plant sap-sucking pest that causes economical damage to cultivated crops as well as wild plants and secretes metabolites containing pathogenic bacteria and viruses into the host plant (Foyer et al. 2016). Stylets, which are specialized mouthparts, are developed in aphids for effective piercing of plant tissues and to feed themselves from phloem tissues. Insects while feeding infect the plants which modulate variety of processes for effective colonization of aphids that activate plant defense signaling pathways involving JA, SA, ABA, ET, and GA (Morkunas et al. 2011; Louis and Shah 2013). The role of RNA silencing in regulation of herbivore-induced plant defense response was investigated by the application of RNA-directed RNA polymerase silencing in *Nicotiana attenuata* that exhibited enhanced susceptibility to *Manduca sexta.* Plants may mediate systemic signaling through siRNAs that are produced from dsRNA by RNA-dependent RNA polymerases (RdRs). The action of miRNAs and RdRs is coordinated. Ta-siRNA biogenesis is initiated by miRNA-guided cleavage of primary transcripts with subsequent RdR6-dependent formation of dsRNA and processing by DCL and forms ta-siRNA which negatively regulates other genes (Allen et al. 2005). The synthesis of nicotine, one of the most substantially induced defense compounds, was decreased in *N. attenuata* with an inverted repeat RdR1 construct irRdR1, which was due to downregulation of arginine decarboxylase (ADC) and ornithine decarboxylase (ODC) as well as nitrate reductase (NR). The study showed the amplification of siRNA biogenesis by RdR activity which is evoked by herbivorous attack. In RdR1-silenced plants, the repressors of alkaloid biosynthesis which are the targets for siRNAs are not degraded effectively; hence irRdR1 plant is unable to generate nicotine rendering the plant susceptible to herbivores. The iRdR1 plants were highly susceptible to herbivore infestation and suggested that RdR1 mediates aphid resistance by fine-tuning of substrates by the smRNAs produced by RdR1 to survive herbivore attack (Pandey and Baldwin 2007). The evaluation of smRNAs in response to oral secretion (OS) of *M. sexta* larvae to wild type and plants in which RdR1 has been silenced (irRdR1) revealed that nine miRNAs (miR157/miR156, miR159, miR396, miR396, miR894, miR168, miR894, miR397) were differentially regulated between the two genotypes. The transcript levels of phytohormone signaling-related genes such as lipoxygenase 2 (LOX2), LOX3, allene oxide synthase (AOS), jasmonate-resistant 4 (JAR4), and ACC synthase 3a (ACS3a) were decreased, while the levels of threonine deaminase (TD), JAR6, and ACC oxidase 3 (ACO3) were increased in irRdR1 plants which suggested the involvement of new smRNAs in OS-elicited irRdR1 plants that differentially regulate the phytohormone signaling genes (Pandey et al. 2008). DCLs are involved in RNA silencing through the production of smRNAs and hence involved in plant-herbivore interaction. DCLs silenced lines of *N. attenuata* (ir-*DCL2*, ir-*DCL3*, ir-*DCL4*) elucidated the regulatory roles of DCL in defense pathway upon herbivore attack. ir-*DCL3* and ir-*DCL4* lines of *N. attenuata* exhibited increased susceptibility to *M. sexta* while ir-*DCL2* did not have any detectable influence. DCL3 and DCL4 were reported to be the chief DCLs that play an instrumental role in plant defense response (Bozorov et al. 2012).

Myzus persicae (Sulzer), the green peach aphid (GPA), is one of the most devastating pests on cultivated crops and also a vector of different plant viruses (Blackman and Eastop 2000). The miRNA pathway involved in imparting resistance against the GPA colonization was suggested from the aphid fecundity on RdR, DCL, and AGO mutants of *Arabidopsis*. GPA production of progeny by GPA on *Arabidopsis* with miRNAs pathway mutant was significantly less. *Arabidopsis* respond to aphid infestation by involving SA, JA, ET, glucosinolate, and camalexin pathway. In *dcl1* mutant plants (*Arabidopsis* with miRNA pathway mutants), phytoal (PAD3), CYP81F2, LOX2, mitogen-activated protein kinase3 (MPK3), PR1, genes involved in camalexin biosynthesis pathway, indolic glucosinolate pathway, JA biosynthesis, defense related, and SA signaling, respectively, were strongly induced which suggested the involvement of these pathways for aphid resistance. Aphid fecundity was significantly higher on *pad3* mutant as compared to *cyp81f2* mutant, which indicated that camalexin plays an important role in aphid resistance. PAD3 is involved in the conversion of dihydrocamalexic acid camalexin, the major phytoalexin in *Arabidopsis* present in the phloem stream. *Arabidopsis* miRNA pathway mutant infected by aphids accumulated more camalexin than those raised on control plants. Lesser progeny was also produced by aphids on artificial diet containing camalexin which implicated the role of camalexin in reducing the reproductive ability of GPA (Kettles et al. 2013). Several miRNAs were also reported to be involved in regulating transcription factor as well as secondary target genes that belong to SA, JA, and ET signaling pathways in the GPA-*Arabidopsis* susceptibility interaction (Barah et al. 2013).

The role of miRNAs in response to *Aphis gossypii* Glover infestation was investigated in resistant and susceptible melon by Sattar et al. (2012). Resistance to aphids is exhibited as antixenosis (non-preference of the aphids) and antibiosis (reduction in the performance of aphids) and plant tolerance (Bohn et al. 1972; Klingler et al. 1998). The resistance of *A. gossypii* in melon is attributed to the presence of a single dominant virus aphid transmission (*Vat*) gene which belongs to the coiled-coil NBS-LRR family of plant resistance gene, expressed primarily in the phloem tissue all through the developmental stages in melon (Garzo et al. 2002). The expression of most conserved miRNA (miR160, miR164, miR165, miR166, miR167, miR169, miR2111, miR2911, miR390, miR393, miR396, miR397, miR398, miR408, and miR894) was differentially expressed during the early and late stages of aphid infestation. In case of resistant interaction (*Vat*$^+$), miR156, miR157, miR159, and miR162 increased in abundance during early stage, whereas miR166 and miR2111 increased in abundance and miR156, miR162, and miR159 decreased in abundance during late stage of aphid infestation. In susceptible interaction (*Vat*$^-$), miR408 increased in abundance during early stage, while miR56, miR166, miR168, miR169, miR171, miR172, and miR396 increased in abundance during the late stage of infestation. During early stage, miRNAs belonging to the family miR159, miR160, miR172, miR168, miR393, and miR397 were significantly decreased in abundance, while two miRNA families, miR408 and miR398, also decreased in abundance during late stages of aphid attack. miRNA expression was enhanced in the *Vat*$^{+}$ interaction, while in *Vat*$^{-}$ there was a decrease in the expression pattern, which might be due to dissimilarity in the transcription or biogenesis of miRNA in the susceptible and resistant melon lines. Sattar et al. (2016) identified 70 target genes of the differentially regulated melon miRNAs through degradome sequencing in resistant melon lines. The target genes were mainly phytohormones that are implicated in plant defenses against aphids as well as those involved in auxin perception and signaling. Auxin accumulates in response to aphid infestation in both resistant and susceptible interaction, in the model proposed for the role of miR167 and miR393 in auxin-regulated gene expression. miR393 increased in abundance in leaves of resistant interaction, wherein it silenced the expression of target genes TIR-1 and AFB2 receptors and hence preventing the formation of ubiquitin ligase complex comprising of SKP1, Cullin, F-box (SCF), and TIR-1 and subsequent degradation of the Aux/IAA proteins through the complex. The concomitant increase in expression of miR167 prevents the expression of ARF6 and ARF8, eventually leading to inhibition of transcription of auxin responsive genes due to unavailability of ARF activators to dimerize with Aux/IAA (Sattar et al. 2016).

Chrysanthemum (*Chrysanthemum morifolium* Ramat), a popular ornamental as well as medicinal plant species, on heavy infestation by the pest *Macrosiphoniella sanbourni* (Gillette) from seedling to flowering stage, adversely affects the quality and yield of the flowers. Many *MIR* genes show high expression in both aphid-infested and mock puncture treatment. *GAMYB-like 2* gene was predicted to be a target for miR159a whose expression was induced upon aphid infestation and mock puncture treatment. *GAMYB-like 2* gene encodes R2R3 MYB domain transcription factors involved in GA signaling pathway. miR159a was suggested to be involved in programmed cell death (PCD) induced by aphid attack. The decrease in abundance of miR160a and miR393a led to the transcript abundance of its targets ARF and TIR, respectively, and hence suggests the association of miRNA-mediated gene regulation in chrysanthemum-aphid interaction (Xia et al. 2015).

10.3 Viral Infection

MiRNAs are known to play important regulatory roles during viral infection in plants by controlling expression of different genes. Upon viral infection in plants, the level of endogenous miRNAs is often altered which is associated with a change in the abundance of its corresponding target mRNA. Next generation sequencing or deep sequencing can deliver insights into virus-induced plant defense mechanism. Interference with miRNA-mediated processes was suggested to be a general feature of pathogenicity, where the miRNA pathways are altered by expression of suppressors of post-transcriptional gene silencing (PTGS) leading to developmental defects in infected plants (Chapman et al. 2004; Chen et al. 2004). Previous studies revealed that post-transcriptional regulation involving silencing suppressor activity led to miRNAs alteration upon viral infection or transgenic expression of viral proteins.

Alteration in the levels of miR156, miR160, miR164, miR166, miR169, and miR171 was correlated with severity in symptom. PTGS suppressors are viral proteins which block one or multiple steps of the PTGS pathway in plants and hence act as a counter-defensive strategy developed by many viruses. Tobacco etch virus (TEV) and Potato virus Y (PVY) were reported to encode strong PTGS suppressor, while Tobacco mosaic virus (TMV), Tomato mosaic virus (ToMV), and Potato virus X (PVX) encode weak PTGS suppressor. The study with five viruses revealed that virus with weak or no suppression activity changed the acquisition of miRNAs, while those that strongly inhibit gene silencing modified miRNAs acquisition to similar levels. Therefore, viral proteins that do not suppress PTGS are effective in alteration of miRNA pathway in infected plants. TMV and ToMV gave rise to severe symptoms on tobacco with altered miRNA accumulation, a common feature of infection, to a considerable degree than TEV and PVY viruses that generated mild symptoms. However, it is still unclear as to how viral infection interferes with miRNA pathway (Bazzini et al. 2007). Further studies revealed the alteration of miRNA pathway at transcriptional level which involves miRNAs promoter after viral infection. It was reported in earlier studies that, after virus infection and GA3 treatment, the transcriptional activity of P-miR164a was enhanced, thus detecting a novel system of plant miRNA biogenesis by viral interference (Bazzini et al. 2009). The change in expression pattern of nine miRNAs and their target mRNAs upon cucumber mosaic virus (CMV) and Tomato aspermy virus (TAV) infection was investigated in tomato flower and fruit development. The study indicated that the level of individual miRNA alteration on infection by different strain varied, suggesting the variation in severity of symptoms phenotypes (Feng et al. 2013). The homeostasis of the most important AGO protein, AGO1, which is responsible for the cleavage or translational inhibition of target mRNA, is controlled by miR168. Several studies have indicated the high abundance of miR168 and elevated expression of AGO1 mRNA in virus-infected plants. The inhibition of AGO1 protein accumulation through virus-mediated induction of miR168 was suggested to be a counter-defensive strategy; moreover virus-occupied region was shown to have enhanced miR168 accumulation which indicated the involvement of virus-derived products in the phenomenon (Varallyay et al. 2010).

Southern rice black-streaked dwarf virus (SRBSDV), a phloem-limited, devastating plant virus, is transmitted by white-backed planthopper [WBPH; *Sogatella furcifera* (Horvath)] that hampers rice production. An insight into miRNA-mediated SRBSDV-rice interaction was furnished using microarray profiling of rice miRNAs upon SRBSDV infection. miR164, R396, R530, and R1846 were reported to target rice genes that suggest the involvement of these miRNAs in the regulation of their respective targets which might be responsible for various symptoms on SRBSDV-infected rice plants. The regulation of three target genes, viz., NAC1, NAM, and SA-induced protein by miR164 in SRBSDV infection, could be responsible for symptoms such as presence of aerial rootlets, branches on stem nodes and excessive tillering, as well as salicylic acid-mediated virus resistance. Upon SRBSDV infection, three transcription factors regulating genes orthologs of growth-regulating factor (AtGRF5) and AtGRF4 that are crucial for cell proliferation in leaf were targeted by osa-miR396 (Xu et al. 2014a). The expression pattern of miRNAs in leaves and roots of healthy as well as RBSDV-infected rice was investigated using high-throughput sequencing. Majority of miRNA were reported to be highly expressed in leaves than in roots, where conserved miRNA distribution pattern was more affected than the non-conserved ones. miR156a-j, miR166abcdf, and miR169hijklm were increased in abundance, while miR408, miR827, and miR1428e decreased in abundance in both leaves and roots upon RBSDV infection. Tissue specific expression pattern was also observed where miR164abf and miR398 increased in abundance in RBSDB-infected leaves but not in roots (Sun et al. 2015).

Rice stripe virus (RSV), which belongs to the genus *Tenuivirus*, causes epidemic disease on rice and hampers the production of rice in many Asian countries. RSV is transmitted by the insect vector small brown planthopper (SBPH), the viruliferous state of which is highly correlated with rice stripe disease epidemics. The previous reports on RSV-infected rice revealed that 7 putative novel pn-miRNAs (5 pn-miRNAs increased in abundance while 2 pn-miRNAs decreased in abundance) were produced in rice whose expression levels altered in RSV-infected rice (Guo et al. 2012). A study also revealed that miRNAs were specifically induced or repressed in specific time point upon RSV infection in rice. A total of 49 identified miRNAs target 399 rice genes associated with signal transduction, transcription factors, differentiation, and diseases resistance suggesting that RSV infection might repress genes that are associated with rice growth and development and also hamper the defense mechanism against virus leading to viral disease symptoms. A total of 15 disease resistance genes were predicted to be targets of two newly identified miRNA (pm4242 and pm5124) as well as other known miRNAs (osa-MIR2118n, osa-MIR2118q, osa-MIR395y, and osa-MIR399g). Members of miR2118 family that target the NBS-LRR containing gene families were decreased in abundance upon RSV infection (Lian et al. 2016).

Mungbean yellow mosaic India virus (MYMIV), which causes the yellow mosaic disease (YMD) in different economically important crop plant species including *Vigna mungo*, is one of the most devastating members of whitefly-transmitted bipartite begomoviruses. The expression pattern of pathogen responsive miRNAs was revealed through high-throughput sRNA-sequencing in the non-model legume plant *V. mungo* when infected with the begomovirus, MYMIV. Enhanced expression of genes involved in SA signaling pathway has been known to confer resistance in *V. mungo* against MYMIV. An increase in miR396 expression was observed in MYMIV infection. miR396 targets LOX, which has a role in the production of JA in plants. An increase in accumulation of miR396 was reported to activate the SA-mediated pathway by suppressing the JA pathway, suggesting its possible role in *V. mungo* immune response. The role of other stress responsive miRNA candidates such as miR159, miR166, miR169, miR319, and miR1514 in conferring resistance in *V. mungo* was also explored. The expression level of miR159, which targets a zinc finger, was reported to be greatly enhanced, while isoforms of miR166 and miR319 were downregulated after MYMIV inoculation. In plants, the recognition of viral effector molecules is mediated by the disease resistance (R) protein receptors that are comprised of LRR-NBS and a putative amino-terminal signaling domain. A moderate level of complementarity was found between miR482 and its corresponding target NB-LRR, suggesting the role of miR482 in regulating MYMIV resistance. Upon MYMIV infection, miR482 transcript level declined with corresponding increase in abundance of NB-LRR. The study also indicated that auxin signaling pathway, which plays a role in plant growth and developmental and physiological aspects, can also regulate plant-pathogen interaction. miR160 and miR393 negatively regulate auxin perception and target auxin response factors (ARFs) family of transcription factors ultimately leading to reduction of normal plant growth and development during MYMIV-infection. Chlorophyll degradation and disruption of photosynthetic apparatus lead to the yellow chlorotic spots phenomenon on leaves of susceptible *V. mungo*. Two miRNAs (vmu-miRn11 and vmu-miRn12) were predicted to target genes involved in photosynthetic machinery, suggesting the destabilization of the photosystem I reaction center and chlorophyll a/b-binding proteins by increase in accumulation of miRn13 and miRn11, respectively (Kundu et al. 2017).

Soybean mosaic virus (SMV), one of the destructive pathogens that hampers soybean production, is a member of the genus *Potyvirus*. A comprehensive study of smRNAs from mock-inoculated and SMV-infected soybean leaves was carried out to explore miRNA families that were involved in SMV infection. Upon SMV infection, miR160 and miR393 were increased in abundance leading to miRNA-mediated suppression of factors of auxin signaling pathways as a defense response (Yin et al. 2013). Developing of next-generation strategies for antiviral resistance in plants requires an understanding of the molecular mechanism of virus-host interaction. Another investigation on the SMV-soybean interaction was carried out using genomic approaches. miRNA profiling in soybean infected by 3 SMV isolates revealed 40 and 15 miRNAs that become abundant or decrease, respectively, in response to SMV infection. Members of miR171 (miR171c/j-5p) and miR156 (miR156t-5p) families increased in abundance upon infection by any of the three SMV isolates, while miR160, miR11510 and, miR2118a/b-5p increased in abundance when infected by isolates G2-L and G2-LRB and miR393 when infected by G2L and G7, respectively. The degradome analysis revealed that the identified miRNAs were found to target transcription factors such as HD-ZIP transcription factor, GRAS transcription factor, and nuclear transcription factor Y as well as disease resistance genes such as LRR protein 1 kinase family genes and TIR-NBS-LRR resistance genes. The transcriptome analysis indicated that upon SMV infection, the expressions of several differentially expressed genes (DEGs) were altered as well as their transcriptional responses were specific to SMV infection with different strains (Chen et al. 2016)

Cotton leaf curl disease (CLCuD) in cotton is caused by a pathogen complex of monopartite begomovirus that is transmitted by the whitefly *Bemisia tabaci* (Gennadius), along with a specific DNA betasatellite (DNA-β) molecule, Cotton leaf curl Multan betasatellite (CLCuMB). Betasatellites encode βC1 which is an inhibitor of PTGS and may intricate in the passage of virus within the plant tissue and in augmenting viral DNA levels in plant (Qazi et al. 2007). Viral dsRNA and single-stranded RNA (ssRNA) are recognized and cleaved by DCL proteins and processed into virus-derived small interfering RNAs (vsiRNAs) upon virus infection in plants, thereby triggering antiviral silencing. During virus infection, two different classes of vsiRNAs are produced, viz., primary siRNAs derived from DCL-mediated cleavage of initial trigger RNA and a secondary siRNAs that involve RdR during their biogenesis. Previous reports predicted that antiviral defenses could be triggered by vsiRNAs through PTGS and utilize the plant RNA silencing apparatus to target host transcripts. CLCuD infection produced a population of 21-nt vsiRNAs which is slightly higher than the 22-nt counterpart suggesting the synergistic interaction of DCL4 and DCL2 in the production of vsiRNAs and also a shift in the size of smRNAs after CLCuD infection. 5′ terminal nucleotides were reported to control their specific sorting into distinct AGO complex. vsiRNA with 5′ terminal U was loaded into AGO1 in CLCuD-infected cotton plants, suggesting that this could be a salient mechanism of RNA silencing against CLCuD. The predicted vsiRNA targets were downregulated suggesting viral-induced gene silencing as the chief regulatory mechanism in CLCuD-infected cotton plants (Wang et al. 2016).

10.4 Fungal Infection

Being sessile organisms, plants have developed a complicated immune response against various fungal pathogens which enable them to survive. Through cell surface and intracellular immune receptors, plants recognize the pathogen invasion and thus initiate defense responses. Plant genome contains several numbers of genes encoding NB-LRR immune receptors, where the uncontrolled activation of these receptors results in autoimmunity responses which effect plant growth and development. Plant smRNAs play a significant role in modulating the expression of genes by PTGS phenomena which is responsible for growth and development and induction of anti-pathogenic responses. miRNAs, being one of the key post-transcriptional gene regulatory molecules in plants that participates in PTI and ETI, respond to microbe infection and alter the gene expression either by suppressing or repressing the target genes (Katiyar-Agarwal and Jin 2010). Thereby, a link has been emerged between miRNAs and fungal infection such as in *Populus szechuanica* species. The expression profiling of miRNAs was studied through high-throughput sequencing in which a total of 90 known miRNAs belonging to 42 families and 378 novel miRNAs were identified on infection with rust fungus, *Melampsora larici-populina* incompatible isolate Sb052, *M. larici-populina* compatible isolate Th053, and uninfected controls. The study revealed that expression of miRNAs was inclusively different after infection with different isolates and most of the miRNAs were suppressed during rust fungal infection. These miRNAs play significant role in modulating genes responsible for disease resistance proteins, serine/threonine kinases, transcription factors, and other related proteins. Among these, 27 miRNAs (7 known miRNAs and 20 novel miRNAs) were found to be relevant in enhancing disease resistivity. miR2118, miR472, miR482, and eight novel miRNAs (novel_mir_11, novel_ mir_166, novel_ mir_211, novel_mir_244, novel_mir_248, novel_mir_250, novel_mir_357, and novel_ mir_403) were found to target disease resistance proteins: resistance to *P. syringae* protein 2 (RPS2), RPS5, and RPM1 [a NB-ARC domain-containing disease resistance protein]. Several novel miRNAs (novel_mir_191, novel_mir_198, novel_mir_368, novel_mir_93) were predicted to be involved in regulating LRR receptor-like serine/threonine kinase, while novel miRNAs, mir290, mir530 were involved in regulation of transcription factors: WRKY and MYC2 which play an important role in plant defense pathway response. The study carried out by Chen and Cao (2015) divulged that most of miRNAs in *P. szechuanica* infection with incompatible Sb052 and compatible Th053 either increased or decreased in abundance. Thirty-two miRNAs were found to be increased in abundance, while 40 miRNAs decreased in abundance in *P. szechuanica* incompatible infection. In case of compatible infection 35 miRNAs were increased abundantly, while 53 miRNAs were found to be decreased. Of this miR1444 showed significant negative regulation of target gene encoding polyphenol oxidase which is an enzyme responsible for resistivity to both biotic and abiotic stresses in incompatible infection (Lu et al. 2008). A decrease in miR1444 level led to high expression of polyphenol oxidase enzymes which enhanced resistivity in tomato against pathogen *P. syringae* (Li and Steffens 2002).

Pythium aphanidermatum, a devastating necrotrophic oomycete, causes *Pythium* soft rot of turmeric (*Curcuma longa* L.). A total of 28 turmeric miRNAs (18 conserved and 3 novel miRNAs) were identified upon infection by *P. aphanidermatum* where most miRNAs decreased in abundance and their expression level was changed in response to fungal stress, which supports the hypothesis of involvement of miRNAs in the turmeric-*P. aphanidermatum* interaction. miR159, miR164, miR393, and miR482 were involved in plant defense against fungal infection. A decrease in abundance of miR164 which resulted in increase in the level of NAC, a protein involved in plant immune response, might prevent hypersensitive response, thereby facilitating fungal infestation while decrease in abundance of miR393 and miR160/miR167 led to the activation of auxin-responsive genes. A novel miRNA, clo-novel-miRx2, significantly increased in abundance in the infected rhizomes when compared to their control while its target pathogen pattern receptor (PPR), a protein which contributes to ROS formation, was repressed resulting in susceptible phenotype (Chand et al. 2016). As plants recognize and develop means to counteract infestation by pathogens, the pathogens require strategies for facilitating infection and to reproduce inside the host plant. A miR1918 (pi-miR1918) in *Phytophthora infestans* was identified and was shown to have sequence similarity with tomato miR1918 (sly-miR1918) which enhances the tomato sensitivity to *P. infestans*. An artificial pi-1918 generated using *Arabidopsis* pre-miR159a as a backbone was used to study the role of miR1918 in tomato-*P. infestans* interaction. A far-reaching disease symptom was developed by transgenic tomato plants overexpressing artificial pi-miR1918 than the wild-type tomato upon infection by *P. infestans*. A decrease in abundance of sly-miR1918 was detected in leaves of tomato inoculated with *P. infestans*. The targets of sly-miR1918 and pi-miR1918, RxLR effector peptide protein, and trafficking protein particle complex subunit decreased in abundance upon fungal infection. *Sly-TG2* encodes a RING finger protein that plays a substantial role in regulation of plant growth and development, hormone signaling, and responses to biotic and abiotic stress in plants. The negative correlation between amiR1918 and its target genes suggested the role of amiR1918 in silencing genes coding for RING finger thereby enhancing the susceptibility of tomato to *P. infestans* infection (Luan et al. 2016).

Blumeria graminis f. sp. *hordei* (Bgh), a natural variant of ascomycete fungus, causes a destructive foliar disease in barley (*Hordeum vulgare* L.) called powdery mildew disease and hampers barley production. A study on *H. vulgare* L.-Bgh, interactions revealed that miR398 mediates the control of disease resistance proteins: *Mla* (*Mildew resistance locus*) and *restoration of Mla resistance 1* (*rom1*) gene expression by suppressing chloroplast copper/zinc HvSOD1. HvSOD1 is involved in barley enhancing effector-induced HR in response to infection of this fungus. Plants anchoring the resistance proteins Mla exhibit HR and accumulation of ROS in *H. vulgare* L.-Bgh interactions (Vanacker et al. 2000). In plants, chloroplasts are one of the important genesis for signaling ROS molecules in which the primary electron transport acceptor (PS I) associated with thylakoid membrane is the finest site for production of superoxide radicals ($O^-{}_2$) (Zurbriggen et al. 2009; Gill and Tuteja 2010). The plastid Cu/Zn SOD is the significant form of SOD (Asada 1999) localized in stromal regions of chloroplast preferring in regulation of $O^-{}_2$ into H_2O_2 (OgawaK et al. 1995). H_2O_2 being more stable easily diffuses through the cell membrane and thereby acts as intercellular signaling molecule to amplify various biological processes and also inducing HR response in plant-pathogen interactions (Liu et al. 2007). miR398, designated as hvmiR398, was reported to regulate hvSOD expression where on infection with fungal pathogen, the hvmiR398 reduced hvSOD accumulation, hence enabling susceptibility to infection (Xu et al. 2014b). Recent studies reported that miR398 directly targets the gene encoding copper chaperone for CCS 1 protein which plays significant role in Cu/Zn SOD stability (Chu et al. 2005). It was reported that hvSOD1 accumulation can be increased by *restoration of Mla resistance 1* (*rom 1*) which induced the downregulation of hv-miR398 expression triggering the resistance activity by upregulation of Mla-mediating genes (Xu et al. 2014b). However, Maekawa et al. (2011) revealed that *Mla* alleles induce rare-specific resistance in barley and *Mla1* in *Arabidopsis*, indicating their conserved nature in CC-NB-LRR-triggered immunity response. Recently, it has been identified that the homolog of *Mla*, *TmMla1*, and the ortholog Sr33 conferred resistivity response in wheat against powdery mildew (*Blumeria graminis* f. sp. *tritici*) and Ug99 stem rust disease implicating the role of *Mla*-mediated control of SOD1 through miR398.

Puccinia cucumerina, a nerotrophic fungus also known as *Fusarium tabacinum* (anamorph *Plectosporium tabacinum*), causes a severe damage to dicot crops as well as *A. thaliana.* A study based on microarray analysis of miRNA regulation in *Arabidopsis-P. cucumerina* interaction revealed the miR396 activity in modulating the target transcription factors, mainly GRFs which enhance the immune response. The GRFs are involved in various developmental processes such as flowering, seed and root development, and plant longevity. In *A. thaliana* miR396 is encoded by two loci: MiR396A and MiR396B. Overexpression of miR396 upon fungal infection revealed a gradient decrease in abundance of its GRFs targets expression which led to developmental effects in plant hence displayed enhanced susceptibility to infection. To avoid such effects, an artificial transgenic line has been constructed which mimics the miR396 targets (MIM396) and conferring a wide resistance to fungal pathogenic infection. Results revealed that MIM396 plants had low level of miR396 on infection with *P. cucumerina* enabling an increase in the target expression to mount accompanied with high accumulation of ROS, callose deposition, and other robust transcriptional reprogramming upon pathogenic challenges. Here, transcriptional reprogramming also includes the alternation in genes involved in cell wall dynamics that induced resistivity in *Arabidopsis* during *P. cucumerina* infection (Hernandez-Blanco et al. 2007; Sanchez-Rodriguez et al. 2009). In addition to generation of stronger response, defensive responses were also enhanced in MIM396 plants which include chromatin remodeling and accumulation of inactive proteins (proteins that are regulated by Ca^{+2} levels, cellular redox state, or phosphorylation processes) that functions in host immune system program. These proteins get activated only on pathogenic infection enabling plants to induce rapid or strong immune responses (Soto-Suarez et al. 2017).

Sclerotinia sclerotiorum, a necrotrophic fungus, causes a white mold disease in oil seed rape (*Brassica napus*). A total of 280 miRNAs families, including 53 novel and 227 known miRNAs, were identified in *B. napus* by high-throughput degradome sequencing upon inoculation of *S. sclerotiorum* and uninoculated leaves. Apart from these a total of 68 miRNAs were analyzed by miRNA microarray, and the expression was exclusively differential after infection at 3 and 12 h post infection (hpi). The diversity in profiling of miRNA expression in response to *S. sclerotiorum* infection interprets the complexity of miRNA-mediated regulation in *B. napus-S. sclerotiorum* interactions. The expression patterns of 68 responsive miRNAs were classified into six groups: Group I miRNAs were highly expressed and constantly lessen in abundance after pathogenic immunization. Group II miRNAs showed exclusive variants in expression where at 3 hpi they were decreased in abundance and then increased abundantly at 12 hpi. The miRNAs in Group III showed no significant difference at 3 hpi and 12 hpi, whereas they were downregulated at 3 hpi. MiRNAs of groups IV and V were distinct to those of groups I and II, respectively. miRNAs in Group VI were observed to regulate the expression lately only at 12 hpi. This expression data analysis implies that the regulations of different miRNAs were involved in early as well as late duration of *S. sclerotiorum-B. napus* interaction. Insights of targets of these responsive miRNAs were significant to study in order to attain their respective role in plant-pathogen interactions. The role of these targets was miscellaneous and involved in variety of cellular and molecular activities such as growth and development metabolisms, transcription, signal transduction, redox homeostasis, and defensive pathway. Among them, about half of the miRNAs targeted genes encoding transcription factors that were involved in plant defense and development. The most important included target genes were the defense/resistance-related genes. The study identified three miRNAs whose target genes were involved in ETI response. These miRNAs were bra-miR1885a which targets TIR-NBS-LRR, a type of *R* gene; bna-miR6030 which targets a At1g12290-like disease resistance proteins; and ath-miR168a_L + 1R_1 which targets a small glutamine-rich tetratricopeptide repeat protein 1 (SGT1a)-phosphatase-like proteins. A decrease in abundance of these miRNAs enhanced the resistivity conferred by the *R* genes-mediated miRNAs during fungal inoculation. Moreover, the two miRNAs such as ath-miR397a and rco-miR397_R + 1 whose level increased after *S. sclerotiorum-B. napus* interaction with decrease in target nitrate reductase (NO-forming) expression played a significant role in defense mechanisms against fungal infection. Four miRNAs viz. ath-miR408 strongly tuned the target expression of L-ascorbate oxidase by decreasing its level upon infection at late interval of 12 hpi; cme-miR166e regulates the expression of its target gene which were copper chaperone of Cu/Zn SOD; ath-miR5021 consistently upregulates its target Cu/Zn SOD gene during infection processes; and the novel miRNA PC-14-5p negatively modulates the redox gene encoding thioredoxin reductase at 12 hpi indicating the reduction status regulation at last stage of fungal infection. Intriguingly, another three miRNAs were identified targeting the two PTGS components: AGO1 and AGO2. ath-miR168a targets the AGO1 components, while both bna-miR403 and aly-miR403a-3p_L + 1 were targeting the AGO2 components, thereby resulting in decrease in level of target expression leading to susceptibility toward the infection. The study collectively reveals a firm synchronization of widely involved miRNAs in *B. napus-S. sclerotiorum* interactions (Cao et al. 2016).

Fusarium oxysporum f. sp. *lycopersici* (Sacc.), a fungal pathogen, causes a vascular wilt disease in tomato (*Solanum lycopersicum* L.). An insight into miRNA regulation was provided through deep-sequencing analysis and revealed that different miRNAs were found to be responsible for tuning their respective targets in the roots of different tomato cultivars such as Moneymaker and Motelle toward fungal infection which demonstrated the significant role of plant miRNAs in defense mechanisms pathway. Two miRNAs, sl-miR482f and sl-miR5300, were shown to target the P-loop region of the genes encoding NB domains. The targets of Sl-miR482f were solyC08g075630 (NB and CC domain) and solyC08g076000 (NB and three LRR domains), while the targets of sl-miR5300 were solyC05g008650 (NB domain and CC motifs) and solyC09g018220. Among these, Moneymaker variety showed susceptibility to fungal infection by decreasing the target's expressions, and the Motelle variety whose target gene expression was high after *F. oxysporum* immunization enhanced the resistance of the cultivar upon infection. The studies revealed that negative modulation of the target genes interprets the resistance activity of the Motelle cultivar toward susceptibility to fungal infection (Ouyang et al. 2014).

10.5 Symbiotic Interactions

Symbiotic nitrogen fixation (SNF) is the mutual interaction phenomenon including both nitrogen (N_2)-fixing rhizobial bacterium and leguminous plants. The symbiotic N_2 fixation which takes place in root evolved with a specialized organ called nodules. The process of nodulation is a complicated process that involves a tight junction between rhizobia and specific legume host through molecular signals. A total of

32 conserved miRNAs belonging to 11 families were isolated in mature nodules, while 20 miRNAs were specific to soybean nodules (Wang et al. 2009). Apparently, 55 families of miRNAs were isolated from the roots of soybean on interaction with rhizobial bacterium and *Bradyrhizobium japonicum* out of which 35 families were found to be novel. The infection strongly induces miR168 and miR172 regulation at 3 hpi, but its modulation decreases by 12 hpi. Several miRNAs such as miR160, miR393, miR164, and miR168 were regulated on rhizobial infection which targets ARFs (ARF10, ARF16, and ARF17), TIR1, NAC1, and AGO1, respectively (Katiyar-Agarwal and Jin 2010; Khraiwesh et al. 2012). Yan et al. (2016) reported the miRNA abundance in root hairs of soybean on rhizobial infection. High-throughput sequencing method identified a total of 48 miRNAs regulating the expression level in root hairs on *B. japonicum* infection. Among these, 38 were conserved miRNAs in which 16 miRNAs were soybean specific. Through degradome sequencing or parallel analysis of RNA ends (PARE) analysis, researchers identified the target of miR166 as HD-ZIP transcription factor Glyma05g30000 which plays a specific role in nodulation where its level gets reduced to two-thirds after rhizobial infection. miR1507 identified in soybean root, stem, leaf and mature green seed showed a remarkable decrease after infection by targeting NB-LRR-type resistance gene, Glyma04g29220 (Yan et al. 2016). Moreover, the same researchers also revealed that the overexpression of two miRNAs in soybean roots resulted in the change of nodule number where gma-miR2606b which regulates the mannosyl-oligosaccharide 1,2-alpha-mannosidase gene and gma-miR4416 regulates the rhizobium-induced peroxidase1 (RIP1)-like peroxidase, GmRIP1. Similarly, miR172 family members were differentially expressed in soybean roots, leaf, and nodules implementing its role in growth and development in a wide range of soybean tissues. Among these, miR172c plays a role as a key regulator in nodulation of soybean and gets up regulated on the onset of infection during nodule development by repressing its target gene, Nodule Number Control1 encoding protein that targets the promoter of nodulin gene, ENOD40 (Wang et al. 2014).

Arbuscular mycorrhizal (AM) is one of the most significant symbiosis implies the alliance of most plants with fungal organisms, *Rhizophagus irregularis* belonging to phylum Glomeromycota. This symbiosis benefits with the exchange of nutrition where the fungus mainly provides phosphorus to the host plant via. penetrating into the roots of the plants and form a structure called arbuscular within the cortical cell branches. Moreover, it also revealed that AM symbiosis enhances plant resistance to both biotic and abiotic stresses. A study of miR171 family was carried out in 2016 demonstrating their regulation in symbiosis interactions. miR171 family comprises of seven members: miR171a, miR171b, miR171c, miR171d, miR171e, miR171f, and miR171h. Previous studies in *M. truncatula* reported that miR171 family was involved in root development as well as AM symbiosis and negatively regulates the target gene *LOST MERISTEM 1* (*LOM1*). *LOM1* positively regulates the AM symbiosis by strongly reducing the root colonization and also the formation of arbuscules by mycorrhizal fungus *Rhizophagus irregularis* and thereby inhibits the AM symbiosis (Lauressergues et al. 2012). The significant role of miR171b among other miR171 family in regulating the mycorrhizal interactions was illustrated in the study. A reporter gene *GUS* was allowed to express conversely under the promoter region of miR171b in the tissues of roots and arbuscule-containing cells of *M. trunctula* upon fungal intercommunication (Formey et al. 2014). While other three miR171 families such as miR171a and miR171c were expressed in all root tissues, miR171e and miR171f were expressed in central cylinder; and the last miR171d was expressed mostly in central cylinder and rarely in root parts when colonized with fungal organism irrespective of their mycorrhizal consequences. This study suggests that these miRNAs can mediate the silencing of the target *LOM1* gene in most of the root cells. In another study, miR171b was reported to be expressed specifically in the root cells and has a conserved mismatch cleavage site that is conserved in mycotrophic plants and hence unable to downregulate the miR171 family target, *LOM1*, leading to establishment of AM symbiosis (Couzigou et al. 2017).

Piriformospora indica, a root-colonizing endophytic fungus of *Sebacinales* which conquer with the roots of *Oncidium* orchid plant also known as "dancing lady orchid" establishes various benefits for the host plant. *Oncidium* is an epiphytic plant that possesses slow growth and development process such as it requires 2–3 years for flowering (Chugh et al. 2009). A perception of study emphasizes the mechanisms underlying the miRNA modulation in growth, phytohormone signaling, cell wall metabolism, and transcription factor regulation during *P. indica* interactions with orchid roots and also induces resistivity responses against biotic and abiotic stresses (Varma et al. 1999; Waller et al. 2005). Ye et al. (2014) reported several conserved and novel miRNAs such as miR160, miR164, miR167, miR393, miR394, and miR5293 which are involved in regulation of auxin signaling factors. Also miR390 was identified which regulates the target gene encoding a LRR-kinase that is responsible in modulating the lateral growth in root. Moreover, some miRNAs such as miR894, miR535a, miR166a, miR166h, miR167, miR168, miR397, and miR2950 were detected to have high expression in root tissues during colonization with *P. indica*. The study shows that the respective miRNAs were related to auxin signaling genes such as TIR1, ARF, and CUC which were recognized to regulate the expression of target genes by either increase or decrease in miRNAs level during different phases of symbiosis. miR159 was found to regulate the MYB factors involved in GA pathway which decreases in expression level during symbiosis. Also, miR397 and miR408 significantly enhanced the anti-oxidative capacity and also enabled resistance against stress by regulating the copper transport. They also had shown convincing role in cell wall metabolisms during growth and development processes when *Oncidium* interaction takes place with *P. indica* (Ye et al. 2014).

Herbaspirillum seropedicae is a diazotrophic bacterial species that have the capability of fixing atmospheric nitrogen (N_2) into useful form such as ammonia (NH_3). These diazotrophic bacteria have been linked with agriculturally important gramineous plants such as maize, rice, and sugarcane, which conquer the intercellular spaces and vascular tissues of the plants without causing any harm to the host plant (James and Olivares 1998). An intuition into miRNA role in regulating differential expression levels in maize during association with diazotrophic bacteria was analyzed by deep-sequencing method. A total of 25 conserved miRNAs families and 15 novel miRNAs were identified in *Zea mays-H. seropedicae* association. Among these four miRNAs, miR397, miR398, miR408, and miR528 were classified as Cu-miRNAs as these microRNAs inhibit their target gene expression level associated with copper homeostasis such as laccase, copper-zinc superoxide dismutase (CSD), and cupredoxin. These miRNAs were highly expressed on inoculation of *H. seropedicae* that resulted in downregulation of the respective target Cu-proteins which enhances during the depletion of defense mechanisms and easily assist in the formation of colonization. Studies show that miR397 was responsible in regulating the nitrogen fixation-associated copper homeostasis in *Lotus japonicus* (De Luis et al. 2012). Study of maize root colonization revealed that miR397 decreased the expression level of the target gene encoding laccase protein which was involved in copper homeostasis (Thiebaut et al. 2014).

10.6 Conclusion

The understanding miRNA-mediated gene regulation is important to unravel the complexity of plant's response to various microorganisms and insects. A plethora of studies of studies have shown the involvement of miRNAs in response to various pathogens as well as beneficial microbes and that alteration in the expression of miRNAs either leads to disease resistance or susceptible. The pathogen-responsive miRNAs induce PTGS through cleavage or translation inhibition of the target mRNA. The involvement of miRNA in stress response has open up an exciting area of research to explore the mechanism of action, biogenesis, and targets of miRNAs involved in plant-microbe interaction.

References

Allen E, Xie Z, Gustafson AM, Carrington JC (2005) microRNA-directed phasing during trans-acting siRNA biogenesis in plants. Cell 121(2):207–221. https://doi.org/10.1016/j.cell.2005.04.004

Asada K (1999) The water-water cycle in chloroplasts: scavenging of active oxygens and dissipation of excess photons. Annu Rev Plant Biol 50(1):601–639

Baldrich P, Campo S, Wu MT, Liu TT, Hsing YI, San Segundo B (2015) MicroRNA-mediated regulation of gene expression in the response of rice plants to fungal elicitors. RNA Biol 12(8):847–863. https://doi.org/10.1080/15476286.2015.1050577

Barah P, Winge P, Kusnierczyk A, Tran DH, Bones AM (2013) Molecular signatures in *Arabidopsis thaliana* in response to insect attack and bacterial infection. PLoS One 8(3):e58987. https://doi.org/10.1371/journal.pone.0058987

Bazzini AA, Hopp HE, Beachy RN, Asurmendi S (2007) Infection and coaccumulation of tobacco mosaic virus proteins alter microRNA levels, correlating with symptom and plant development. Proc Natl Acad Sci USA 104(29):12157–12162. https://doi.org/10.1073/pnas.0705114104

Bazzini AA, Almasia NI, Manacorda CA, Mongelli VC, Conti G, Maroniche GA, Rodriguez MC, Distefano AJ, Hopp HE, del Vas M, Asurmendi S (2009) Virus infection elevates transcriptional activity of miR164a promoter in plants. BMC Plant Biol 9:152. https://doi.org/10.1186/1471-2229-9-152

Blackman RL, Eastop VF (2000) Aphids on the World's crops. In: An identification and information guide, 2nd edn. John Wiley & Sons, Chichester, p 414, 59 figs, 51 plates

Bohn GW, Kishaba AN, Toba HH (1972) Mechanisms of resistance to melon aphid in a muskmelon line. Hort Sci. 7:281–282

Boke H, Ozhuner E, Turktas M, Parmaksiz I, Ozcan S, Unver T (2015) Regulation of the alkaloid biosynthesis by miRNA in opium poppy. Plant Biotechnol J 13(3):409–420. https://doi.org/10.1111/pbi.12346

Bové JM (2006) Huanglongbing: a destructive, newly-emerging, century-old disease of citrus. J Plant Pathol 88(1):7–37

Bozorov TA, Pandey SP, Dinh ST, Kim SG, Heinrich M, Gase K, Baldwin IT (2012) DICER-like proteins and their role in plant-herbivore interactions in Nicotiana attenuata. J Integr Plant Biol 54(3):189–206. https://doi.org/10.1111/j.1744-7909.2012.01104.x

Cao JY, Xu YP, Zhao L, Li SS, Cai XZ (2016) Tight regulation of the interaction between Brassica napus and Sclerotinia sclerotiorum at the microRNA level. Plant Mol Biol 92(1–2):39–55. https://doi.org/10.1007/s11103-016-0494-3

Chakraborty S, Devi KJ, Deb B, Rajwanshi R (2016) Identification and characterization of novel microRNAs and their targets in *Cucumis melo* L.: an *insilico* approach. Focus Sci 2(1)

Chand SK, Nanda S, Rout E, Mohanty J, Mishra R, Joshi RK (2016) Identification and characterization of microRNAs in turmeric (Curcuma longa L.) responsive to infection with the pathogenic fungus Pythium aphanidermatum. Physiol Mol Plant Pathol 93:119–128

Chapman EJ, Prokhnevsky AI, Gopinath K, Dolja VV, Carrington JC (2004) Viral RNA silencing suppressors inhibit the microRNA pathway at an intermediate step. Genes Dev 18(10):1179–1186. https://doi.org/10.1101/gad.1201204

Chen M, Cao Z (2015) Genome-wide expression profiling of microRNAs in poplar upon infection with the foliar rust fungus Melampsora larici-populina. BMC Genomics 16:696. https://doi.org/10.1186/s12864-015-1891-8

Chen J, Li WX, Xie D, Peng JR, Ding SW (2004) Viral virulence protein suppresses RNA silencing-mediated defense but upregulates the role of microrna in host gene expression. Plant Cell 16(5):1302–1313. https://doi.org/10.1105/tpc.018986

Chen H, Arsovski AA, Yu K, Wang A (2016) Genome-wide investigation using sRNA-Seq, Degradome-Seq and Transcriptome-Seq reveals regulatory networks of microRNAs and their target genes in soybean during soybean mosaic virus infection. PLoS One 11(3):e0150582. https://doi.org/10.1371/journal.pone.0150582

Chu C-C, Lee W-C, Guo W-Y, Pan S-M, Chen L-J, H-m L, Jinn T-L (2005) A copper chaperone for superoxide dismutase that confers three types of copper/zinc superoxide dismutase activity in *Arabidopsis*. Plant Physiol 139(1):425–436

Chugh S, Guha S, Rao IU (2009) Micropropagation of orchids: a review on the potential of different explants. Sci Hortic 122(4):507–520

Couzigou JM, Lauressergues D, Andre O, Gutjahr C, Guillotin B, Becard G, Combier JP (2017) Positive gene regulation by a natural protective miRNA enables arbuscular mycorrhizal symbiosis. Cell Host Microbe 21(1):106–112. https://doi.org/10.1016/j.chom.2016.12.001

De Luis A, Markmann K, Cognat V, Holt DB, Charpentier M, Parniske M, Stougaard J, Voinnet O (2012) Two microRNAs linked to nodule infection and nitrogen-fixing ability in the legume Lotus japonicus. Plant Physiol 160(4):2137–2154. https://doi.org/10.1104/pp.112.204883

Devi KJ, Chakraborty S, Deb B, Rajwanshi R (2016) Computational identification and functional annotation of microRNAs and their targets from expressed sequence tags (ESTs) and genome survey sequence (GSSs) of *Coffea arabica* L. Plant Gene 6:30–42

Devi KJ, Saha P, Chakraborty S, Rajwanshi R (2018) Computational identification and functional annotation of microRNAs and their targets in three species of kiwifruit (*Actinidia* spp.). Indian J Plant Physiol 23(1):179–191

El-Sharkawy MA (2004) Cassava biology and physiology. Plant Mol Biol 56(4):481–501

Fahlgren N, Howell MD, Kasschau KD, Chapman EJ, Sullivan CM, Cumbie JS, Givan SA, Law TF, Grant SR, Dangl JL, Carrington JC (2007) High-throughput sequencing of *Arabidopsis* microRNAs: evidence for frequent birth and death of miRNA genes. PLoS One 2:e219

Feng J, Lin R, Chen J (2013) Alteration of tomato microRNAs expression during fruit development upon cucumber mosaic virus and Tomato aspermy virus infection. Mol Biol Rep 40(5):3713–3722. https://doi.org/10.1007/s11033-012-2447-5

Formey D, Sallet E, Lelandais-Brière C, Ben C, Bustos-Sanmamed P, Niebel A, Frugier F, Combier JP, Debellé F, Hartmann C (2014) The small RNA diversity from Medicago truncatula roots under biotic interactions evidences the environmental plasticity of the miRNAome. Genome Biol 15(9):457

Foyer CH, Rasool B, Davey JW, Hancock RD (2016) Cross-tolerance to biotic and abiotic stresses in plants: a focus on resistance to aphid infestation. J Exp Bot 67(7):2025–2037. https://doi.org/10.1093/jxb/erw079

Garzo E, Soria C, Gómez-Guillamón ML, Fereres A (2002) Feeding behavior of Aphis gossypii on resistant accessions of different melon genotypes (Cucumis melo). Phytoparasitica 30(2):129–140. https://doi.org/10.1007/bf02979695

Gill SS, Tuteja N (2010) Reactive oxygen species and antioxidant machinery in abiotic stress tolerance in crop plants. Plant Physiol Biochem 48(12):909–930

Guo W, Wu G, Yan F, Lu Y, Zheng H, Lin L, Chen H, Chen J (2012) Identification of novel Oryza sativa miRNAs in deep sequencing-based small RNA libraries of rice infected with Rice stripe virus. PLoS One 7(10):e46443. https://doi.org/10.1371/journal.pone.0046443

Hernandez-Blanco C, Feng DX, Hu J, Sanchez-Vallet A, Deslandes L, Llorente F, Berrocal-Lobo M, Keller H, Barlet X, Sanchez-Rodriguez C, Anderson LK, Somerville S, Marco Y, Molina A (2007) Impairment of cellulose synthases required for *Arabidopsis* secondary cell wall formation enhances disease resistance. Plant Cell 19(3):890–903. https://doi.org/10.1105/tpc.106.048058

Jagadeeswaran G, Saini A, Sunkar R (2009) Biotic and abiotic stress down-regulate miR398 expression in *Arabidopsis*. Planta 229(4):1009–1014. https://doi.org/10.1007/s00425-009-0889-3

James EK, Olivares FL (1998) Infection and colonization of sugar cane and other graminaceous plants by endophytic diazotrophs. Crit Rev Plant Sci 17(1):77–119

Katiyar-Agarwal S, Jin H (2010) Role of small RNAs in host-microbe interactions. Annu Rev Phytopathol 48:225–246. https://doi.org/10.1146/annurev-phyto-073009-114457

Kettles GJ, Drurey C, Schoonbeek HJ, Maule AJ, Hogenhout SA (2013) Resistance of *Arabidopsis thaliana* to the green peach aphid, Myzus persicae, involves camalexin and is regulated by microRNAs. New Phytol 198(4):1178–1190. https://doi.org/10.1111/nph.12218

Khraiwesh B, Zhu JK, Zhu J (2012) Role of miRNAs and siRNAs in biotic and abiotic stress responses of plants. Biochim Biophys Acta 1819(2):137–148. https://doi.org/10.1016/j.bbagrm.2011.05.001

Klingler J, Powell G, Thompson GA, Isaacs R (1998) Phloem specific aphid resistance in Cucumis melo line AR 5: effects on feeding behaviour and performance of Aphis gossypii. Entomol Exp Appl 86(1):79–88. https://doi.org/10.1046/j.1570-7458.1998.00267.x

Kumar A, Sarma Y (2012) Characterization of Ralstonia solanacearum causing bacterial wilt in ginger. Indian Phytopathol 57:12–17

Kumar D, Dutta S, Singh D, Prabhu KV, Kumar M, Mukhopadhyay K (2017) Uncovering leaf rust responsive miRNAs in wheat (Triticum aestivum L.) using high-throughput sequencing and prediction of their targets through degradome analysis. Planta 245(1):161–182. https://doi.org/10.1007/s00425-016-2600-9

Kundu A, Paul S, Dey A, Pal A (2017) High throughput sequencing reveals modulation of microRNAs in Vigna mungo upon Mungbean yellow mosaic India virus inoculation highlighting stress regulation. Plant Sci 257:96–105. https://doi.org/10.1016/j.plantsci.2017.01.016

Lauressergues D, Delaux PM, Formey D, Lelandais-Brière C, Fort S, Cottaz S, Bécard G, Niebel A, Roux C, Combier JP (2012) The microRNA miR171h modulates arbuscular mycorrhizal colonization of Medicago truncatula by targeting NSP2. Plant J 72(3):512–522

Li L, Steffens JC (2002) Overexpression of polyphenol oxidase in transgenic tomato plants results in enhanced bacterial disease resistance. Planta 215(2):239–247. https://doi.org/10.1007/s00425-002-0750-4

Li Y, Zhang Q, Zhang J, Wu L, Qi Y, Zhou JM (2010) Identification of microRNAs involved in pathogen-associated molecular pattern-triggered plant innate immunity. Plant Physiol 152(4):2222–2231. https://doi.org/10.1104/pp.109.151803

Lian S, Cho WK, Kim SM, Choi H, Kim KH (2016) Time-course small RNA profiling reveals rice miRNAs and their target genes in response to rice stripe virus infection. PLoS One 11(9):e0162319. https://doi.org/10.1371/journal.pone.0162319

Liu Y, Ren D, Pike S, Pallardy S, Gassmann W, Zhang S (2007) Chloroplast-generated reactive oxygen species are involved in hypersensitive response-like cell death mediated by a mitogen-activated protein kinase cascade. Plant J 51(6):941–954

Louis J, Shah J (2013) *Arabidopsis thaliana-Myzus persicae* interaction: shaping the understanding of plant defense against phloem-feeding aphids. Front Plant Sci 4:213. https://doi.org/10.3389/fpls.2013.00213

Lu S, Sun YH, Chiang VL (2008) Stress-responsive microRNAs in populus. Plant J 55(1):131–151. https://doi.org/10.1111/j.1365-313X.2008.03497.x

Luan Y, Cui J, Zhai J, Li J, Han L, Meng J (2015) High-throughput sequencing reveals differential expression of miRNAs in tomato inoculated with Phytophthora infestans. Planta 241(6):1405–1416. https://doi.org/10.1007/s00425-015-2267-7

Luan Y, Cui J, Wang W, Meng J (2016) MiR1918 enhances tomato sensitivity to Phytophthora infestans infection. Sci Rep 6:35858. https://doi.org/10.1038/srep35858

Maekawa T, Cheng W, Spiridon LN, Töller A, Lukasik E, Saijo Y, Liu P, Shen QH, Micluta MA, Somssich IE, Takken FLW, Petrescu AJ, Chai J, Schulze-Lefert P (2011) Coiled-coil domain-dependent Homodimerization of intracellular barley immune receptors defines a minimal functional module for triggering cell death. Cell Host Microbe 9(3):187–199

Morkunas I, Mai VC, Gabryś B (2011) Phytohormonal signaling in plant responses to aphid feeding. Acta Physiol Plant 33(6):2057–2073. https://doi.org/10.1007/s11738-011-0751-7

Navarro L, Dunoyer P, Jay F, Arnold B, Dharmasiri N, Estelle M, Voinnet O, Jones JD (2006) A plant miRNA contributes to antibacterial resistance by repressing auxin signaling. Science 312(5772):436–439. https://doi.org/10.1126/science.1126088

Naya L, Paul S, Valdes-Lopez O, Mendoza-Soto AB, Nova-Franco B, Sosa-Valencia G, Reyes JL, Hernandez G (2014) Regulation of copper homeostasis and biotic interactions by microRNA 398b in common bean. PLoS One 9(1):e84416. https://doi.org/10.1371/journal.pone.0084416

Nejat N, Mantri N (2017) Plant immune system: crosstalk between responses to biotic and Abiotic stresses the missing link in understanding plant defence. Curr Issues Mol Biol 23:1–16. https://doi.org/10.21775/cimb.023.001

Niu D, Lii YE, Chellappan P, Lei L, Peralta K, Jiang C, Guo J, Coaker G, Jin H (2016) miRNA863-3p sequentially targets negative immune regulator ARLPKs and positive regulator SERRATE upon bacterial infection. Nat Commun 7:11324. https://doi.org/10.1038/ncomms11324

Nova-Franco B, Iniguez LP, Valdes-Lopez O, Alvarado-Affantranger X, Leija A, Fuentes SI, Ramirez M, Paul S, Reyes JL, Girard L, Hernandez G (2015) The micro-RNA72c-APETALA2-1 node as a key regulator of the common bean-Rhizobium etli nitrogen fixation symbiosis. Plant Physiol 168(1):273–291. https://doi.org/10.1104/pp.114.255547

OgawaK KS, Takabe K, Asada S (1995) Attachment of CuZn-superoxide dismutase to thylakoid membranes at the site of superoxide generation (PSI) in spinach chloroplasts: detection by immunogold labeling after rapid freezing and substitution method. Plant Cell Physiol 36:565–573

Ouyang S, Park G, Atamian HS, Han CS, Stajich JE, Kaloshian I, Borkovich KA (2014) MicroRNAs suppress NB domain genes in tomato that confer resistance to Fusarium oxysporum. PLoS Pathog 10(10):e1004464. https://doi.org/10.1371/journal.ppat.1004464

Pandey SP, Baldwin IT (2007) RNA-directed RNA polymerase 1 (RdR1) mediates the resistance of Nicotiana attenuata to herbivore attack in nature. Plant J 50(1):40–53. https://doi.org/10.1111/j.1365-313X.2007.03030.x

Pandey SP, Shahi P, Gase K, Baldwin IT (2008) Herbivory-induced changes in the small-RNA transcriptome and phytohormone signaling in Nicotiana attenuata. Proc Natl Acad Sci USA 105(12):4559–4564. https://doi.org/10.1073/pnas.0711363105

Park YJ, Lee HJ, Kwak KJ, Lee K, Hong SW, Kang H (2014) MicroRNA400-guided cleavage of pentatricopeptide repeat protein mRNAs renders *Arabidopsis thaliana* more susceptible to pathogenic bacteria and fungi. Plant Cell Physiol 55(9):1660–1668. https://doi.org/10.1093/pcp/pcu096

Pérez-Quintero ÁL, Quintero A, Urrego O, Vanegas P, López C (2012) Bioinformatic identification of cassava miRNAs differentially expressed in response to infection by Xanthomonas axonopodis pv. manihotis. BMC Plant Biol 12(1):29. https://doi.org/10.1186/1471-2229-12-29

Prasath D, Karthika R, Habeeba NT, Suraby EJ, Rosana OB, Shaji A, Eapen SJ, Deshpande U, Anandaraj M (2014) Comparison of the transcriptomes of ginger (Zingiber officinale Rosc.) and mango ginger (Curcuma amada Roxb.) in response to the bacterial wilt infection. PLoS One 9(6):e99731. https://doi.org/10.1371/journal.pone.0099731

Qazi J, Amin I, Mansoor S, Iqbal MJ, Briddon RW (2007) Contribution of the satellite encoded gene βC1 to cotton leaf curl disease symptoms. Virus Res 128(1–2):135–139

Rajwanshi R, Chakraborty S, Jayanandi K, Deb B, Lightfoot DA (2014) Orthologous plant microRNAs: microregulators with great potential for improving stress tolerance in plants. Theor Appl Genet 127(12):2525–2543. https://doi.org/10.1007/s00122-014-2391-y

Rawat N, Kiran SP, Du D, Gmitter FG Jr, Deng Z (2015) Comprehensive meta-analysis, co-expression, and miRNA nested network analysis identifies gene candidates in citrus against huanglongbing disease. BMC Plant Biol 15:184. https://doi.org/10.1186/s12870-015-0568-4

Sanchez-Rodriguez C, Estevez JM, Llorente F, Hernandez-Blanco C, Jorda L, Pagan I, Berrocal M, Marco Y, Somerville S, Molina A (2009) The ERECTA receptor-like kinase regulates cell wall-mediated resistance to pathogens in *Arabidopsis thaliana*. Mol Plant-Microbe Interact 22(8):953–963. https://doi.org/10.1094/MPMI-22-8-0953

Sattar S, Song Y, Anstead JA, Sunkar R, Thompson GA (2012) Cucumis melo microRNA expression profile during aphid herbivory in a resistant and susceptible interaction. Mol Plant-Microbe Interact 25(6):839–848. https://doi.org/10.1094/MPMI-09-11-0252

Sattar S, Addo-Quaye C, Thompson GA (2016) miRNA-mediated auxin signalling repression during Vat-mediated aphid resistance in Cucumis melo. Plant Cell Environ 39(6):1216–1227. https://doi.org/10.1111/pce.12645

Shiu S-H, Karlowski WM, Pan R, Tzeng Y-H, Mayer KF, Li W-H (2004) Comparative analysis of the receptor-like kinase family in *Arabidopsis* and rice. Plant Cell 16(5):1220–1234

Singh N, Srivastava S, Shasany AK, Sharma A (2016) Identification of miRNAs and their targets involved in the secondary metabolic pathways of Mentha spp. Comput Biol Chem 64:154–162. https://doi.org/10.1016/j.compbiolchem.2016.06.004

Soto-Suarez M, Baldrich P, Weigel D, Rubio-Somoza I, San Segundo B (2017) The *Arabidopsis* miR396 mediates pathogen-associated molecular pattern-triggered immune responses against fungal pathogens. Sci Rep 7:44898. https://doi.org/10.1038/srep44898

Sun Z, He Y, Li J, Wang X, Chen J (2015) Genome-wide characterization of rice black streaked dwarf virus-responsive microRNAs in rice leaves and roots by small RNA and degradome sequencing. Plant Cell Physiol 56(4):688–699. https://doi.org/10.1093/pcp/pcu213

Thiebaut F, Rojas CA, Grativol C, Motta MR, Vieira T, Regulski M, Martienssen RA, Farinelli L, Hemerly AS, Ferreira PC (2014) Genome-wide identification of microRNA and siRNA responsive to endophytic beneficial diazotrophic bacteria in maize. BMC Genomics 15:766. https://doi.org/10.1186/1471-2164-15-766

Vanacker H, Carver TL, Foyer CH (2000) Early H2O2 accumulation in mesophyll cells leads to induction of glutathione during the hyper-sensitive response in the barley-powdery mildew interaction. Plant Physiol 123(4):1289–1300

Varallyay E, Valoczi A, Agyi A, Burgyan J, Havelda Z (2010) Plant virus-mediated induction of miR168 is associated with repression of ARGONAUTE1 accumulation. EMBO J 29(20):3507–3519. https://doi.org/10.1038/emboj.2010.215

Varma A, Verma S, Sahay N, Bütehorn B, Franken P (1999) Piriformospora indica, a cultivable plant-growth-promoting root endophyte. Appl Environ Microbiol 65(6):2741–2744

Waller F, Achatz B, Baltruschat H, Fodor J, Becker K, Fischer M, Heier T, Hückelhoven R, Neumann C, von Wettstein D (2005) The endophytic fungus Piriformospora indica reprograms barley to salt-stress tolerance, disease resistance, and higher yield. Proc Natl Acad Sci USA 102(38):13386–13391

Wang Y, Li P, Cao X, Wang X, Zhang A, Li X (2009) Identification and expression analysis of miRNAs from nitrogen-fixing soybean nodules. Biochem Biophys Res Commun 378(4):799–803. https://doi.org/10.1016/j.bbrc.2008.11.140

Wang Y, Wang L, Zou Y, Chen L, Cai Z, Zhang S, Zhao F, Tian Y, Jiang Q, Ferguson BJ, Gresshoff PM, Li X (2014) Soybean miR172c targets the repressive AP2 transcription factor NNC1 to activate ENOD40 expression and regulate nodule initiation. Plant Cell 26(12):4782–4801. https://doi.org/10.1105/tpc.114.131607

Wang J, Tang Y, Yang Y, Ma N, Ling X, Kan J, He Z, Zhang B (2016) Cotton leaf curl multan virus-derived viral small RNAs can target cotton genes to promote viral infection. Front Plant Sci 7:1162. https://doi.org/10.3389/fpls.2016.01162

Xia X, Shao Y, Jiang J, Du X, Sheng L, Chen F, Fang W, Guan Z, Chen S (2015) MicroRNA expression profile during Aphid feeding in Chrysanthemum (Chrysanthemum morifolium). PLoS One 10(12):e0143720. https://doi.org/10.1371/journal.pone.0143720

Xu W, Meng Y, Wise RP (2014a) Mla- and Rom1-mediated control of microRNA398 and chloroplast copper/zinc superoxide dismutase regulates cell death in response to the barley powdery mildew fungus. New Phytol 201(4):1396–1412. https://doi.org/10.1111/nph.12598

Xu D, Mou G, Wang K, Zhou G (2014b) MicroRNAs responding to southern rice black-streaked dwarf virus infection and their target genes associated with symptom development in rice. Virus Res 190:60–68. https://doi.org/10.1016/j.virusres.2014.07.007

Yan Z, Hossain MS, Valdes-Lopez O, Hoang NT, Zhai J, Wang J, Libault M, Brechenmacher L, Findley S, Joshi T, Qiu L, Sherrier DJ, Ji T, Meyers BC, Xu D, Stacey G (2016) Identification and functional characterization of soybean root hair microRNAs expressed in response to Bradyrhizobium japonicum infection. Plant Biotechnol J 14(1):332–341. https://doi.org/10.1111/pbi.12387

Yang T, Xue L, An L (2007) Functional diversity of miRNA in plants. Plant Sci 172(3):423–432. https://doi.org/10.1016/j.plantsci.2006.10.009

Yang L, Mu X, Liu C, Cai J, Shi K, Zhu W, Yang Q (2015) Overexpression of potato miR482e enhanced plant sensitivity to Verticillium dahliae infection. J Integr Plant Biol 57(12):1078–1088. https://doi.org/10.1111/jipb.12348

Ye W, Shen CH, Lin Y, Chen PJ, Xu X, Oelmuller R, Yeh KW, Lai Z (2014) Growth promotion-related miRNAs in Oncidium orchid roots colonized by the endophytic fungus Piriformospora indica. PLoS One 9(1):e84920. https://doi.org/10.1371/journal.pone.0084920

Yin X, Wang J, Cheng H, Wang X, Yu D (2013) Detection and evolutionary analysis of soybean miRNAs responsive to soybean mosaic virus. Planta 237(5):1213–1225. https://doi.org/10.1007/s00425-012-1835-3

Zhang W, Gao S, Zhou X, Chellappan P, Chen Z, Zhou X, Zhang X, Fromuth N, Coutino G, Coffey M, Jin H (2011) Bacteria-responsive microRNAs regulate plant innate immunity by modulating plant hormone networks. Plant Mol Biol 75(1–2):93–105. https://doi.org/10.1007/s11103-010-9710-8

Zhang Q, Li Y, Zhang Y, Wu C, Wang S, Hao L, Li T (2017) Md-miR156ab and Md-miR395 target WRKY transcription factors to influence apple resistance to leaf spot disease. Front Plant Sci 8:526. https://doi.org/10.3389/fpls.2017.00526

Zhao H, Sun R, Albrecht U, Padmanabhan C, Wang A, Coffey MD, Girke T, Wang Z, Close TJ, Roose M, Yokomi RK, Folimonova S, Vidalakis G, Rouse R, Bowman KD, Jin H (2013) Small RNA profiling reveals phosphorus deficiency as a contributing factor in symptom expression for citrus huanglongbing disease. Mol Plant 6(2):301–310. https://doi.org/10.1093/mp/sst002

Zurbriggen MD, Carrillo N, Tognetti VB, Melzer M, Peisker M, Hause B, Hajirezaei MR (2009) Chloroplast-generated reactive oxygen species play a major role in localized cell death during the non-host interaction between tobacco and Xanthomonas campestris pv. vesicatoria. Plant J 60(6):962–973

11 Stress Protectant Secondary Metabolites and their Metabolic Engineering to Enhance Abiotic Stress Tolerance in Plants

Gurminder Kaur and Deepak Ganjewala

Abstract

Plants produce huge numbers of secondary metabolites, which play many essential roles in their adaptation to the changing environment and during abiotic stresses, such as drought, temperature, salinity, etc. Major classes of secondary metabolites, such as isoprenoids, carotenoids, and flavonoids, exhibited potential of abiotic stress tolerance. Synthesis and accumulation of secondary metabolites markedly increased during abiotic stress to cope with harsh abiotic stress conditions, suggesting a tight link between secondary metabolite accumulation and tolerance of plants to abiotic stresses. Therefore, abiotic stress tolerance of plants can be enhanced by manipulating the synthesis and accumulation of secondary metabolites. Metabolic engineering approaches have been very useful to manipulate biosynthesis and accumulation of secondary metabolites such as isoprenoids, carotenoids, and flavonoids and plant hormones, such as abscisic acid. In the past few years, metabolic engineering of secondary metabolic pathways to confer abiotic stress tolerance has been successfully performed in Arabidopsis (*Arabidopsis thaliana*), tobacco (*Nicotiana tabacum*), sweet potato (*Ipomoea batatas*), alfalfa (*Medicago sativa* L.), *S. europaea*, and Brassica (*Brassica napus*). The present chapter highlights roles of abiotic stress-protecting secondary metabolites and discusses progresses and prospects of metabolic engineering of secondary metabolite biosynthesis to enhance abiotic stress tolerance of plants.

G. Kaur
Faculty of Bioscience, Institute of Bioscience and Technology,
Shri Ramswaroop Memorial University, Barabanki, Uttar Pradesh, India

D. Ganjewala (✉)
Amity Institute of Biotechnology, Amity University, Noida, Uttar Pradesh, India

M. Kumar et al. (eds.), *In vitro Plant Breeding towards Novel Agronomic Traits*,
https://doi.org/10.1007/978-981-32-9824-8_11

Keywords
Abiotic stress · Metabolic engineering · Isoprenoid · Carotenoid · Flavonoid · Abscisic acid

11.1 Introduction

Abiotic stress can be defined as any negative impact on living organisms, plants, animals, and microbes caused by a nonliving factor in the environment. Abiotic stresses and their adverse effects on plant productivity have gained attention of researchers and as well as large and small agro-biotech companies. Our knowledge and deeper understanding of abiotic stresses and plant's response to abiotic stresses is massive, and it is being rapidly increasing by intervention of newer advanced techniques of molecular biology. Molecular biology techniques, particularly genetic and metabolic engineering, have been proven to be excellent tools in providing deeper insight into plants performance under abiotic stresses and rather conferring abiotic stress tolerance in a number of plant species. Abiotic stress tolerance is a highly complex mechanism underlying many processes, which yet need to be completely elucidated and understood. There are several abiotic factors, e.g., light, temperature, drought, soil salinity, air pollution, and mechanical damage, that may cause abiotic stress, which directly affect plants performance and productivity (Vickers et al. 2011). Among all these, the most threatening abiotic stresses are drought (low water availability), salt (high salinity), and heat (high temperatures) stresses. Only this three abiotic stress factors are responsible for substantial loss (up to 80%) in the productivity of the plants (Bromham et al. 2013). In plants, abiotic stresses mainly stimulate increased production of reactive oxygen species (ROS), which adversely affect antioxidant defenses of plants resulting in difficulty in survival of plants under stressed condition (Miller et al. 2010). In fact, abiotic stress factors can potentially influence almost all physiological, biochemical, and molecular processes in plants from the early stage of seed germination to maturity and eventually cause severe losses in the economic yield of crop plants. As per reports abiotic stresses drastically affect (70%) the productivity of staple food crops (Kaur et al. 2008; Mantri et al. 2012). Our studies on drought stress in tomato (*Lycopersicon esculentum*) have revealed that drought stress can be alleviated by silicon supplementation (Malhotra et al. 2016a, b). Silicon supplementation has been responsible for modifying antioxidant potential of the tomato which is linked with abiotic stress tolerance (Malhotra et al. 2017). Literature survey revealed that approximately 20% of the irrigated soils worldwide are suffering from salt stress (Zhao et al. 2013). Salt stress is reported to severely limit the productivity and distribution of major plants worldwide (Tuteja 2007). Under salinity, levels of $Na+$ are excessively very high which poses several harmful effects on plants, including osmotic stress, ionic toxicity, and related oxidative stress. Finally, all these stresses decline the rate of photosynthesis and productivity (Zhu 2002; Oh et al. 2000). It is reported that some plants have adapted to salt and drought stress conditions by evolving counter-mechanisms, which comprise of growth and development regulation, detoxification, ion homeostasis, and osmotic adjustment (Bohnert et al. 1995; Zhu 2001, 2002). However, still large number of commercial plants remains to be made abiotic stress tolerable.

In view of the great impact of the abiotic stress directly on plant productivity, deep understanding of mechanism of abiotic stresses and counterresponse of plants to such stresses are very interesting and desirable. Studies of abiotic stresses are fundamental for the development of new plants with enhanced abiotic stress tolerance or to modulate plants response to abiotic stresses. Currently, the researchers have understood the changes in cellular, biochemical, and molecular process under stress condition. Also, many genes induced under stress circumstances have been identified, characterized, and cloned (Bhatnagar-Mathur et al. 2008). The review in genetic engireeing provids a sound background towards the concurrent approaches in crop improvent with provided stress factors at biotic and abiotic levels (Bhatnagar-Mathur et al. 2008). This information has laid a solid background to engineer plants with enhanced abiotic stress tolerance through modern molecular biology techniques. In this context, genetic and metabolic engineering approaches have been very useful to improve abiotic stress tolerance and productivity of plants (Varshney et al. 2012). A number of reports indicated that the molecular manipulation of genes, such as those encoding antioxidant enzymes (Natwar et al. 2014; Wu et al. 2014; Zhang et al. 2014), transcription factors (Tang et al. 2014; Xu et al. 2014; Min et al. 2013; Sun et al. 2014), and ion transporters (Muhammad et al. 2014; Yarra et al. 2012), offers great promises to overcome multiple limitations to plant productivity by developing abiotic stress-tolerant transgenic plants. Also, limited numbers of reports have advocated metabolic engineering of the secondary metabolic pathways leading to the synthesis of isoprenoids, carotenoids, phenolics, and flavonoids, which have shown stress-protecting potential. With the cloning of the genes of the secondary metabolic pathways, scope of metabolic engineering of secondary metabolic pathways in plants for enhancement of abiotic stress tolerance is widening. To feed the burgeoning population, the abiotic stress problem in plants needs to be solved, and this would be a major challenge for plant biotechnologist. Certainly, metabolic engineering approaches have paved the way for researchers to overcome abiotic stress problem in plants. Previously, a plenty of reviews published have discussed molecular approaches for manipulation of genes involved in the synthesis of mannitol, glycine betaine, and heat shock proteins (Chen and Murata, 2002; Wani et al. 2013). Recently, Pandey and Arora (2017) have elaborately discussed molecular strategies that have been employed to develop transgenic crops resistant to specific abiotic stresses in plants. However, there is no chapter that has been published till date covering recent progress on metabolic engineering of secondary metabolite biosynthesis in plants for abiotic stress tolerance. Here, this chapter provides information on secondary metabolites with abiotic stress-protecting potential and progress on metabolic engineering approaches to manipulate secondary metabolite synthesis in order to improve abiotic stress tolerance in plants.

11.2 Metabolic Pathways Involved in Abiotic Stress Tolerance

Plants have numerous biochemical pathways leading to tens of thousands of primary and secondary metabolites (Nascimento and Fett-Neto 2010). Primary metabolic pathways produce sugars (trehalose, sucrose, and fructan), amino acids (tryptophan and proline), and ammonium compounds (polyamines and glycine betaine), which serve as osmoprotectants under stressed condition. The biosynthesis of osmoprotectants is reported to be enhanced through genetic and metabolic engineering approaches to improve abiotic stress tolerance in crop plants (Rathinasabapathi 2000; Rontein et al. 2002). Molecular biology approaches including genetic and metabolic engineering of the betaines and other compatible solutes for enhancement of abiotic stress tolerance in plans have been discussed previously (Chen and Murata 2002; Lawlor 2013; Wani et al. 2013). Similarly, genetic engineering of metabolic pathways of antioxidant enzymes impairing ROS can enhance tolerance to abiotic stresses. On the other hand, the secondary metabolic pathways produce isoprenoids, carotenoids, and flavonoids and plant hormones, such as abscisic acid (ABA), which exhibited stress-protecting potential. Secondary metabolic pathways are induced by oxidative stress caused by heat, salt, drought, flooding, UV light, or temperature extremes. In spite of the complexity of secondary metabolism, important advances have been achieved, leading to success stories in engineering this diverse reservoir of useful molecules (Nascimento and Fett-Neto 2010). The pathways involved in the biosynthesis of isoprenoids, carotenoids, flavonoids, and plant hormones such as abscisic acid (ABA) are also attractive target to improve abiotic stress tolerance in plants (Fig. 11.1). Biosynthetic pathway, leading

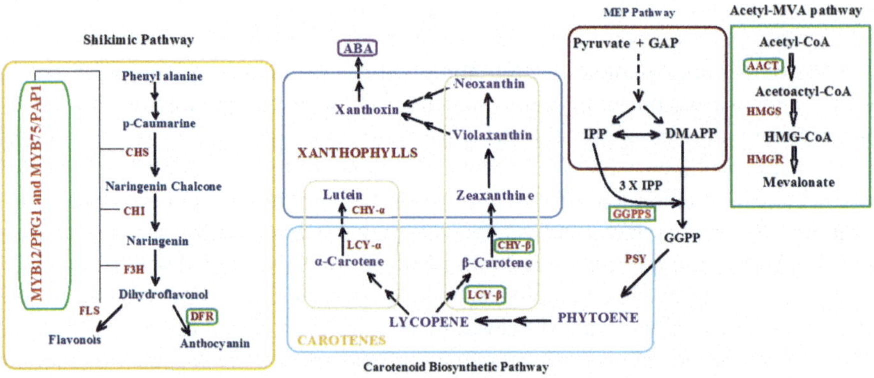

Fig. 11.1 Schematic diagram of secondary metabolic pathways induced during abiotic stress condition in plants. Overexpression of genes depicted in green box, AACT (acetoacetyl-CoA thiolase), GGPS (geranylgeranyl pyrophosphate synthase), LCY-β (lycopene cyclase), CHY-β (carotene hydroxylase), and DFR (dihydroflavonol4-reductase) results in the enhancement of abiotic stress tolerance. Overexpression of regulatory gene *MYB12* also upregulate the genes involved in the flavonoid and ABA biosynthesis

to cellulose and suberin, also plays important role in stress adaptation (Wang et al. 2011; Franke et al. 2012). Jain (2013) highlighted emerging role of metabolic pathways in abiotic stress tolerance. Recently, Baksh and Hussain (2015) have discussed about recent advances made in the development of transgenic crop plants of commercial importance with enhanced tolerance to abiotic stress. Here we have discussed only biochemical pathways, which produce secondary metabolites capable of effectively responding to stress situations imposed by abiotic factors. However, secondary metabolic pathways, often recruited from primary metabolic pathways upon initial gene duplication, are frequently restricted to specific taxonomic groups and play a major role in the plant and environment interaction (Mazid et al. 2011).

11.3 Stress Protectant Secondary Metabolites Produced During Abiotic Stresses

Plants produce myriads of secondary metabolites, which play many essential roles in plant's life such as in adaptation of plants to the changing environment and in abiotic stress conditions. There are four major classes of the secondary metabolites, namely, isoprenoids, alkaloids, phenolics, and flavonoids. However, phenolics, flavonoids, and carotenoids have been considered as the most important stress-protective secondary metabolites owing to their tremendous antioxidant potential. Many review and research papers have been published on antioxidant properties of phenolics and flavonoids (Ferdinando et al. 2012; Mierziak et al. 2014; Ganjewala 2016; Yogendra and Nirmalajit 2017) and their roles in protection from the bioactive stresses in plants. In tobacco a brown pigment produced by condensation of chlorogenoquinone with proteins curbs the spread of stress-induced tissue damage. Polyamines and phenylamides in tobacco and bean accumulated as a result of water stress and heat shocks, respectively, have potential ROS-scavenging ability which helps in abiotic stress management. Accumulation of anthocyanins in leaves of cotton suffering Na/K imbalance also confers a high ROS-scavenging capacity. The drought tolerance in cotton has been correlated with the level of ROS-scavenging polyphenol compounds. The protective effect of a gaseous secondary metabolite, isoprene, against ozone fumigation and heat shock has been reported (Edreva et al. 2008). Isoprene also demonstrated an ability to scavenge singlet oxygen. Isoprene emitted by plants provides a protective effect under abiotic stress conditions (high light, drought, ozone, etc.), but most agricultural plants do not produce isoprene. Emission of isoprene has been associated with stress responses in a naturally emitting species. It is reported that the isoprene shows stress-protective role by alteration of the reactive oxygen species (ROS) response system (Vickers et al. 2009a, b). Therefore, engineering of non-isoprene-emitting plants might provide protection against abiotic stress. Transgenic tobacco plants engineered to produce isoprene show resistance to ozone stress (Vickers et al. 2009a); similarly, transgenic isoprene-emitting Arabidopsis plants show resistance to thermal stress (Loivamaki et al. 2007; Sasaki et al. 2007). The transgenic tobacco plants were engineered with a single constitutively expressed IspS gene, and they showed classical isoprene emission patterns with no obvious phenotypic effects (Vickers et al. 2009a, 2011). Besides IspS, other genes might also have protective effects, particularly under specific abiotic stress conditions. For example, in alfalfa an enzyme acetoacetyl-CoA thiolase has been reported to regulate the acetate-MVA pathway during adaptation to abiotic stress. Here, we have discussed some of the secondary metabolites and their stress-protective roles.

11.4 Isoprenoids

Isoprenoids represent one of the largest classes of secondary metabolites. In plants, isoprenoids are biosynthesized by two pathways (i) the cytosolic acetate-MVA pathway and (ii) the plastidic methylerythritol-4-phosphate (MEP) pathway. The 3-hydroxy-3-methylglutaryl-CoA reductase (HMGR), the enzyme that catalyzes the third step of the acetate-MVA pathway, has been described as the regulatory step. Several experiments suggest that HMGR activity is posttranscriptionally and posttranslationally regulated in plants (Kirby and Keasling 2009). The 1-deoxy-D-xylulose-5-phosphate synthase (DXS), the first enzyme of the MEP pathway, plays a major role in the overall regulation of this route (Cordoba et al. 2009).

Several published reports have revealed that isoprenoids have an antioxidant function under biotic and abiotic stress in planta (Bajda et al. 2009; Vickers et al. 2011). However, many molecular aspects of this function remain unclear. Volatiles in the isoprenoid class have also been reported to provide protection under abiotic stress conditions (Vickers et al. 2009b; Loreto and Schnitzler 2010). These important roles in both biotic and abiotic response physiology represent high potential for biotechnological exploitation (Aharoni et al. 2005). The most obvious biotechnological application is modification of these emissions in plants to augment or repress their activities to achieve specific outcomes, such as deterring herbivores, influencing tritrophic interactions, protecting against pathogens and abiotic stress, and improving fragrances for fruits and flowers.

11.5 Flavonoids

Flavonoids are representative group of polyphenolic compounds containing a 15-carbon skeleton (C6-C3-C6), which consists of a heterocyclic ring and two phenyl rings. The heterocyclic benzopyran ring is referred to as the C ring, the fused aromaticring as the A ring, and the phenyl constituent as the B ring. They differ structurally according to the nature of the stereochemistry of the asymmetric carbons on the C rings and the number of hydroxyl groups on the B rings. Flavonoids have been divided into subgroups such as flavonols, (iso)flavones, flavanones, flavan-3-ols, proanthocyanidins, and anthocyanins. Recently, flavonoids have been named as "specialized metabolites" because plants synthesize species-specific flavonoids (Harborne and Williams 2000). Flavonoids possess strong antioxidant properties which makes them excellent dietary supplements for beneficial effects in humans. Besides many useful bioactive properties of flavonoids, they are potentially able to confer protection against various abiotic and biotic stresses (Buer et al. 2010; Dixon et al. 2013). Studies conducted in model plant species have shown that several transcription factors (TFs), including MYB, bHLH, and WD40, played important roles in flavonoid biosynthesis, leading to its accumulation (Grotewold 2006; Lepiniec et al. 2006; Feller et al. 2011; Qiu et al. 2014). Reports published till date indicated that flavonoids, including anthocyanins, protect tissues against oxidative stress from various abiotic stressors.

11.6 Carotenoids

Carotenoids represent the largest class of pigments in nature that are widely distributed in plants, algae, fungi, and cyanobacteria (Cunningham and Gantt 1998). Carotenoids are fat-soluble terpenoids with a number of conjugated double bonds that contribute to their characteristic colors in the yellow to red range (Bramley 1997). Carotenoids are biosynthesized from the isopentenyl diphosphate (IPP), which is derived from the MEP pathway present in the chloroplasts of photosynthetic tissues and chromoplasts of fruits and flowers in plants (Galpaz et al. 2006). Further, condensation of the IPP units gives rise to geranylgeranyl pyrophosphate (GGPP), which condenses to produce a linear C_{40} backbone for the formation of different structures of carotenoids.

Carotenoids provide potent nutritional benefits to human and animal health because animals and humans are unable to synthesize vitamin A (Fraser and Bramley 2004). They are major accessory pigments for light-harvesting system and protective agents for scavenging the ROS produced in the photosynthetic apparatus (Tao et al. 2007). Carotenoids participate in light-harvesting processes and protect the photosynthetic machinery from photooxidative damage by quenching triplet chlorophyll and singlet oxygen derived from excess light energy (Dall'Osto et al. 2007; Andrade-Souza et al. 2011). Recently, many reports have demonstrated that increased carotenoid content in plants can improve tolerance to abiotic stresses such as high light conditions, UV irradiation, and salt stress. Carotenoids play critical roles in the stress tolerance of many plants, indicating their value as powerful antioxidants and important nutrients. High carotenoid content increases resistance to abiotic stresses such as dehydration and salt by scavenging reactive oxygen species (ROS) or by acting synergistically with other antioxidants such as vitamins E and C. Krauss et al. (2005) reported that in tomato, high salt concentrations increased vitamin C, lycopene, and b-carotene content by up to 35% and enhance antioxidant capacity (Kim et al. 2012). Carotenoids are precursors for the synthesis of the hormone abscisic acid (ABA), which functions in plants as an important signal in a variety of developmental processes and in adaptive stress responses to environmental stimuli (Nambara and Marion-Poll, 2005; Taylor et al. 2005). Carotenoid-derived phytohormone abscisic acid (ABA) regulates numerous processes in plants including seed dormancy and the plant responses to low water availability. Under the salinity and drought stress, plants produce and accumulate numerous reactive oxygen species, which cause lipid peroxidation and destabilize membrane structure. Carotenoids provide protection for plants against oxidative stress as nonenzymatic antioxidants, by scavenging ROS generated due to excess excitation energy from chlorophyll during photosynthesis. The antioxidant nature of carotenoids helps to maintain the redox state of the cell and facilitates proper functioning of the cell under abiotic stress.

11.7 Metabolic Engineering of Secondary Metabolites to Improve Abiotic Stress Tolerance

Engineering of the metabolic pathways has been emerging as a useful approach for enhancement of abiotic stress tolerance in plants. In the past several years, an impressive progress has been achieved on metabolic engineering of the abiotic stress tolerance in plants. Metabolic engineering approaches have been a great success in development of transgenic plants with eminent abiotic stress tolerance with high productivity. Metabolic engineering is defined as the modification of endogenous metabolic pathways in order to alter and/or increase the production of valuable molecules or produce new molecules (Aharoni et al. 2005). For the success of metabolic engineering, knowledge of the various biochemical pathways and their genes and enzymes along with their roles in biosynthesis and regulation of an end product is highly essential. Till date, a number of genes encoding enzymes of the primary biochemical pathways leading to the synthesis of abiotic stress-protectant metabolites have been cloned and characterized from plants, and the number of genes related to abiotic stress tolerance is still increasing. Besides, several transcriptional factors with their specific roles in abiotic stress tolerance have been identified. Most of the metabolic engineering performed till date for abiotic stress tolerance has been focused on altering expression profile of genes, which encode enzymes of the biosynthesis of stress-protecting compounds, such as mannitol, glycine betaine, and heat shock proteins. Also, expressions of various transcription factors like DREB1 (dehydration-responsive element-binding protein), MAPK (mitogen-activated protein kinases), WRKY, etc. have been modified to confer abiotic stress tolerance in plants. The WRKY name has been coined from the highly conserved 60 amino acid long WRKY domains of the TFs, which contain a conserved heptapeptide motif WRKYGQK at the N-terminus and a novel zinc finger-like motif at the C-terminus. Both heptapeptide sequence and zinc finger-like motif are required for the high binding affinity of WRKY TFs to the consensus *cis*-acting elements termed W box (TTGACT/C). WRKY transcription factors are one of the largest families of transcriptional regulators found exclusively in plants. They have diverse biological functions in plant disease resistance, abiotic stress responses, nutrient deprivation, senescence, seed and trichome development, embryogenesis, as well as additional developmental and hormone-controlled processes (Bakshi and Oelmuller 2014). A number of studies revealed that apart from the primary metabolites, accumulation of the secondary metabolites (i.e., isoprenoids, alkaloids, carotenoids, and flavonoids) markedly changes in response to abiotic stress factors in plants. Identification, cloning, and characterization of genes encoding enzymes of the secondary metabolic pathways, such as acetate-MVA and MEP pathways of isoprenoids/terpenoids and carotenoids, shikimic acid pathway of alkaloids, and phenyl propanoid pathway of flavonoid biosynthesis, will have substantial importance in improvement of abiotic stress tolerance through metabolic engineering (Fig. 11.1). Cloned genes and/or cDNAs have been proven to be excellent tools of metabolic engineering. Transgenic plants of Arabidopsis (*Arabidopsis thaliana*), tobacco (*Nicotiana tabacum*), sweet potato (*Ipomoea batatas*), alfalfa (*Medicago sativa* L.), *S. europaea* and Brassica (*Brassica napus*) generated by altering accumulation of isoprenoid, carotenoid, and flavonoid through metabolic engineering have displayed promising abiotic stress tolerance (Davison et al. 2002, Gotz et al. 2002, Wagner et al. 2002, Han et al. 2008; Chen et al. 2011). Genes and regulatory factors involved in the abiotic stress tolerance are presented in Table 11.1.

In plants, many important agronomic traits are not regulated by a single gene but multiple genes, and many long and complex metabolic pathways might be involved in the synthesis and regulation of a primary and secondary metabolite. Earlier, transformation of plants with such multigene controlled and regulated traits was quite difficult that has certainly adversely hampered development of transgenic plants to cope with abiotic stress conditions. But, now with the modification/advancement in genetic engineering approaches, the difficulties with multigene transformation have been circumvented. Multigene transformation has become a conventional approach to generate transgenic plants with more determined phenotypes, including more complex examples of metabolic engineering (Naqvi et al. 2009). Several recently evolved molecular biology techniques, such as antisense gene, RNAi, and CRISPR/Cas9, have entirely changed our approaches for development of transgenic plants. RNAi has been successfully used in ornamental plants for modification of flower color, increasing shelf life and reducing postharvest losses whereas CRISPR/Cas9 in tomato, petunia, citrus, grape, potato, and apple for gene mutation, repression, activation, and epigenome editing. Very recently, Jain (2013) highlighted the emerging applications and future avenues of CRISPR-Cas9 system to understand the biology of abiotic stress tolerance in plants. Owing to the simplicity of programming, CRISPR-Cas9 system has opened a plethora of options for genome editing in various biological contexts. Outcome of previously conducted studies on genetic and metabolic engineering of the isoprenoids, carotenoids, and flavonoids biosynthesis has provided state-of-the-art enhancement of plants performances and productivity of industrially important metabolites under stressed conditions. The knowledge gathered can also be utilized for improvement of performance of plants to regulate abiotic stress responses at the initial stage or prior to the presence of the abiotic stress in a timely and quantitative manner. In the following section, we have deliberately discussed manipulation of secondary metabolites, such as isoprenoids, flavonoids, and carotenoids, in plants through metabolic engineering approaches.

Table 11.1 Summary of metabolic engineering of secondary metabolic pathways and important genes for improvement of abiotic stress tolerance

Pathway	Gene	Source plant	Transformed plant	Stress-protectant metabolite	Expression	Stress tolerance	Reference
Flavonoid biosynthetic pathway	*MYB12*	*Arabidopsis thaliana*	*Arabidopsis*	Flavonoids and ABA	Overexpression	Drought and salt	Wang et al. (2016a, b)
Flavonoid biosynthetic pathway	*MYB12/PFG1* or *MYB75/PAP1*	*Arabidopsis thaliana*	*Arabidopsis*	Flavonoids and ABA	Overexpression	Oxidative and drought	Nakabayashi et al. (2014)
Flavonoid biosynthetic pathway	*DFR-OX B*	*Arabidopsis thaliana*	*Brassica napus*	Anthocyanin	Overexpression	Drought and salt stress	Kim et al. (2012)
Carotenoid biosynthetic pathway	*β-LCY1*	*Nicotiana tabacum*	*Nicotiana tabacum*	Carotenoids	Overexpression	Drought and salt stress	
Carotenoid biosynthetic pathway	*Or*	*Ipomoea batatas* cv. Sinhwangmi	*Ipomoea batatas* cv. Yulmi (Ym)	β-Carotene, lutein, carotenoids	Overexpression	Salt stress	Kim et al. (2012)
Carotenoid biosynthetic pathway	*β-LCY1*	*Salicornia europaea* L.	*Arabidopsis thaliana and Nicotiana benthamiana*	Carotenoids	Inhibition	Salt stress	Chen et al. (2011)
Carotenoid biosynthetic pathway	*CHY-β*	*Ipomoea batatas* cv. Sinhwangmi	*Ipomoea batatas* cv. Yulmi (Ym)	β-Cryptoxanthin and zeaxanthin, β-carotene	Inhibition (downregulation)	Salt stress	Kim et al. (2012)
IPP biosynthetic pathway	*GGPS*	*Ipomoea batatas*	*Arabidopsis thaliana*	Carotenoid	Overexpression	Osmotic stress	

11.8 Improvement of Carotenoids Production

In plants, roles of isoprenoids in physiological adaptation to low temperature and high salt condition have been demonstrated. Alfalfa is a particularly abiotic stress-sensitive crop. In the case of osmotic stress, alfalfa yield is reduced by 50% at 100 mMNaCl. A study by Soto et al. has revealed that acetoacetyl-CoA thiolase regulates the biosynthesis of isoprenoid during abiotic stress adaptation in alfalfa, which has role in physiological adaptation to low temperature and high salinity. Acetoacetyl-CoA thiolase also known as thiolase-II has been cloned from alfalfa, which is referred to as MsAACT1. Expression of MsAACT1 rapidly increased in roots and leaves under cold and salinity stress. Transgenic alfalfa plants (overexpressing MsAACT1) showed enhanced salinity tolerance and production of squalene without altering the 3-hydroxy-3-methylglutaryl-CoA reductase (HMGR) activity in salt stress conditions. Further, applications of substrates, such as mevastatin (a potent inhibitor of acetate-MVA pathway) and vitamin C (a potent antioxidant), have suggested a critical function of the acetate-MVA pathway in abiotic stress-inducible antioxidant defense in plants.

11.9 Improvement of Flavonoids Production

Metabolic engineering of the flavonoid biosynthetic pathways has been performed to confer abiotic stress tolerance in plants (Butelli et al. 2008, 2012; Nakabayashi et al. 2014). Previously, researchers have talked about interaction between flavonoid accumulation and plant's tolerance to abiotic stresses, such as UV-B irradiation (Stracke et al. 2010a; Kusano et al. 2011); however, functions of flavonoid in stresses remain to be determined experimentally. The biosynthesis of flavonoid in *Arabidopsis* and other plants can be manipulated by altering expressions of flavonoid biosynthetic genes (Butelli et al. 2008, 2012). There are 35 genes encoding biosynthetic enzymes or transcription factors that are involved in the processing and production of flavonoids and anthocyanins (Yonekura-Sakakibara et al. 2008). The flavonol regulators MYB12/PFG1 (PRODUCTION OF FLAVONOL GLYCOSIDES1), MYB11/PFG2, and MYB111/PFG3 regulate expression of the five early biosynthetic genes, namely, chalcone synthase (CHS), chalcone isomerase (CHI), flavanone 3-hydroxylase (F3H), flavonoid 3′-hydroxylase (F3′H), and flavonol synthase (FLS) (Stracke et al. 2010b), whereas the anthocyanin regulators MYB75/PAP1 (PRODUCTION OF ANTHOCYANIN PIGMENT1), MYB90/PAP2, MYB113, and MYB114 control the expression of the late biosynthetic genes dihydroflavonol reductase (DFR) and leucoanthocyanidin dioxygenase/anthocyanidin synthase (LDOX/ANS) (Gonzalez et al. 2008; Dubos et al. 2010). Nakabayashi et al. (2014) have reported enhancement of oxidative and drought tolerance in *A. thaliana* by overaccumulation of flavonoids with radical scavenging activity, which mitigates oxidative and drought stress. Analysis of metabolome and transcriptome profiles under the oxidative and drought stress of wild-type, single overexpressors of MYB12/PFG1 (PRODUCTION OF FLAVONOL GLYCOSIDES1) or MYB75/PAP1 (PRODUCTION OF

ANTHOCYANIN PIGMENT1), double overexpressors of MYB12 and PAP1, transparent testa4 (tt4) as a flavonoid-deficient mutant, and flavonoid-deficient MYB12 or PAP1 overexpressing lines (obtained by crossing tt4 and the individual MYB overexpressor) has revealed overaccumulation of flavonoid, which was a key for enhanced tolerance to such stresses. Antioxidant assay has been performed for further confirmation of overaccumulation of flavonoids and anthocyanins (Nakabayashi et al. 2014). Regulators of flavonoid biosynthesis (i.e., MYB12/PFG1 (PRODUCTION OF FLAVONOL GLYCOSIDES1), MYB11/PFG2, and MYB111/PFG3) play essential role in abiotic stress tolerance in plants including *A. thaliana*. The *AtMYB12* has been found to regulate flavonoids accumulation and abiotic stress tolerance in *A. thaliana* (Wang et al. (2016a, b)). In particular, upregulation of the *AtMYB12* gene has been shown to increase the accumulation of flavonoids in several plant species. Wang et al. (2016a, b) have chemically synthesized the *AtMYB12* gene through the PTDS (PCR-based two-step DNA synthesis) method. They overexpressed *AtMYB12* in transgenic *Arabidopsis* and observed a marked increase in accumulation of flavonoids and salt and drought tolerance. These studies have revealed that regulator of flavonoid biosynthesis like *MYB12* gene is a promising target for increasing the accumulation of flavonoids and enhancing tolerance to abiotic stresses. In plants, high accumulation of anthocyanins is also linked with salt stress tolerance. In *B. napus* high accumulation of anthocyanins via the overexpression of *A. thaliana* dihydroflavonol-4-reductase (At*DFR*) confers significant salt stress tolerance. In view of increasing significance of *B. napus* and its cultivation area worldwide, development of stress-resistant Brassica plants is highly desirable. Manipulation of anthocyanin accumulation via *AtDFR* gene can also be a useful strategy to modulate and/or confer salinity and drought stress tolerance in other commercially important plants. However, intensified research efforts are needed to gain detailed insight into mechanisms of flavonoid accumulation during abiotic stress for convenience, which is at present scarcely understood.

11.10 Improvement of Isoprenoids Production

In view of the specialized roles of carotenoid in mediating abiotic stresses, its metabolic engineering has gained tremendous momentum in recent years. Metabolic engineering offers great promises to improve carotenoids synthesis and abiotic stress tolerance in plants. Plenty of research papers are available on metabolic engineering of primary (e.g., carbohydrates, lipids, and amino acids) and secondary metabolic pathways (e.g., alkaloids, terpenoids, flavonoids, lignins, quinones, and other benzoic acid derivatives) in plants (Gomez et al. 2007). Both primary and secondary metabolic pathways produce a number of metabolites, which are functionally important during abiotic stress conditions supporting plant in adaptation, and also aid on food quality. Most of the work on metabolic engineering of carotenoid biosynthesis anticipating enhancing abiotic stress tolerance mainly has been carried out in Arabidopsis, tobacco, and sweet potato. In plants, production of carotenoid can be improved through regulated expression of a few genes, such as β-carotene hydroxylase (*chy-β*) and lycopene *β*-cyclase (*lcy-β*) (Kim et al. 2012, 2013a, b). The LCY-β functions at the branch point of the carotenoid biosynthesis pathway, catalyzing the cyclization of lycopene. A downregulation of *chy-β* gene was observed to increase the contents of β-carotene and total carotenoids in cultured transgenic sweet potato cells. Besides, downregulation of *lcy-β* gene increased carotenoid synthesis and enhanced salt stress tolerance in transgenic calli of sweet potato (Kim et al. 2013a). Overexpression of *lcy-β* gene in *Arabidopsis* has been found to increase the level of lutein and the rate of non-photochemical quenching induction under high light (Pogson and Rissler, 2000). Engineering of a bacterial *chy-β* gene in tobacco and *Arabidopsis* has enhanced tolerance to UV and high light stress, respectively (Davison et al. 2002, Gotz et al. 2002). The improvement in stress tolerance of tobacco and *Arabidopsis* has been correlated with the biosynthesis of xanthophylls, the level of which has been significantly increased due to bacterial *chy-β*. Similarly, overexpression of a phytoene synthase gene (Se*PSY*) in transgenic *Arabidopsis* conferred tolerance against salt and oxidative stresses (Han et al. 2008). The phytoene synthase gene (*PSY*) was cloned from *S. europaea* thus named as Se*PSY*. Inhibition of the lycopene cyclization of the carotenoid synthesis pathway affects the salt suitability in *S. europaea* (Chen et al. 2011).

Most elaborated work on metabolic engineering of the carotenoid biosynthesis to confer abiotic stress tolerance has been attempted in sweet potato (Chen et al. 2011; Kim et al. 2012). Kang et al. (2017) in their recent review have discussed about the regulation of biosynthesis, accumulation, and catabolism of carotenoids in sweet potato along with the prospects for metabolic engineering of carotenoids in terms of molecular breeding for higher levels of antioxidants and stress tolerance. Sweet potato is a valuable industrial crop and source of antioxidants carotenoids. A β-carotene hydroxylase (CHY-β) is a key regulatory enzyme in the β-β-branch of carotenoid biosynthesis. It catalyzes hydroxylation into both β-carotene to β-cryptoxanthin and β-cryptoxanthin to zeaxanthin (Kim et al. 2012). Kim et al. (2012) have cloned partial cDNA encoding CHY-β from the storage roots of orange-fleshed sweet potato (cv. Shinhwangmi). Further, they generated an RNA interference-Ib*chy*-β construct and introduced into cultured cells of white-fleshed sweet potato (cv. Yulmi). The construct inhibited the activity of *chy*-β gene as a result of which β-carotene and total carotenoids content increased in transgenic cultured cells of sweet potato. The increase in carotenoids content has been attributed to enhanced salt stress tolerance (Kim et al. 2012). The major carotenoid found in sweet potato is violaxanthin, which mediates plant's response to abiotic stress. Violaxanthin is downstream of β-carotene and is a good candidate to control the upstream pathway. Metabolic engineering of violaxanthin biosynthesis has been performed in white-fleshed sweet potato cv. Yulmi (Ym) callus. Transgenic cells produced could be distinguished by variation of color from white to yellow, which corresponded to the amount of carotenoids (Kim et al. 2012). Another study by reported cloning and characterization of an Orange gene (Or) that increases carotenoid accumulation and salt stress tolerance in transgenic sweet potato cultures. The Or gene is responsible for the accumulation of carotenoids in plants. They isolated the Or gene from storage roots of orange-fleshed sweet potato cv.

Sinhwangmi and named as IbOr, which is an open reading frame in the 942 bp cDNA encoding a 313-amino acid protein containing a cysteine-rich zinc finger domain. Overexpression of IbOr gene in transgenic white-fleshed sweet potato cv. Yulmi enhanced the contents of b-carotene, lutein, and total carotenoids which in turn increased antioxidant acidity and tolerance to salt stress. A gene IbGGPS encoding geranylgeranyl pyrophosphate synthase has also been cloned from the storage roots of sweet potato. The enzyme GGPS catalyzes formation of geranylgeranyl pyrophosphate (GGPP), which is believed to be a key step in biosynthetic pathway of carotenoids and many other terpenes. The expression of GPPS has been found to be associated with resistance of plants to different stresses, for example, the expression of GGPS in *A. thaliana* protects it from photooxidative stress (Xie et al. 2012). In sweet potato, IbGGPS favors biosynthesis of carotenoids which is likely associated with osmotic stress tolerance. When the IbGGPS was overexpressed in *A. thaliana*, total carotenoids and osmotic stress tolerance have been significantly improved. In *Arabidopsis* and tobacco, salt stress tolerance has been conferred by expression of *lcy-β* gene from *S. europaea*. Observation of ectopic expression of Se*lcy-β* and At*lcy-β* in *Nicotiana benthamiana* by the pea early browning virus (PEBV) system was carried out to compare functional difference between them. No significant differences have been observed between Se*lcy-β* and At*lcy-β* transformed plants. As reported earlier, enhanced salt stress tolerance of transgenic *Arabidopsis* has been credited to carotenoids accumulation, which impairs reactive oxygen species and protects photosynthesis system under salt stress (Chen et al. 2011). Shi et al. (2009) have also cloned and characterized a Nt*lcy-β* gene from *N. tabacum*. The transcript expression levels of the Nt*lcy-β* gene markedly varied with developmental stages, in response to salt, drought, and ABA treatment (Shi et al. 2009). The overexpression of Nt*lcy-β* gene in *N. tabacum* enhanced accumulation of carotenoids and tolerance to salt and drought stress. On the other hand, in RNAi transgenic plants, where expression of Nt*lcy-β* was suppressed, no carotenoid, chlorophyll, and abscisic acid accumulation was observed, and plants were highly prone to salt and drought stress.

Abscisic acid (ABA) which is derived from the carotenoid has been known to regulate a number of processes, such as seed dormancy and plant responses to drought condition. Previous studies have revealed that ABA binds with small soluble protein receptors, namely, PYR/PYL/RCAR (pyrabactin resistance1/PYR1-like/regulatory component of ABA receptor), which are found in the cytoplasm and nucleus of the plant cell. The names of proteins are from pyrabactin resistance/pyrabactin-like or regulatory components of ABA receptor. Pyrabactin is a synthetic ABA that helps plant to cope with drought condition by selectively inhibiting seed germination. In their nonbonded state, the ABA receptor proteins exist in dimeric form. There are six members of the PYR/PYL/RCAR family of proteins (PYR1/RCAR11, PYL1/RCAR12, PYL2/RCAR14, PYL3/RCAR13, PYL8/RCAR3, PYL9/RCAR1), which binds a protein phosphatase 2C in the presence of ABA. Applications of pyrabactin have been very useful in the identification of the PYR1/PYL/RCAR family (PYL) of proteins as the ABA receptor. Based on the structures of ABA receptors, they have been kept under the START superfamily of ligand-binding proteins. Studies have revealed that constitutive overexpression of ABA receptors in plants could be an excellent approach to improve drought tolerance. However, studies have shown that the overexpression of ABA may have some negative impact on yield under non-stress conditions. Thus, the precise regulation of the activity of individual or multiple ABA receptors is very important in achieving enhanced drought tolerance without compromising plant yield. A drought-resistant *Arabidopsis* was developed using an engineered PYR1 ABA receptor that can be activated by an existing non-herbicidal agrochemical that is not a natural inducer of ABA responses. Some chemicals (herbicides) in plants are reported to activate or repress the receptors that sense the stress or the signaling pathways triggered by hormones which mediate the corresponding stress responses. Applications of such chemicals could be an alternative approach to actively control tolerance to abiotic stress in plants.

11.11 Conclusion

Metabolic engineering is now rapidly being used for development of abiotic stress-resistant transgenic plants via manipulation of the genes involved in the synthesis of stress-protectant metabolites and transcription factors. Abiotic stress factors such as drought, salinity, and extreme temperatures, light, radiation, and CO_2 levels markedly influence the pathways of several secondary metabolites in plants. Here, we have underlined importance of stress-protectant secondary metabolites and their metabolic engineering of their biosynthesis for enhancing abiotic stress tolerance in plants. The chapter provides information on manipulation of secondary metabolites, such as isoprenoids, carotenoids, and flavonoids, to confer abiotic stress tolerance in plants. Based on analysis of published reports, it has been realized that instead of single gene, it would be more advantageous to engineer multiple enzymatic reactions of the same or different metabolic pathways to generate abiotic stress-tolerant plants. In view of increasing significance of metabolic engineering, there is an immense necessity of methodical thorough investigations of genes and metabolic pathways to understand their precise role during abiotic stresses. In this regard, model plants like Arabidopsis may be very useful as a model system to unravel the genes for improvement of abiotic stress tolerance in plants. In this chapter several examples of metabolic engineering have been discussed. Most extensive work on metabolic engineering has been carried out in sweet potato. Transgenic sweet potato plants with an overexpressed IbGGPS gene showed better osmotic stress tolerance, while transgenic *Arabidopsis* with overexpression of *AtMYB12* exhibited increased levels of flavonoids and enhanced salt and drought tolerance. Similarly, transgenic tobacco overexpressing Nt*b*-LCY1 gene showed tolerance to drought and salt stresses, whereas transgenic alfalfa overexpressing thiolase II performs better under abiotic stress conditions. These studies have clearly indicated that secondary metabolic pathways may be engineered to increase tolerance to abiotic stresses in plants. Scientific advances in molecular biology over the past years have facelifted our understanding of secondary metabolite biosynthesis and regulation in plants which is crucial for metabolic engineering. Also, recent techniques in gene editing CRSPER-Cas9 can also be useful for expression or inhibition of the genes. Gene stacking approaches may also be useful for generation of transgenic plants. Currently, this approach is being used to generate abiotic stress tolerance of transgenic sweet potato varieties. Metabolic engineering is of course the most straightforward approach to manipulate gene expressions. At the same time, transgenic plants have several complex issues of concerns associated with population. Nevertheless, metabolic engineering of secondary metabolic pathways may be very useful to confer abiotic stress tolerance in major crop plants.

Acknowledgment Corresponding author of this article is grateful to Dr. Ashok Kumar Chauhan, Founder President, and Atul Chauhan, Chancellor, Amity University, Uttar Pradesh, Noida, India, for providing necessary support and facilities.

References

Aharoni A, Jongsma MA, Bouwmeester HJ (2005) Volatile science? Metabolic engineering of terpenoids in plants. Trends Plant Sci 10:594–602

Andrade-Souza V, Costa MGC, Chen CX, Gmitter FG Jr, Costa MA (2011) Physical location of the carotenoid biosynthesis genes Psy and b-Lcy in Capsicum annuum (Solanaceae) using heterologous probes from *Citrus sinensis* (Rutaceae). Genet Mol Res 10:404–409

Bajda A, Konopka-Postupolska D, Krzymowska M et al (2009) Role of polyisoprenoids in tobacco resistance against biotic stresses. Physiol Plant 135:351–364

Bakhsh A, Hussain T (2015) Engineering crop plants against abiotic stress: current achievements and prospects. Emir J Food Agric 27:24–39

Bakshi M, Oelmüller R (2014) WRKY transcription factors: Jack of many trades in plants. Plant Signal Behav 9:e27700

Bhatnagar-Mathur P, Vadez V, Sharma KK (2008) Transgenic approaches for abiotic stress tolerance in plants: retrospect and prospects. Plant Cell Rep 27:411–424

Bohnert HJ, Nelson DE, Jensen RG (1995) Adaptations to environmental stresses. Plant Cell 7:1099–1111

Bramley PM (1997) The regulation and genetic manipulation of carotenoid biosynthesis in tomato fruit. Pure Appl Chem 69:2159–2162

Bromham L, Saslis-Lagoudakis CH, Bennett TH, Flowers TJ (2013) Soil alkalinity and salt tolerance: adapting to multiple stresses. Biol Lett 9:1744

Buer CS, Imin N, Djordjevic MA (2010) Flavonoids: new roles for old molecules. J Integr Plant Biol 52:98–111

Butelli E, Titta L, Giorgio M et al (2008) Enrichment of tomato fruit with health-promoting anthocyanins by expression of select transcription factors. Nat Biotechnol 26:1301–1308

Butelli E, Licciardello C, Zhang Y, Liu J, Mackay S, Bailey P, Reforgiato-Recupero G, Martin C (2012) Retrotransposons control fruit-specific, cold-dependent accumulation of anthocyanins in blood oranges. Plant Cell 24:1242–1255

Chen Tony HH, Murata N (2002) Enhancement of tolerance of abiotic stress by metabolic engineering of betaines and other compatible solutes. Curr Opin Plant Biol 5:250–257

Chen X, Han H, Jiang P, Nie L, Bao H, Fan P, Lv S, Feng J, Li Y (2011) Transformation of beta-lycopene cyclase genes from Salicornia europaea and Arabidopsis conferred salt tolerance in Arabidopsis and tobacco. Plant Cell Physiol 52:909–921

Cordoba E, Salmi M, Leon P (2009) Unravelling the regulatory mechanisms that modulate the MEP pathway in higher plants. J Exp Bot 60:2935–2943

Cunningham FX, Gantt E (1998) Genes and enzymes of carotenoid biosynthesis in plants. Annu Rev Plant Physiol Plant Mol Biol 49:557–583

Dall'Osto L, Fiore A, Cazzaniga S, Giuliano G, Bassi R (2007) Different roles of - and -branch xanthophylls in photosystem assembly and photoprotection. J Biol Chem 282:35056–35068

Davison PA, Hunter CN, Horton P (2002) Overexpression of beta-carotene hydroxylase enhances stress tolerance in Arabidopsis. Nature 418:203–206

Dixon RA, Liu C, Jun JH (2013) Metabolic engineering of anthocyanins and condensed tannins in plants. Curr Opin Biotechnol 24:329–335

Dubos C, Stracke R, Grotewold E, Weisshaar B, Martin C, Lepiniec L (2010) MYB transcription factors in Arabidopsis. Trends Plant Sci 15:573–581

Edreva A, Velikova V, Tsonev T, Dagnon A, Gurel L, Gesheva AE (2008) Stress-protective role of secondary metabolites: diversity of functions and mechanisms. Gen Appl Plant Physiol 34:67–78

Feller A, Machemer K, Braun EL, Grotewold E (2011) Evolutionary and comparative analysis of MYB and bHLH plant transcription factors. Plant J 66:94–116

Ferdinando MD, Brunetti C, Fini A, Tattini M (2012) Flavonoids as antioxidants in plants under abiotic stresses. In: Ahmad P, Prasad M (eds) Abiotic stress responses in plants. Springer, New York

FL, Schnitzler JP (2010) Abiotic stresses and induced BVOCs. Trends Plant Sci 15(3):154–166

Franke RB, Dombrink I, Schreiber L (2012) Suberin goes genomics: use of a short living plant to investigate a long lasting polymer. Front Plant Sci 3:4–5

Fraser PD, Bramley PM (2004) The biosynthesis and nutritional uses of carotenoids. Prog Lipid Res 43:228–265

Galpaz N, Ronen G, Khalfa Z, Zamir D, Hirschberg J (2006) A chromoplast-specific carotenoid biosynthesis pathway is revealed by cloning of the tomato white-flower locus. Plant Cell 18:1947–1960

Ganjewala D (2016) Secondary metabolite credentials and biological properties of *Litchi chinensis*. In: Kumar M, Kumar V, Prasad R, Varma A (eds) The Litchi: biotechnology. Springer, Heidelberg, pp 213–242

Ganjewala D, Kumar S, Luthra R (2009) An account of cloned genes of methyl-erythritol-4-phosphate pathway of isoprenoid biosynthesis in plants. Curr Issues Mol Biol S1:34–44

Gomez Galera S, Pelacho AM, Gene A, Capell T, Christou P (2007) The genetic manipulation of medicinal and aromatic plants. Plant Cell Rep 26:1689–1715

Gonzalez A, Zhao M, Leavitt JM, Lloyd AM (2008) Regulation of the anthocyanin biosynthetic pathway by the TTG1/bHLH/Myb transcriptional complex in Arabidopsis seedlings. Plant J 53:814–827

Gotz T, Sandmann G, Romer S (2002) Expression of a bacterial carotene hydroxylase gene (crtZ) enhances UV tolerance in tobacco. Plant Mol Biol 50:129–142

Grotewold E (2006) The genetics and biochemistry of floral pigments. Annu Rev Plant Biol 57:761–780

Han H, Li Y, Zhou S (2008) Overexpression of phytoene synthase gene from *Salicornia europaea* alters response to reactive oxygen species under salt stress in transgenic Arabidopsis. Biotechnol Lett 30:1501–1507

Harborne JB, Williams CA (2000) Advances in flavonoid research since 1992. Phytochemistry 55:481–504

Jain M (2013) Emerging role of metabolic pathways in abiotic stress tolerance. J Plant Biochem Physiol 1:2–3

Kang L, Park SC, Ji CY, Kim HS, Lee HS, Kwak SS (2017) Metabolic engineering of carotenoids in transgenic sweet potato. Breed Sci 67:27–34

Kaur G, Kumar S, Nayyar H, Upadhyaya HD (2008) Cold stress injury during the pod-filling phase in chickpea (*Cicer arietinum* L.): effects on quantitative and qualitative components of seeds. J Agron Crop Sci 194:457–464

Kim SH, Ahn YO, Ahn MJ, Lee HS, Kwak SS (2012) Down-regulation of β-carotene hydroxylase increasesβ-carotene and total carotenoids enhancing salt stress tolerance in transgenic cultured cells of sweet potato. Phytochemistry 74:69–78

Kim SH, Kim YH, Ahn YO, Ahn MJ, Jeong JC, Lee HS et al (2013a) Down regulation of the lycopene ϵ-cyclase gene increases carotenoid synthesis via the β-branch-specific pathway and enhances salt-stress tolerance in sweet potato transgenic calli. Physiol Plant 147:432–442

Kim SH, Ahn YO, Ahn MJ, Jeong JC, Lee HS, Kwak SS (2013b) Cloning and characterization of an Orange gene that increases carotenoid accumulation and salt stress tolerance in transgenic sweet potato cultures. Plant Physiol Biochem 70:445–454

Kirby J, Keasling JD (2009) Biosynthesis of plant isoprenoids: perspectives for microbial engineering. Annu Rev Plant Biol 60:335–355

Krauss S, Schnitzler WH, Grassmann J, Woitke M (2005) The influence of different electrical conductivity values in a simplified recirculating soilless system on inner and outer fruit quality characteristics of tomato. J Agric Food Chem 54:441–448

Kusano M, Tohge T, Fukushima A, Kobayashi M, Hayashi N, Otsuki H, Kondou Y, Goto H, Kawashima M, Matsuda F, Niida R, Matsui M, Saito K, Fernie AR (2011) Metabolomics reveals comprehensive reprogramming involving two independent metabolic responses of Arabidopsis to UV-B light. Plant J 67:354–369

Lawlor DW (2013) Genetic engineering to improve plant performance under drought: physiological evaluation of achievements, limitations, and possibilities. J Exp Bot 64:83–108

Lepiniec L, Debeaujon I, Routaboul JM, Baudry A, Pourcel L et al (2006) Genetics and biochemistry of seed flavonoids. Annu Rev Plant Biol 57:405–430

Loivamäki M, Gilmer F, Fischbach RJ, Sorgel C, Bachl A, Walter A, Schnitzler JP (2007) Arabidopsis, a model to study biological functions of isoprene emission? Plant Physiol 144:1066–1078

Malhotra C, Kapoor RT, Ganjewala D (2016a) Alleviation of abiotic and biotic stresses in plants by silicon supplementation. Sci Agric 13:59–73

Malhotra C, Kapoor RT, Ganjewala D (2016b) Protective role of sodium silicate against water stress in *Lycopersicon esculentum* mill. Int J Pharm Biosci 7:909–917

Malhotra C, Kapoor RT, Ganjewala D, Singh NB (2017) Sodium silicate mediated response of antioxidative defense system in *Lycopersicon esculentum* mill. under water stress. Int J Phytomedicine 9:364–378

Mantri N, Patade V, Penna S, Ford R (2012) Abiotic stress responses in plants: present and future. In: Ahmad P, Prasad MNV (eds) Abiotic stress responses in plants. Springer, New York, pp 1–19

Mazid M, Khan TA, Mohammad F (2011) Effect of abiotic stress on synthesis of secondary plant products: a critical review. Agric Rev 32:172–182

Mierziak J, Kostyn K, Kulma A (2014) Flavonoids as important molecules of plant interactions with the environment. Molecules 19:16240–16265

Miller G, Suzuki N, Ciftci-Yilmaz S, Mittler R (2010) Reactive oxygen species homeostasis and signaling during drought and salinity stresses. Plant Cell Environ 33:453–467

Min DH, Zhao Y, Huo DY, Li LC, Chen M, Xu ZS, Ma YZ (2013) Isolation and identification of a wheat gene encoding a zinc finger protein (TaZnFP) responsive to abiotic stresses. Acta Physiol Plant 35:1597–1604

Muhammad R, Khurram S, Rashid A, Moddassir A, Imran H, Shahid M, Gerald AB, Asir AS (2014) Cloning and characterization of Na+/H+ antiporter (LfNHX1) gene from a halophyte grass *Leptochloa fusca* for drought and salt tolerance. Mol Biol Rep 41:1669–1682

Nakabayashi R, Yonekura-Sakakibara K, Urano K, Suzuki M, Yamada Y, Nishizawa T, Matsuda F, Kojima M, Sakakibara H, Shinozaki K, Michael AJ, Tohge T, Yamazaki M, Saito K (2014) Enhancement of oxidative and drought tolerance in Arabidopsis by over accumulation of antioxidant flavonoids. Plant J 77:367–379

Nambara E, Marion-Poll A (2005) Abscisic acid biosynthesis and catabolism. Annu Rev Plant Biol 56:165–185

Naqvi S, Zhu C, Farre G, Ramessar K, Bassie L, Breitenbach J et al (2009) Transgenic multivitamin corn through biofortification of endosperm with three vitamins representing three distinct metabolic pathways. Proc Natl Acad Sci U S A 106:7762–7767

Nascimento NC, Fett-Neto AG (2010) Plant secondary metabolism and challenges in modifying its operation: an overview. In: Fett-Neto A (ed) Plant secondary metabolism engineering. Methods in molecular biology (methods and protocols), vol 643. Humana Press, Totowa

Natwar S, Avinash M, Bhavanath J (2014) Over-expression of the peroxisomalascorbate peroxidase (SbpAPX) gene cloned from *Halophyte Salicornia* brachiata confers salt and drought stress tolerance in transgenic tobacco. Mar Biotechnol 16:321–332

Oh SK, Kim IJ, Shin DH, Yang J, Kang H, Han KH (2000) Cloning, characterization, and heterologous expression of a functional geranylgeranyl pyrophosphate synthase from sunflower (*Helianthus annuus* L.). J Plant Physiol 157:535–542

Pande A, Arora S (2017) Molecular strategies for development of abiotic stress tolerance in plants. Cell Cellular Life Sci J 2:000112

Pogson BJ, Rissler HM (2000) Genetic manipulation of carotenoid biosynthesis and photoprotection. Philos Trans R Soc Lond B 355:1395–1403

Qiu J, Sun S, Luo S, Zhang J, Xiao X et al (2014) Arabidopsis AtPAP1 transcription factor induces anthocyanin production in transgenic *Taraxacum brevicorniculatum*. Plant Cell Rep 33:669–680

Rathinasabapathi B (2000) Metabolic engineering for stress tolerance: installing osmoprotectant synthesis pathways. Ann Bot 86:709–716

Rontein D, Basset G, Hanson AD (2002) Metabolic engineering of osmoprotectant accumulation in plants. Metab Eng 4:49–56

Sasaki K, Saito T, Lamsa M, Oksman-Caldentey KM, Suzuki M, Ohyama K, Muranaka T, Ohara K, Yazaki K (2007) Plants utilize isoprene emission as a thermotolerance mechanism. Plant Cell Physiol 48:1254–1262

Stracke R, Favory JJ, Gruber H, Bartelniewoehner L, Bartels S, Binkert M, Funk M, Weisshaar B, Ulm R (2010a) The Arabidopsis bZIP transcription factor HY5 regulates expression of the PFG1/MYB12 gene in response to light and ultraviolet-B radiation. Plant Cell Environ 33:88–103

Stracke R, Jahns O, Keck M, Tohge T, Niehaus K, Fernie AR, Weisshaar B (2010b) Analysis of production of flavonol glycosides-dependent flavonol glycoside accumulation in *Arabidopsis thaliana* plants reveals MYB11-, MYB12- and MYB111-independent flavonol glycoside accumulation. New Phytol 188:985–1000

Sun PP, Zhu XF, Huang XS, Liu JH (2014) Overexpression of a stress-responsive MYB transcription factor of *Poncirus trifoliata* confers enhanced dehydration tolerance and increases polyamine biosynthesis. Plant Physiol Biochem 78:71–79

Shi GR, Cai QS, Liu QQ, Wu L (2009) Salicylic acid-mediated alleviation of cadmium toxicity in hemp plants in relation to cadmium uptake, photosynthesis, and antioxidant enzymes. Acta Physiol Plant 31:969–977

Tang LL, Cai H, Zhai H, Luo X, Wang ZY, Cui L, Bai X (2014) Overexpression of *Glycine soja* WRKY20 enhances both drought and salt tolerance in transgenic alfalfa (*Medicago sativa* L.). Plant Cell Tissue Organ Cult 118:77–86

Tao N, Hu Z, Liu Q, Xu J, Cheng Y, Guo L et al (2007) Expression of phytoene synthase gene (Psy) is enhanced during fruit ripening of Cara Cara navel orange (*Citrus sinensis* Osbeck). Plant Cell Rep 26:837–843

Taylor IB, Sonneveld T, Bugg TDH, Thompson AJ (2005) Regulation and manipulation of the biosynthesis of abscisic acid, including the supply of xanthophyll precursors. J Plant Growth Regul 24:253–273

Tuteja N (2007) Abscisic acid and abiotic stress signaling. Plant Signal Behav 2:135–138

Varshney RK, Ribaut JM, Buckler ES, Tuberosa R, Rafalski JA, Langridge P (2012) Can genomics boost productivity of orphan crops? Nat Biotechnol 30:1172–1176

Vickers CE, Gershenzon J, Lerdau MT, Loreto F (2009a) A unified mechanism of action for volatile isoprenoids in plant abiotic stress. Nat Chem Biol 5:283–291

Vickers CE, Possell M, Cojocariu CI, Velikova VB, Laothawornkitkul J, Ryan A, Mullineaux PM, Nicholas Hewitt C (2009b) Isoprene synthesis protects transgenic tobacco plants from oxidative stress. Plant Cell Environ 32:520–531

Vickers CE, Possell M, Laothawornkitku J, Ryan AC, Hewitt CN, Mullineaux PM (2011) Isoprene synthesis in plants: lessons from a transgenic tobacco model. Plant Cell Environ 34:1043–1053

Wagner T, Windhovel U, Romer S (2002) Bansformation of tobacco with a mutated cyanobacterial phytoene desaturase gene confers resistance to bleaching herbicides. Z Naturforsch C 57:671–679

Wang ZY, Xiong L, Li W, Zhu JK, Zhu J (2011) The plant cuticle is required for osmotic stress regulation of abscisic acid biosynthesis and osmotic stress tolerance in Arabidopsis. Plant Cell 23:1971–1984

Wang H, Fan W, Li H, Yang J, Huang J, Zhang P (2013) Functional characterization of dihydroflavonol-4-reductase in anthocyanin biosynthesis of purple sweet potato underlies the direct evidence of anthocyanins function against abiotic stresses. PLoS One 8:e78484

Wang FB, Tong WJ, Zhu H, Kong WL, Peng RH et al (2016a) A novel Cys2/His2 zinc finger protein gene from sweet potato, IbZFP1, is involved in salt and drought tolerance in transgenic Arabidopsis. Planta 243:783–797

Wang FB, Zhu H, Chen DH, Li ZJ, Peng RH, Yao QH (2016b) A grape bHLH transcription factor gene, VvbHLH1, increases the accumulation of flavonoids and enhances salt and drought tolerance in transgenic *Arabidopsis thaliana*. Plant Cell Tissue Org Cult 125:387–398

Wani SH, Singh NB, Haribhushan A, Mir JI (2013) Compatible solute engineering in plants for abiotic stress tolerance-role of glycine betaine. Curr Genomics 14:157–165

Wu G, Wang G, Ji J, Gao H, Guan W, Wu J, Guan C, Wang Y (2014) Cloning of a cytosolic ascorbate peroxidase gene from *Lycium chinense* Mill. and enhanced salt tolerance by overexpressing in tobacco. Gene 543:85–92

Xie Y, Xu D, Cui W, Shen W (2012) Mutation of Arabidopsis Hy1 causes Uv-C hypersensitivity by impairing carotenoid and flavonoid biosynthesis and the down-regulation of antioxidant defence. J Exp Bot 63:3869–3883

Xu Q, He Q, Li S, Tian Z (2014) Molecular characterization of StNAC2 in potato and its overexpression confers drought and salt tolerance. Acta Physiol Plant 36:1841–1851

Yarra R, He SJ, Abbagani S, Ma B, Bulle M, Zhang WK (2012) Overexpression of a wheat Na+/H+ antiporter gene (TaNHX2) enhances tolerance to salt stress in transgenic tomato plants (*Solanum lycopersicum* L.). Plant Cell Tissue Org Cult 111:49–57

Yogendra KM, Nirmaljit K (2017) Antioxidants: environmental stress mitigating metabolites. Curr Trends Biomed Eng Biosci 5:555660

Yonekura-Sakakibara K, Tohge T, Matsuda F, Nakabayashi R, Takayama H, Niida R, Watanabe-Takahashi A, Inoue E, Saito K (2008) Comprehensive flavonol profiling and transcriptome coexpression analysis leading to decoding gene-metabolite correlations in Arabidopsis. Plant Cell 20:2160–2176

Zhang DY, Yang HL, Li XS, Li HY, Wang YC (2014) Overexpression of *Tamarix albiflonum* TaMnSOD increases drought tolerance in transgenic cotton. Mol Breed 34:1–11

Zhao Q, Zhang H, Wang T, Chen SX, Dai SJ (2013) Proteomics based investigation of salt-responsive mechanisms in plant roots. J Proteome 82:230–253

Zhu JK (2001) Plant salt tolerance. Trends Plant Sci 6:66–71

Zhu JK (2002) Salt and drought stress signal transduction in plants. Annu Rev Plant Biol 53:247–273

VAM-Assisted Adaptive Response and Tolerance Mechanism of Plants Under Heavy Metal Stress: Prospects for Bioremediation

12

Ritu Chaturvedi and Garima Malik

Abstract

Heavy metal (HM) toxicity is an imperative abiotic stress component of the environment that is negatively influencing the agricultural productivity throughout the world. Due to the unprecedented population explosion mainly in developing countries, land resources that are shrinking and therefore utilizing any and every space for cultivation have become a necessity to meet the expanding demands for food, fuel, and fodder. Accumulation of HMs in soil inflicts detrimental effect on various morphological, physiological, and metabolic pathways leading to various other stresses and anomalies in plants. However, plants are equipped with a range of inherent mechanisms to neutralize HM toxicity. Additionally, the phytomicrobiome-microorganisms that colonize plants can allow them to better tolerate soil pollutants. Considering that even contaminated soils are a potential asset for agricultural production, it is crucial to develop remediation techniques involving plants and associated microbiota for sustainable cleanup of defiled land resources so as to avoid biomagnification of toxic trace elements in the food chain. Fungi, both symbiotic and nonsymbiotic, are an important component of soil microbiota. Vesicular arbuscular mycorrhizal fungi (VAMF) are known to enhance nutrient acquisition from soil, promote growth and development, boost reproductive success, modulate secondary metabolic pathways, and confer biotic and abiotic stress tolerance to its host plants. In this chapter, we present the discussion on fungal symbionts and role of VAMF in combating HM stress that interferes with plant development and affects crop productivity. A detailed update on recent research has also been covered which contributes to a better understanding of the potential of this unique mutualistic relationship that can be

R. Chaturvedi
St. John's College, Agra, India

G. Malik (✉)
R.G. (PG) College, C.C.S University, Meerut, Uttar Pradesh, India

M. Kumar et al. (eds.), *In vitro Plant Breeding towards Novel Agronomic Traits*,
https://doi.org/10.1007/978-981-32-9824-8_12

engaged in the sustainable production of agricultural and other economically important plants in a changing world.

Keywords
Heavy metal · Stress · Arbuscular mycorrhizal fungi · Sustainable development

12.1 Introduction

Higher plants are closely associated with the microbial community present in their microenvironments, the phytomicrobiome. The presence of microbes in and on plants is nature's rule rather than the exception as plants have likely had allied microorganisms ever since they colonized the terrestrial ecosystem (Partida-Martínez and Heil 2011; Knack et al. 2015). Plants interact and provide precise biotic and abiotic conditions for microbial existence. The natural world of plant-microbe interactions ranges from symbiosis to pathogenicity. This depends on various abiotic and biotic factors like the genotypes of plants and microbes, ecological environment, and the active network of interactions within the plant biome (Hardoim et al. 2015). The rhizosphere, the soil surrounding the plant roots, offers a unique ecological niche for a variety of microorganisms. Microorganisms found in the rhizosphere include virus, archaea, bacteria, cyanobacteria, fungi, oomycetes, and protozoa. In comparison to the other plant microenvironments (e.g., phyllosphere), the rhizosphere portrays high microbial abundance and activities (Berg et al. 2014). Plant root exudates influence the growth of microorganisms living in the vicinity of plant roots and have a profound effect on the composition of the rhizosphere microbiome (Grayston et al. 1998; Chaparro et al. 2014). Rhizosphere microbiome diversity, in turn, may affect plant growth, health, and development by regulating nutrient uptake via immobilization of nutrients into microbial biomass, promoting stress resistance, and providing protection against soilborne pathogens (Weller et al. 2002; Rodriguez et al. 2004; Barrow et al. 2008; Lugtenberg and Kamilova 2009; Yang et al. 2009). Several microorganisms grow within living plant tissue and are hence known as endophytes (*endo*-within, *phyte*-plant). Orlando Petrini expanded the use of the term endophytes in 1991 and included all organisms inhabiting plant organs that at some time in their life cycle can colonize internal plant tissues without causing evident damage to their host in endophytes (Petrini 1991; Hardoim et al. 2015). Endophytic fungal alliance with plants helps it to adapt and endure in harsh biotic and abiotic stress situation.

Plant-fungus association is ancient and provided the foundation for colonization of land by land plants (Redecker et al. 2000). Fossil evidences indicate that plant-fungi association goes back to the early Devonian period (more than 400 million years ago), and this beneficial relationship played a crucial role in the adaptation, expansion, and evolution of terrestrial plants (Pirozynski and Malloch 1975; Krings et al. 2007). Mycorrhiza is one of the finest examples of mutualistic association between plants and fungi. The term "mycorrhiza" was coined by Albert Bernhard Frank derived from Greek *mycos* (meaning fungus) and *rhiza* (meaning roots).

Mycorrhizas (fungus roots) are highly evolved, interdependent, symbiotic associations between the roots of higher plants and soil fungi where the host plant receives mineral nutrients while the fungus obtains photosynthetically fixed carbon compounds (Harley 1989). In nature, more than 80% of all vascular land plants live in some association with mycorrhizal fungi. Mycorrhizae increase the surface area of the plant root system because hyphae, which are long and narrow extensions of the fungus, can penetrate and spread into the small soil pores beyond the nutrient depletion zone and allow access to nutrients otherwise unavailable to the plant. The fungus partners in this association are members of the kingdom *Zygomycetes*, *Ascomycetes*, and *Basidiomycetes* (Harley and Smith 1983; Kendrick 1985). Several different types of mycorrhizal associations are found in nature, involving different groups of fungi and host plants and diverse morphology patterns (Brundrett 1991). However, the two most abundant and widespread mycorrhizal associations are as follows: (a) Ectomycorrhizas (ECM) are present in 3% of vascular plants where *Basidiomycetes* and *Ascomycetes* fungi form an extensive dense sheath around the roots (mantle) and a Hartig net (hyphal network) between the root cortical cells, and (b) vesicular arbuscular mycorrhizae (VAM) are present in 80% of vascular plants where the hyphae of the zygomycetous fungi not only grow inside the root of the plant but also penetrate the root cortical cells and form clusters of finely divided hyphae which eventually develops into arbuscules (small ellipsoidal structures) and vesicles (Brundrett 1991).

12.2 Vesicular Arbuscular Mycorrhizal (VAM) Fungi

Mycorrhizal fungi are chief constituents of the soil microflora in various ecosystems (Harley 1971). Fossil record investigations and discovery of 450-million-year-old fossilized spores alike to the spores of present AM fungi revealed that mycorrhiza evolved along with land plants as VAM associations and played an imperative role in facilitating the adaption of plants to the terrestrial environment (Pirozynski and Dalpé 1989; Redecker 2000). However, it is still not clear whether the first AM fungi were already mutualistic symbionts or it evolved from pathogenic forms. The AM association is formed in a large number of plant species (mainly herbaceous vegetation, grasses, and tropical trees and some woody temperate wood plants) and a relative limited diversity of fungal species. Earlier, approximately 200 species of the AM fungi belonging to three families and six genera, viz., *Acaulospora* and *Entrophospora* (*Acaulosporaceae*), *Glomus* and *Sclerocystis* (*Glomeraceae*), and *Gigaspora* and *Scutellospora* (*Gigasporaceae*), were recognized (Morton and Benny 1990). They were classified on the basis of their asexual spore morphology and spore formation characteristics and were placed within one order, *Glomerales*, of the phylum *Zygomycota*. This fungal group has recently been removed from the polyphyletic phylum *Zygomycota* and elevated to the rank of a new monophyletic phylum *Glomeromycota* based on their rDNA sequences analysis (Schüßler et al. 2001). Now, the AM-forming fungi are organized into 3 classes (*Archaeosporomycetes*, *Glomeromycetes*, and

Paraglomeromycetes), 5 orders (*Archaeosporales*, *Diversisporales*, *Gigasporales*, *Glomerales*, and *Paraglomerales*), 14 families, 29 genera, and around 230 species (Morton and Redecker 2001; Schüßler et al. 2001; Oehl and Sieverding 2004; Walker and Schüßler 2004; Sieverding and Oehl 2006; Spain et al. 2006; Oehl et al. 2008, 2011a, b, c, d, e, Palenzuela et al. 2008).

AM fungi are obligate biotrophs, and completion of their life cycle relies on their capability to colonize a host plant. These fungi have aseptate (lack cross walls) and coenocytic hyphae. There is no evidence of sexual reproduction in them, and they do not stay alive without the presence of plant roots. Genetic studies using a molecular marker have detected only low levels or no genetic recombination at all (Kuhn et al. 2001). Glomeromycotan fungi produce large size multinucleate/coenocytic spores with layered walls, which may be formed either singly/in clusters/in distinctive fruitbodies – "sporocarps" (Morton 1988; Bécard and Pfeffer 1993). The symbiotic association between plant roots and fungi induces many vital physiological modifications in both the partners leading to reciprocal assistance from each other. At the molecular level, signals are exchanged between the partners, which in turn led to modulation of gene expression pattern and hormonal signaling. This complex molecular dialogue may possibly trigger the morphological and physiological changes required for the union of two partners into one association (Harrier 2001; Gutjahr and Parniske 2013). The AMF symbiosis begins with the germination of fungal spores. The establishment of symbiosis includes hyphal branching, appressorium production, colonization, development of intracellular arbuscules, and simultaneous formation of extraradical mycelium (Akiyama et al. 2005).

The VAM symbiosis is of tremendous worth for both natural and man-made ecosystems. Apart from nutrient exchange, the VAM symbiosis is associated with an array of added benefits for the plant such as acquisition of vital mineral nutrients (nitrogen and phosphorus), resistance to biotic stresses (soil and root borne pathogens), various abiotic stresses (drought, alkalinity, salinity, heavy/toxic metals, mineral imbalance), soil stability, and fertility, influencing the bacterial community composition in the mycorrhizosphere and synergistic interactions with other beneficial microorganisms. Recent advances in scientific research over the past decades on AM fungi have provided us greater insight into the multiple roles and influence of mycorrhizae on the ecosystem.

12.3 Heavy Metal Stress and Its Impact on Crops

Heavy metal (HM) contamination of soils through natural and/or anthropogenic activities is of serious concern worldwide for its possible consequences on the environment and all living organisms. Heavy metals (HMs) enter into the body of living organisms through the food chain, thereby raising the risk of metal toxicity on the well-being of the entire food web. HMs are imprecisely defined as a group of non-biodegradable, persistent inorganic chemical substances or metallic elements with atomic mass over 20 (excluding alkaline metals) and specific mass/density higher than 5 g/cm^3 (Rascio and Navari-Izzo 2011; Emamverdian et al. 2015). The essential elements play an important role in the growth, metabolism, and development of plants; however, their concentration in excess can lead to diminution and inhibition of growth in plants. 53 out of 90 naturally occurring elements are HMs, but not all of them are biologically important. HM occurs in colloidal, particulate, or dissolved phases, and their toxicity depends upon their solubility/bioavailability, which is further governed by physiochemical properties of soil such as pH. HMs at toxic levels disrupt normal plant functioning and act as a hindrance to plant's metabolic processes. Based on their solubility under physiological conditions, 17 HMs may be available for living cells and are of importance for living organisms. Among these metals, Zn, Cu, Fe, Mo, Mn, and Ni are referred to as essential micronutrients required for normal plant growth and development. V, Co, W, and Cr are toxic elements with high or low importance as trace elements. As, Ag, Cd, Hg, Pb, Sb, and Se are nonessential elements, have unknown biological function, and seem to be more or less toxic to plants and microorganisms.

Land used for crop production has been reported to be contaminated by HMs in many parts of the world (Singh et al. 2010; Foucault et al. 2013; Xiong et al. 2014). In India, most of the urban and densely populated cities with large industrial waste generation have been found to have contaminated soil. According to the European Environment Agency (EEA), 32 member countries of the European Union have reported around three million sites to be contaminated with HMs, of which more than 8% (or nearly 250,000 sites) are under highly contaminated category and need immediate remediation (Guimaraes et al. 2010). Shahid et al. (2015) reported on recent research that the total number of contaminated sites needing remediation may increase by more than 50% in 2025. Subsequently, the crops growing on contaminated soils have been shown to contain HMs beyond permissible limits, thereby posing a great threat to the food chain and also ultimately leading to reduction in crop productivity (Jarup 2003; Foucault et al. 2013).

Soil contaminated with HM can lead to uptake and accumulation of these toxic elements in plant parts, disrupting several physiological and biochemical pathways in a variety of ways. For example, disruption of protein structure and stability, reduced functionality of essential biomolecules such as enzymes and pigments, repression of major events in plant life such as chlorophyll synthesis, photosynthesis and respiration, disturbed water and mineral uptake, interrupted nitrogen metabolism, reduced root and shoot growth, chlorosis, and necrosis have been shown to occur in plants after excessive uptake of HMs (Hall 2002; Gill and Tuteja 2010; Cambrolle et al. 2011; Pourrut et al. 2011, 2013; Ali et al. 2013; Austruy et al. 2014; Emamverdian et al. 2015). Besides these direct impacts, elevated levels of HMs also lead to excessive accumulation of reactive oxygen species (ROS), non-free radical species (singlet oxygen and hydrogen peroxide), and cell toxicity that disturb the redox balance within the plant cells (Sytar et al. 2013). This results in oxidative stress, which in turn results in damage of protein and lipid molecules, ion leakage, and disruption of cell membrane, eventually ensuing the activation of programmed cell death (PCD) pathways (Flora et al. 2008; Nagajyoti et al. 2010; Sharma et al. 2012). To combat oxidative stress, cells are equipped with various defense mechanisms, and production of enzymatic/nonenzymatic antioxidants is one such mechanism. While enzymatic antioxidant defense machinery includes superoxide dismutase (SOD), catalase (CAT), ascorbate peroxidase (APX), glutathione reductase (GR), monodehydroascorbate reductase (MDHAR), dehydroascorbate reductase (DHAR), glutathione peroxidase (GPX), guaiacol peroxidase (GOPX), and glutathione *S*-transferase (GST), nonenzymatic antioxidants include glutathione (GSH), ascorbic acid (AsA), α-tocopherol, proline, and carotenoids (Gill and Tuteja 2010). Enzymatic antioxidants efficiently convert superoxide radicals into hydrogen peroxide and subsequently to water and oxygen, whereas nonenzymatic antioxidants are low molecular weight quenchers that directly detoxify ROS.

The efficiency of plants to survive under HM stress depends upon their ROS management machinery which is impeccably affected by rhizospheric microorganisms.

12.4 Mycorrhizoremediation: An Enhanced Form of Bioremediation

The restoration and reclamation of soil polluted with HMs require their characterization and remediation. The remediation of HM-contaminated soil is a difficult task as they are highly persistent and nondegradable elements. The appropriate and environment-friendly remedial techniques should be applied to clean up contaminated soils in order to avoid or lessen the risk of spreading of metal pollution to other ecosystems. Bioremediation is one of the most effective and innovative technologies that uses biological systems/processes, either natural or genetically engineered, for treatments and reduction or complete elimination of various kinds of toxic contaminants present in our ecosystem. It has emerged as an ecologically and economically sound solution for pollutant control in today's scenario. Bioremediation is a broad term that covers all those processes that utilize microorganisms, fungi (*mycoremediation*), green plants (*phytoremediation*), or their enzymes, to return the natural environment altered by contaminants to its original state, i.e., concept of *using nature to cleanse nature* (Etim 2012).

Various researches revealed that plants employ a range of inherent defense mechanisms to remove toxic HMs from the contaminated soil including *phytoextraction* (plants absorb contaminants from soil and accumulate them into their shoot tissue), *phytostabilization* (pollutant is being immobilized in the root zone), *phytovolatilization* (plants extract metals from soil and release them into the atmosphere by volatilization), *rhizofiltration* (plant roots absorb or adsorb metals from aqueous solution), and *phytostimulation* or *rhizodegradation* (breakdown of organic pollutants in the soil by utilizing rhizospheric associations with symbiotic soil microbes) (Garbisu and Alkorta 2001; Berti and Cunningham 2000; Thangavel and Subhuram 2004; Pilon-Smits 2005). The use of plants in conjunction with associated soil microbiota offers tremendous potential for rhizoremediation. The involvement of the rhizomicrobial population in remediation process is termed as rhizoremediation (Anderson et al. 1993; Schwab and Banks 1994; Kuiper et al. 2004). Most plants growing on HM-contaminated soils are reported to possess arbuscular mycorrhizae signifying that these AMF might have evolved high tolerance limit against HMs and thus can play crucial role in decontamination of soil polluted with HMs termed as *mycorrhizoremediation* (Khan 2006). Several research experiments conducted to evaluate the potential of AMF to extract HMs from contaminated soil establish the position of AMF as a potential bioremediation tool in the restoration of contaminated soil (Giasson 2005, 2006; Marques et al. 2007; Dudhane et al. 2012; Jarrah et al. 2014).

12.5 Unraveling the Role of VAM in Plant Adaptation to Environmental Stresses

The global environmental change, poor land management, and non-sustainable agricultural practices present a serious threat to soil fertility and have been anticipated to largely affect plant distribution, yield, and productivity in the twenty-first century. Since plant yield and productivity are closely connected to soil well-being, understanding the roles of soil microbiota can help in restoring degraded lands and soil. Arbuscular mycorrhizal fungi (AMF) are an integral part of the microbial soil community. AMF play a very significant role in the ecological improvement of rhizosphere and act as a living conduit between soil and plants (Azcón-Aguilar et al. 2003).

Arbuscular mycorrhiza is a mutually valuable biological alliance between the roots of higher plants and various fungal species in the phylum *Glomeromycota*. This mutualistic association is probably the most ecologically and agriculturally important symbiosis in terrestrial ecosystems. *Glomeromycota*, one of the oldest fungal lineages, is believed to advance the development of land plants from rootless gametophytes to rooted sporophytes amid the Paleozoic era (Remy et al. 1994; Redecker et al. 2000). They form an arbuscular mycorrhizal symbiotic association with the vast majority of plant species, including several members of important crop plant families such as Gramineae and Leguminosae. AMF help plants to adapt to antagonistic environmental conditions and increase plant tolerance to various abiotic stresses including mineral deficiency or imbalance, damage by pathogens, soil alkalinity or acidity, and toxicities associated with HMs. The extraradical mycelium of the symbiont acts as an extension of the root system, efficiently penetrates into the soil particles, and increases the uptake of water and minerals than the plant roots. There are repeated reports that AMF promote the uptake of key nutrients, particularly phosphorus, which is relatively inaccessible to the plants and also secretes organic acids and phosphatase enzymes that amplify the concentration of phosphorus available to plants (Miller 2000; Bucher 2007; Elbon and Whalen 2014). In addition, this interaction aids plant establishment and growth due to better nutrition, water availability, and soil aggregation properties despite elevated levels of HMs in the soil (Enkhtuya et al. 2002; Feng et al. 2003; Taylor and Harrier 2001; Auge 2001; Rillig and Steinberg 2002). The importance of AM symbiosis in stress tolerance has been reported by many recent researches and proposed as one of the many mechanisms of plant HM tolerance. It includes immobilization of metal compounds within the plant root tissues that contain the fungal cells, precipitation of polyphosphate granules in the soil, adsorption to fungal chitin cell walls, chelation of heavy metals inside the fungus cytosol, and sequestration in the vacuolar compartment (Hall 2002; Hildebrandt et al. 2007; Joachim et al. 2009). The fine fungal hyphae binding in the soil are an important sink for heavy metals due to their large surface area.

Recent studies have emphasized on the use of VAM fungi as potential biocontrol agents in modern sustainable agriculture as they increase the resistance of host plant against soilborne plant pathogen (Whipps 2004; Jung et al. 2012). Different mechanisms have been proposed to play a role in the biocontrol effect of AMF: (a) creating a mechanical barrier to pathogen penetration and subsequent spread as in case of sheathing mycorrhiza; (b) thickening of cell wall through lignification and the production of other polysaccharides, which in turn hinder the entry of root pathogens (Dehne and Schoenbeck 1979); (c) stimulation of the host roots to produce and accumulate sufficient concentration of metabolites, e.g., terpenes, phenols, flavonoids, antibiotics, etc., which impart resistance to the host tissue against pathogen invasion (Krupa et al. 1973; Sampangi 1989; Strobel and Sinclair 1991); and (d) competing with the pathogens for the uptake of essential nutrients in the rhizosphere and at the root surface (Reid 1990). Studies conducted on VAM fungi and onion and tomato plants during the last few decades revealed that VAM can also be used as potential biofertilizer which may reduce the needs of current levels of chemical fertilizer in agriculture industry and thus help in restoration of disturbed ecosystems in the near future (Gray and Gerdemann 1969; Kim et al. 1997).

12.6 The Protective Role of VAM Against HM Stress

Several recent studies indicate that the fitness benefits conferred by colonization of roots with AMF contribute in alleviating the heavy metal-induced stress in the host plants (Tao 1997; Göhre and Paszkowski 2006; Lingua et al. 2008; Cornejo et al. 2013; Tamayo et al. 2014; Meier et al. 2015). For example, *Glycine max* (soybean) plants inoculated with AMF showed a reduced concentration of Zn, Cd, and Mn in plant leaves grown in soil with high levels of these HMs as compared to control plants (Heggo et al. 1990). The roots of a fern, *Pteridium aquilinum*, collected from Cd-treated soil demonstrated greater accumulation of Cd, Ti, and Ba in the cytoplasm of mycorrhizal fungi than that of the host cells of the fern (Turnau et al. 1993). Similarly, one of the experimental studies with *Zea mays* (maize) grown in soil contaminated with Cd, Cu, Zn, and Mn revealed that the concentrations of heavy metals were considerably lower in the mycorrhizal plants than in the non-mycorrhizal plants, indicating a protective effect of mycorrhiza against HM toxicity (Weissenhorn et al. 1995). Examination of *Euphorbia cyparissias* plants collected from Zn waste sites showed that crystalloids deposited within cortical cells and around fungal hyphae contained higher concentrations of Zn than root cell wall and fungal structures (Turnau 1998). The translocation of Cd, Ni, and Zn from soil to different parts of the plant was significantly reduced in ryegrass (*Lolium perenne* L.) plants with mycorrhizal symbiosis as compared to control plants due to immobilization of HMs in soil by AMF (Takács et al. 2001). Similar results were obtained with green gram (*Vigna radiata*) grown in soil contaminated with Zn (Shivakumar et al. 2011). Various other experimental studies have also proved that arbuscular mycorrhizal plants may display higher plant biomass and much lower concentrations of Zn in shoots as compared to non-mycorrhizal controls in soil contaminated with Zn, suggesting the protective role of mycorrhizas against Zn toxicity on the host plants (Diaz et al. 1996; Liu et al. 2000; Li and Christie 2001). VAM symbiosis has been shown to improve gas exchange capacity of host plants by a variety of mechanisms such as maintaining open stomata, increasing transpiration flux, and reducing stomatal resistance. Also VAM symbiosis helped *Robinia* plants to cope with Pb stress by increasing antioxidant enzyme activities (Yang et al. 2015).

Several physical and biological mechanisms have been proposed to explain metal tolerance of AMF and its contribution in annulling the impacts of HM stress on plants:

(a) AMF can act as a physical barrier when plants are faced with HM stress (Emamverdian et al. 2015).
(b) By depositing metals within cortical cells of the root containing the fungal structures/arbuscules (Turnau 1998; Kaldorf et al. 1999).
(c) Immobilization of metals in the fungal biomass or adsorption of metals to the fungal cell wall as chitin has an important metal-binding capacity or sequestering them in their vacuole or compartmentalization into plastids or other membrane-rich organelles (Turnau et al. 1993; Joner and Leyval 1997; Zhou 1999; Joner et al. 2000a, Zhu et al. 2001; Hall 2002).
(d) Immobilization in the rhizosphere through precipitation or chelation of metals in the soil matrix by producing glycoprotein, siderophores, metallothioneins, etc. or due to intracellular precipitation of metallic cations with PO_4^- making phosphate-metal complexes inside the hyphae (Joner and Leyval 1997; Joner et al. 2000b; Tullio et al. 2003; Aloui et al. 2011).
(e) By encoding proteins involved in HM stress tolerance such as heat-shock protein and glutathione (Hildebrandt et al. 2007).
(f) Diluting the strength of metals by increased root and shoot growth, i.e., by higher biomass production (Mohammadi et al. 2011).
(g) By indirect effect of AM fungi on rhizosphere characteristics such as changes in soil pH, microbial communities, and root exudation patterns (Li et al. 1991; Olsson et al. 1998; Laheurte et al. 1990).

A pictorial representation regarding effect of VAM on plants is demonstrated in Fig. 12.1.

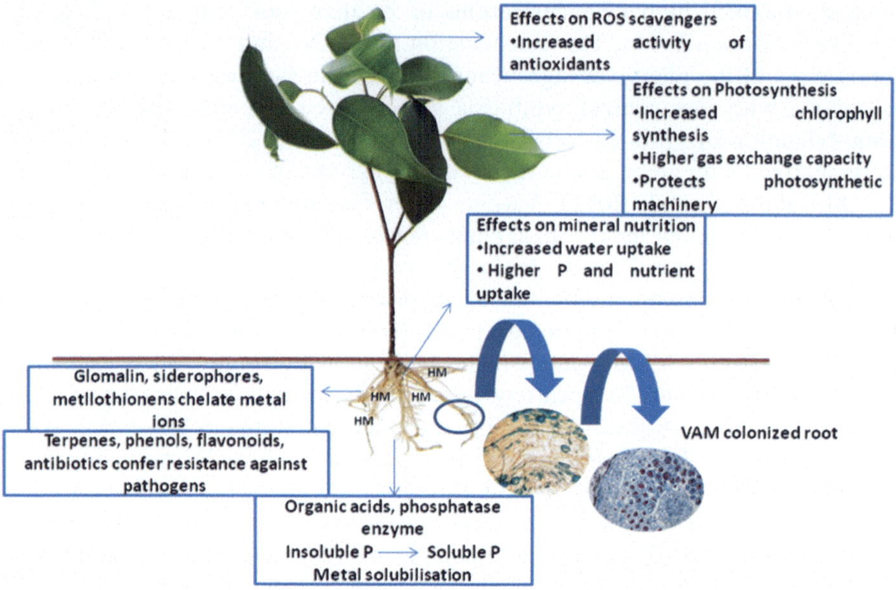

Fig. 12.1 Effects of VAM inoculation on plants grown under heavy metal stress. Mycorrhizal plants show better nutrient uptake, higher photosynthesis, and antioxidant activities compared to non-mycorrhizal plants. VAM inoculation leads to secretion of variety of molecules that confer resistance to plants and chelate metal ions to phytostablize them. (References: Auge 2001; Feng et al. 2003; Yang et al. 2015; Image Source: Bittman et al. 2004; Mark Perkins-Arbuscular mycorrhizal fungus (AMF) ~100×)

12.7 Glomalin: A Natural Product from VAM

An organic compound "glomalin" was discovered by the US Department of Agriculture (USDA) soil scientist Sara F. Wright in 1996. Glomalin is a stable and persistent glycoprotein that is produced by the hyphae of *Glomerales* AMF and has been detected in large amounts in numerous soils (Wright and Upadhyaya 1998; Nichols 2003). The glomalin structure has not been fully characterized, but it appears to be a complex of repeated monomeric structures bound together by hydrophobic interactions (Nichols 2003). Glomalin has a significant beneficial effect on soil structure. It contains substantial carbon and enhances soil aggregation and soil aeration which in turn lead to improved microbial activity and soil fertility. AMF can bind HMs beyond the plant rhizosphere by releasing glomalin (Gonzalez-Chavez et al. 2004; Gohre and Paszkowski 2006). Glomalin has been defined as an alkaline-soluble protein material associated to AMF and quantified in soil as glomalin-related soil protein (GRSP) (Rillig 2004; Nichols and Wright 2006). Gonzalez-Chavez et al. (2004) showed that glomalin pool (GRSP) in the soil has the capability to sequestrate potentially toxic elements (Cu, Cd, and Pb) and suggested that glomalin could be effective for heavy metal biostabilization in polluted soil. Cornejo et al. (2008) reported that GRSP binds to about 27% Cu and 5.8% of Zn in a soil highly polluted with these heavy metals and suggests that the higher the concentration of the pollutant, the higher is the ability of GRSP to bind to the pollutants and make them unavailable. Malekzadeh et al. (2016) examined the influence of Cd concentrations on glomalin production and its role in the sequestration of Cd. The results revealed that Cd sequestrated by hyphal glomalin was significantly increased as the levels of Cd increased indicating that the stimulated production of glomalin by AM fungus under Cd-induced stress could be a defensive mechanism against the toxic effect of Cd. All the above studies indicate that glomalin may be of great utility to reduce the bioavailability of toxic elements and could be an asset in bioremediation of HM-contaminated soils.

12.8 Genomics of VAM

Despite being ubiquitous in nature, it took almost a century to document the ecological and agronomical importance of VAM. VAM's obligate biotrophic nature makes it difficult to culture it in the laboratory. Also, its direct genetic studies are difficult owing to its asexual nature, unknown ploidy level, and lack of manipulability (Malbreil et al. 2014). Similarly, the aseptate nature led to a limited morphological description of VAM, restricted up to mycelium and spore organization only. The first evidence of their mutualistic nature came out in the mid of the nineteenth century, when Barbara Mosse obtained mycorrhized plants de novo and provided a gateway to further research by optimizing in vitro culture of the AM fungus *Funneliformis mossae* (Mosse 1953; Mosse and Hepper 1975). This proved to be a milestone since it fulfilled the requirement of biological material produced in vitro for genomic studies. It led to the description of developmental studies in well-sporulating isolate DAOM 197198 of *Rhizophagus irregularis* (formerly *Glomus intraradices*) around 1990s. The last few years have evidenced enormous physiological, taxonomic, phylogenic, and genetic accounts of this symbiosis. Generation of molecular and transcriptomic data has led to the taxonomic repositioning of AM fungi into an independent group; *Glomeromycota* concerning around 29 genera and 230 species with some taxa are still under revision (Schußler et al. 2001; Redecker et al. 2013).

However, despite various studies the genetic status of *Glomeromycetes* is still unclear, and conflicting reports have been presented. Identification of multiple variants of genetic markers within each spore has led to an array of heterokaryotic, polyploidy, and gene duplication hypotheses pertaining to homogeneity of all nuclei in a spore (Hijri and Sanders 2005; Pawlowska 2005). The estimation of genome size of *Glomeromycota* ranges from 0.18 to 1.08 pg of DNA per nucleus, i.e., approximately 176 MB to 1GB which is largest among the fungi, of which the genome size of *Rhizophagus irregularis* isolate DAOM 197198 remains among the lowest (Hosny 1998). Since, it is the most studied AM fungus and can sporulate heavily in vitro, it was chosen for genome sequencing in the framework of the poplar mesocosm analysis and hence became the model organism (Martin et al. 2004). The genome sequencing program was supported by the Joint Genome Initiative and funded by the US Department of Energy. By virtue of the latest bioinformatic tools and next-generation sequencing technique (suitable for sequencing low biomass producing discrete organisms), the first version of the genome was released in April 2013 (Tisserant et al. 2013).

Gene regulation study of AM fungi is constrained by features specific to it. Similarly, the lack of possibility to obtain stable transformants constrains the applicability of forward and reverse genetics. Helber et al. (2011) performed host-induced gene silencing of a candidate *Rhizophagus* gene for its functional validation. Such tools being utilized after the release of the *Rhizophagus* gene repertoire are acting as a gateway to functional genomics. Several new fungal genes involved in symbiosis, fungal metabolism, nutrient acquisition, and heavy metal homeostasis are being identified. The main benefit that plants obtain by associating with AM fungi is nitrogen and phosphate acquisition, which is mainly due to the extensive network of extraradical mycelium of the fungi. Two high-affinity nitrogen transporters have been identified in *R. irregularis*, namely, GinAMT1 and GinAMT2 (Lopez-Pedrosa et al. 2006; Pérez-Tienda et al. 2011). Similarly, candidate genes responsible for alkaline phosphatase activity have also been determined. The first one to be identified was in *Glomus versiforme* (Harrison and van Buuren 1995). At least four kinds of putative phosphatases are found to be expressed in *R. irregularis* (Tisserant et al. 2013).

Metal transporters play a major role in heavy metal homeostasis. Several putative genes encoding for Cu, Fe, and Zn transporters have been recently identified in *Rhizophagus irregularis* through genome-wide analysis (Tamayo et al. 2014). Lanfranco et al. (2002) reported upregulation of a metallothionein gene of *Gigaspora margarita* (BEG 34) in symbiotic mycelia under Cu exposure. Similarly, increased transcript levels of a putative Zn transporter gene (*GintZnT1*) were observed in the mycelium of *G. intraradices* under short- and longtime exposure to Zn, indicating a possible function of this gene in protection against Zn stress (González-Guerrero et al. 2005). González-Guerrero et al. (2006) reported the Cd- and Cu- dependent upregulation of a putative ABC transporter gene (*GintABC1*) in the extraradical mycelium of *G. intraradices*, indicating *GintABC1* role in Cd and Cu detoxification in the extraradical mycelia of *G. intraradices*. The considerably enhanced expression of metallothionein and metal transporter gene, particularly under HM stress, indicates their role in HM tolerance. Various researches have reported a noteworthy effect of AMF on gene expression pattern of several plants by coding for proteins apparently involved in HM tolerance and detoxification (Repetto et al. 2003; Rivera-Becerril et al. 2005; Hildebrandt et al. 2007). Hildebrandt et al. 2006 reported the HM-dependent induction of genes encoding heat-shock protein and glutathione S-transferase in the mycelium of *G. intraradices* Sy167, well-known for their role in alleviating HM stress in plants.

12.9 Conclusion and Future Prospects

Anthropogenic activities and unsustainable exploitation of earth's biological resources have accelerated the accumulation of metals in the soil. As a result, plants are now exposed to high level of HM toxicity than ever before. Therefore, it is necessary to look out for plants and associated microbiota that can tolerate HMs and accumulate high concentrations of metals in their tissues and, thus, can be used for soil remediation. AMF regulate plant growth and development and contribute genes and natural products that facilitate plants to adapt and survive in stressed environments. The protection and enhanced capability of uptake of water and minerals provided by the AMF result in greater plant biomass, a prerequisite for successful use of plants in bioremediation. Identification and isolation of indigenous AMF existing naturally in heavy metal-polluted soils and understanding their genomics could lead to the discovery of valuable symbionts that can contribute in improvement of plant capabilities growing under HM stress conditions and in turn can be used in mainstream soil remediation practices. Thus, in order to guarantee the effectiveness of AM symbiosis in cleanup of metal-contaminated soils, it is very important to screen indigenous and heavy metal-tolerant isolates. Careful selection of compatible host/fungus/substrate combinations would allow more appropriate management of mycorrhizae in poor soils and would allow a substantial reduction in the amount of metals present in the soil leading to better soil management. Since AM fungi are asexual organisms and refractory to transformation, transgenic approaches cannot be undertaken to improve fungal phytoremediation properties. Thus it is essential to understand physio-biochemical and molecular events during interaction of host and AM symbionts to effectively employ them for phytoremediation and to unravel novel fungal genes suitable for introduction into plant backgrounds in order to tolerate HM stress.

References

Akiyama K, Matsuzaki K, Hayashi H (2005) Plant sesquiterpenes induce hyphal branching in arbuscular mycorrhizal fungi. Nature 435:824–827

Ali H, Khan E, Sajad MA (2013) Phytoremediation of heavy metals—concepts and applications. Chemosphere 91:869–881

Aloui A, Recorbet G, Robert F et al (2011) Arbuscular mycorrhizal symbiosis elicits shoot proteome changes that are modified during cadmium stress alleviation in *Medicago truncatula*. BMC Plant Biol. https://doi.org/10.1186/1471-2229-11-75

Anderson TA, Guthrie EA, Walton BT (1993) Bioremediation in the rhizosphere. Environ Sci Technol 27:2630–2636

Auge RM (2001) Water relations, drought and vesicular-arbuscular mycorrhizal symbiosis. Mycorrhiza 11:3–42

Austruy A, Shahid M, Xiong T et al (2014) Mechanisms of metal-phosphates formation in the rhizosphere soils of pea and tomato: environmental and sanitary consequences. J Soils Sediments 14:666–678

Azcón-Aguilar C, Palenzuela J, Roldán A et al (2003) Analysis of the mycorrhizal potential in the rhizosphere of representative plant species from desertification- threatened Mediterranean shrublands. Appl Soil Ecol 22:29–37
Barrow JR, Lucero ME, Reyes-Vera I et al (2008) Do symbiotic microbes have a role in plant evolution, performance and response to stress? Commun Integr Biol 1:69–73
Bécard G, Pfeffer PE (1993) Status of nuclear division in arbuscular mycorrhizal fungi during in-vitro development. Protoplasma 174:62–68
Berg G, Grube M, Schloter M (2014) Unraveling the plant microbiome: looking back and future perspectives. Front Microbiol 5:148–152
Berti WR, Cunningham SD (2000) Phytostabilization of metals. In: Raskin I, Ensley BD (eds) Phytoremediation of toxic metals: using plants to clean-up the environment. Wiley, New York, pp 71–88
Bittman S, Hunt D, Kowalenko CG et al (2004) Early phosphorus nutrition in corn and the role of mycorrhizae. In: Bittman S, Kowalenko CG (eds) Advanced silage corn management: a production guide for coastal British Columbia and the Pacific Northwest
Brundrett MC (1991) Mycorrhizas in natural ecosystems. In: Macfayden A, Begon M, Fitter AH (eds) Advances in ecological research, vol 21. Academic, London, pp 171–313
Bucher M (2007) Functional biology of plant phosphate uptake at root and mycorrhiza interfaces. New Phytol 173(1):11–26
Cambrolle J, Mateos-Naranjo E, Redondo-Gomez S et al (2011) Growth, reproductive and photosynthetic responses to copper in the yellow-horned poppy, *Glaucium flavum* Crantz. Environ Exp Bot 71:57–64
Chaparro JM, Badri VD, Vivanco MJ (2014) Rhizosphere microbiome assemblage is affected by plant development. ISME J 8(4):790–803. https://doi.org/10.1038/ismej.2013.196
Cornejo P, Meier S, Borie G et al (2008) Glomalin-related soil protein in a Mediterranean ecosystem affected by a copper smelter and its contribution to Cu and Zn sequestration. Sci Total Environ 406:154–160
Cornejo P, Pérez-Tienda J, Meier S et al (2013) Copper compartmentalization in spores as a survival strategy of arbuscular mycorrhizal fungi in Cu-polluted environments. Soil Biol Biochem 57:25–928
Dehne HW, Schoenbeck F (1979) The influence of endotrophic mycorrhizae on plant disease. I. Colonization of tomato plants by *Fusarium oxysporum* f.sp. *lycopersici*. Phytopathology 95:105–110
Diaz G, Azcón-Aguilar C, Honrubia M (1996) Influence of arbuscular mycorrhizae on heavy metal (Zn and Pb) uptake and growth of *Lygeum spartum* and *Anthyllis cytisoides*. Plant Soil 180:241–249
Dudhane M, Borde M, Jite PK (2012) Effect of aluminium toxicity on growth responses and antioxidant activities in *Gmelina arborea* Roxb inoculated with AM Fungi. Int J Phytoremediation 14:643–655
Elbon A, Whalen JK (2014) Phosphorus supply to vegetable crops from arbuscular mycorrhizal fungi: a review. Biol Agric Hortic 31:73–90
Emamverdian A, Ding Y, Mokhberdoran F et al (2015) Heavy metal stress and some mechanisms of plant defense response. Sci World J 2015:1–18. https://doi.org/10.1155/2015/756120
Enkhtuya B, Rydlová J, Vosátka M (2002) Effectiveness of indigenous and non-indigenous isolates of arbuscular mycorrhizal fungi in soils from degraded ecosystems and man-made habitats. Appl Soil Ecol 14:201–211
Etim EE (2012) Phytoremediation and its mechanism: a review. Int J Environ Bioener 2:120–136
Feng G, Song YC, Li XL et al (2003) Contribution of arbuscular mycorrhizal fungi to utilization of organic sources of phosphorus by red clover in a calcareous soil. Appl Soil Ecol 22:139–148
Flora SJS, Mittal M, Mehta A (2008) Heavy metal induced oxidative stress and its possible reversal by chelation therapy. Indian J Med Res 128:501–523
Foucault Y, Leveque T, Xiong T et al (2013) Green manure plants for remediation of soils polluted by metals and metalloids: ecotoxicity and human bioavailability assessment. Chemosphere 93:1430–1435

Garbisu C, Alkorta I (2001) Phytoextraction: a cost-effective plant-based technology for the removal of metals from the environment. Bioresour Technol 77:229–236

Giasson P, Jaouich A, Gagné S et al (2005) Phytoremediation of zinc and cadmium: a study of arbuscular mycorrhizal hyphae. Remediation 15:113–122

Giasson P, Jaouich A, Cayer P et al (2006) Enhanced phytoremediation: a study of heavy mycorrhizoremediation metal–contaminated soil. Remediation 17:97–110

Gill SS, Tuteja N (2010) Reactive oxygen species and antioxidant machinery in abiotic stress tolerance in crop plants. Plant Physiol Biochem 48:909–930

Göhre V, Paszkowski U (2006) Contribution of the arbuscular mycorrhizal symbiosis to heavy metal phytoremediation. Planta 223:1115–1122

González-Chávez C, Carrillo-González R, Wright SF et al (2004) The role of glomalin, a protein produced by arbuscular mycorrhizal fungi in sequestering potentially toxic elements. Environ Pollut 130:317–323

González-Guerrero M, Azcón-Aguilar C, Mooney M et al (2005) Characterization of a *Glomus intraradices* gene encoding a putative Zn transporter of the cation diffusion facilitator family. Fungal Genet Biol 42:130–140

González-Guerrero M, Azcón-Aguilar C, Ferrol N (2006) GintABC1 and GintMT1 are involved in Cu and Cd homeostasis in *Glomus intraradices*. In: Abstracts of the 5th international conference on mycorrhiza, Granada, Spain

Gray LE, Gerdemann JW (1969) Uptake of phosphorus-32 by vesicular-arbuscular mycorrhizae. Plant Soil 30:415–422. https://doi.org/10.1007/BF01881967

Grayston SJ, Wang S, Campbell CD et al (1998) Selective influence of plant species on microbial diversity in the rhizosphere. Soil Biol Biochem 30:369–378

Guimaraes BCM, Arends JBA, van der Ha D et al (2010) Microbial services and their management: recent progresses in soil bioremediation technology. Appl Soil Ecol 46:157–167

Gutjahr C, Parniske M (2013) Cell and developmental biology of arbuscular mycorrhiza symbiosis. Annu Rev Cell Dev Biol 29:593–617

Hall JL (2002) Cellular mechanisms for heavy metal detoxification and tolerance. J Exp Bot 53:1–11

Hardoim PR, van Overbeek LS, Berg G et al (2015) The hidden world within plants: ecological and evolutionary considerations for defining functioning of microbial endophytes. Microbiol Mol Biol Rev 79:293–320. https://doi.org/10.1128/MMBR.00050-14

Harley JL (1971) Fungi in ecosystems. J Appl Ecol 8:627–642

Harley JL (1989) The significance of mycorrhiza. Mycol Res 92:129–139

Harley JL, Smith SE (1983) Mycorrhizal symbiosis. Academic Press, Toronto, pp 112–115

Harrier L (2001) The arbuscular mycorrhizal symbiosis: a molecular review of the fungal dimension. J Exp Bot 52:469–478

Harrison MJ, van Buuren ML (1995) A phosphate transporter from the mycorrhizal fungus *Glomus versiforme*. Nature 378:626–629

Heggo A, Angle JS, Chaney RL (1990) Effects of vesicular arbuscular mycorrhizal fungi on heavy-metal uptake by soybeans. Soil Biol Biochem 22:865–869

Helber N, Wippel K, Sauer N et al (2011) A versatile monosaccharide transporter that operates in the arbuscular mycorrhizal fungus Glomus sp is crucial for the symbiotic relationship with plants. Plant Cell 23:3812–3823

Hijri M, Sanders IR (2005) Low gene copy number shows that arbuscular mycorrhizal fungi inherit genetically different nuclei. Nature 433:160–163

Hildebrandt U, Ouziad F, Marner FJ, Bothe H (2006) The bacterium *Paenibacillus validus* stimulates growth of the arbuscular mycorrhizal fungus *Glomus intraradices* up to the formation of fertile spores. FEMS Microbiol Lett 254:258–267

Hildebrandt U, Regvar M, Bothe H (2007) Arbuscular mycorrhiza and heavy metal tolerance. Phytochemistry 681:139–146

Hosny M, Gianinazzi-Pearson V, Dulieu H (1998) Nuclear DNA content of 11 fungal species in glomales. Genome 41:422–428

Jarrah M, Ghasemi-Fasaei R, Karimian N et al (2014) Investigation of *Arbuscular mycorrhizal* fungus and EDTA efficiencies on lead phytoremediation by sunflower in a calcareous soil. Biorem J 18:71–79

Jarup L (2003) Hazards of heavy metal contamination. Br Med Bull 68:167–182

Joachim HJ, Makoi R, Ndakidemi PA (2009) The agronomic potential of vesicular-arbuscular mycorrhiza (VAM) in cereals-legume mixtures in Africa. Afr J Microbiol Res 3:664–675

Joner EJ, Leyval C (1997) Uptake of ^{109}Cd by roots and hyphae of a *Glomus mosseae/Trifolium subterraneum* mycorrhiza from soil amended with high and low concentrations of cadmium. New Phytol 135:353–360

Joner EJ, Leyval C, Briones R (2000a) Metal binding capacity of arbuscular mycorrhizal mycelium. Biol Fertil Soils 226:227–234

Joner EJ, Ravnskov S, Jakobsen I (2000b) Arbuscular mycorrhizal phosphate transport under monoxenic conditions using radiolabeled inorganic and organic phosphate. Biotechnol Lett 22:1705–1708

Jung SC, Martinez-Medina A, Lopez-Raez JA et al (2012) Mycorrhiza-induced resistance and priming of plant defenses. J Chem Ecol 38(6):651–664

Kaldorf M, Kuhn AJ, Schroder WH et al (1999) Selective element deposits in maize colonized by a heavy metal tolerance conferring arbuscular mycorrhizal fungus. J Plant Physiol 154:718–728

Kendrick B (1985) The fifth kingdom. Mycologue Publications, Waterloo

Khan AG (2006) Mycorrhizoremediation-an enhanced form of phytoremediation. J Zhejiang Univ Sc B 7:503–514

Kim KY, Jordan D, McDonald GA (1997) Effect of phosphate-solubilizing bacteria and vesicular-arbuscular mycorrhizae on tomato growth and soil microbial activity. Biol Fertil Soils 26:79–87

Knack JJ, Wilcox LW, Ané JM et al (2015) Microbiomes of streptophyte algae and bryophytes suggest that a functional suite of microbiota fostered plant colonization of land. Int J Plant Sci 176:405–420. https://doi.org/10.1086/681161

Krings M, Taylor TN, Hass H et al (2007) Fungal endophytes in a 400 million-yr-old land plant: infection pathways, spatial distribution and host responses. New Phytol 174:648–657

Krupa S, Anderson J, Marx DH (1973) Studies on ectomycorrhizae of pine. IV. Volatile organic compounds in mycorrhizal and nonmycorrhizal root systems of *Pinus echinata* Mill. Eur J For Pathol 3:194–200

Kuhn G, Hijri M, Sanders IR (2001) Evidence for the evolution of multiple genomes in arbuscular mycorrhizal fungi. Nature 414:745–748

Kuiper I, Lagendijk EL, Bloemberg GV et al (2004) Rhizoremediation: a beneficial plant-microbe interaction. Mol Plant Microb Interact 17:6–15

Laheurte F, Leyval C, Berthelin J (1990) Root exudates of maize, pine and beech seedlings influenced by mycorrhizal and bacterial inoculation. Symbiosis 9:111–116

Lanfranco L, Bolchi A, Ros EC et al (2002) Differential expression of a metallothionein gene during the presymbiotic versus the symbiotic phase of an arbuscular mycorrhizal fungus. Plant Physiol 130:58–67

Li XL, Christie P (2001) Changes in soil solution Zn and pH and uptake of Zn by arbuscular mycorrhizal red clover in Zn-contaminated soil. Chemosphere 42:201–207

Li XL, George E, Marschner H (1991) Phosphorus depletion and pH decrease at the root-soil and hyphae-soil interfaces of VA mycorrhizal white clover fertilized with ammonium. New Phytol 119:397–404

Lingua G, Franchin C, Todeschini V et al (2008) Arbuscular mycorrhizal fungi differentially affect the response to high zinc concentrations of two registered poplar clones. Environ Pollut 153:137–147

Liu A, Hamel C, Hamilton RI et al (2000) Acquisition of Cu, Zn, Mn and Fe by mycorrhizal maize (*Zea mays* L.) grown in soil at different P and micronutrient levels. Mycorrhiza 9:331–336

López-Pedrosa A, González-Guerrero M, Valderas A (2006) GintAMT1 encodes a functional high-affinity ammonium transporter that is expressed in the extraradical mycelium of *Glomus intraradices*. Fungal Genet Biol 43:102–110

Lugtenberg B, Kamilova F (2009) Plant-growth promoting rhizobacteria. Annu Rev Microbiol 63:541–556. https://doi.org/10.1146/annurev.micro.62.081307.162918
Malbreil M, Tisserant E, Martin F et al (2014) Genomics of arbuscular mycorrhizal fungi: out of the shadows. Adv Bot Res 70:259–290
Malekzadeh E, Aliasgharzad N, Majidi J et al (2016) Cd-induced production of glomalin by arbuscular mycorrhizal fungus (*Rhizophagus irregularis*) as estimated by monoclonal antibody assay. Environ Sci Pollut Res 23(20):20711–20718. https://doi.org/10.1007/s11356-016-7283-z
Marques AP, Oliveira RS, Samardjieva KA et al (2007) *Solanum nigrum* grown in contaminated soils. Effects of AMF on zinc accumulation and histolocalisation. Environ Pollut 145:691–699
Martin F, Tuskan GA, DiFazio SP et al (2004) Symbiotic sequencing for the Populus mesocosm. New Phytol 161:330–335
Meier S, Cornejo P, Cartes P et al (2015) Interactive effect between Cu-adapted arbuscular mycorrhizal fungi and biotreated agrowaste residue to improve the nutritional status of *Oenothera picensis* growing in Cu-polluted soils. J Plant Nutr Soil Sci 178:126–135
Miller MH (2000) Arbuscular mycorrhizae and the phosphorus nutrition of maize: a review of Guelph studies. Can J Plant Pathol Sci 80(1):47–52
Mohammadi K, Khalesro S, Sohrabi Y et al (2011) A review: beneficial effects of the mycorrhizal fungi for plant growth. J Appl Environ Biol Sci 9:310–319
Morton JB (1988) Taxonomy of VA mycorrhizal fungi: classification, nomenclature and identification. Mycotaxon 32:267–324
Morton JB, Benny GL (1990) Revised classification of arbuscular mycorrhizal fungi (Zygomycetes): a new order, Glomales, two new suborders Glomineae and Gigasporineae and two new families Acaulosporaceae and Gigasporaceae with an emendation of Glomaceae. Mycotaxon 37:471–491
Morton JB, Redecker D (2001) Two new families of *Glomales*, *Archaeosporaceae* and *Paraglomaceae*, with two new genera *Archaeospora* and *Paraglomus*, based on concordant molecular and morphological characters. Mycologia 93:181–195
Mosse B (1953) Fructifications associated with mycorrhizal strawberry roots. Nature 171:974
Mosse B, Hepper C (1975) Vesicular-arbuscular mycorrhizal infections in root organ cultures. Physiol Plant Pathol 5:215–223
Nagajyoti PC, Lee KD, Sreekanth TVM (2010) Heavy metals, occurrence and toxicity for plants: a review. Environ Chem Lett 8:199–216
Nichols K (2003) Characterization of glomalin-a glycoprotein produced by arbuscular mycorrhizal fungi. Ph. D Dissertation. University of Maryland, College Park, Maryland
Nichols KA, Wright SF (2006) Carbon and nitrogen in operationally defined soil organic matter pools. Biol Fertil Soils 43:215–220
Oehl F, Sieverding E (2004) *Pacispora*, a new vesicular-arbuscular mycorrhizal fungal genus in the *Glomeromycetes*. J Appl Bot Food Qual 78:72–82
Oehl F, Souza FA, Sieverding E (2008) Revision of *Scutellospora* and description of five new genera and three new families in the arbuscular mycorrhizal-forming *Glomeromycetes*. Mycotaxon 106:311–360
Oehl F, Silva GA, Goto BT et al (2011a) *Glomeromycetes*: three new genera and glomoid species reorganized. Mycotaxon 116:75–120
Oehl F, Silva DKA, Maia LC et al (2011b) *Orbispora* gen. nov., ancestral in the *Scutellosporaceae* (*Glomeromycetes*). Mycotaxon 116:161–169
Oehl F, Silva GA, Goto BT et al (2011c) *Glomeromycota*: two new classes and a new order. Mycotaxon 116:365–379
Oehl F, Silva GA, Sanchez-Castro I et al (2011d) Revision of *Glomeromycetes* with entrophosporoid and glomoid spore formation with three new genera. Mycotaxon 117:297–316
Oehl F, Sieverding E, Palenzuela J et al (2011e) Advances in *Glomeromycota* taxonomy and classification. IMA Fungus: The Global Mycological Journal 2(2):191–199
Olsson PA, Francis R, Read DJ et al (1998) Growth of arbuscular mycorrhizal mycelium in calcareous dune sand and its interaction with other soil microorganisms as estimated by measurement of specific fatty acids. Plant Soil 201:9–16

Palenzuela J, Ferrol N, Boller T et al (2008) *Otospora bareai*, a new fungal species in the Glomeromycetes from a dolomitic shrub land in Sierra de Baza National Park (Granada, Spain). Mycologia 100(2):296–305

Partida-Martínez LP, Heil M (2011) The microbe-free plant: factor artifact? Front Plant Sci 2:100. https://doi.org/10.3389/fpls.2011.00100

Pawlowska TE (2005) Genetic processes in arbuscular mycorrhizal fungi. FEMS Microbiol Lett 251:85–192

Pérez-Tienda J, Testillano PS, Balestrini R et al (2011) GintAMT2, a new member of the ammonium transporter family in the arbuscular mycorrhizal fungus *Glomus intraradices*. Fungal Genet Biol 48:1044–1055

Petrini O (1991) Fungal endophytes of tree leaves. In: Andrews JH, Hirano SS (eds) Microbial ecology of leaves. Springer, New York, pp 179–197. https://doi.org/10.1007/978-1-4612-3168-4_9

Pilon-Smits E (2005) Phytoremediation. Annu Rev Plant Biol 56:15–39

Pirozynski KA, Dalpé Y (1989) Geological history of the Glomaceae with particular reference to mycorrhizal symbiosis. Symbiosis 7:1–36

Pirozynski KA, Malloch DW (1975) The origin of land plants a matter of mycotrophism. Biosystems 6:153–164

Pourrut B, Jean S, Silvestre J et al (2011) Lead-induced DNA damage in *Vicia faba* root cells: potential involvement of oxidative stress. Mutat Res 726:123–128

Pourrut B, Shahid M, Douay F et al (2013) Molecular mechanisms involved in lead uptake, toxicity and detoxification in higher plants. In: Gupta DK, Corpas FJ, Palma JM (eds) Heavy metal stress in plants. Springer, Berlin, pp 121–147

Rascio N, Navari-Izzo F (2011) Heavy metal hyperaccumulating plants: how and why do they do it? And what makes them so interesting? Plant Sci 180:169–181

Redecker D, Kodner R, Graham LE (2000) Glomalean fungi from the Ordovician. Science 289(5486):1920–1921

Redecker D, Schussler A, Stockinger H et al (2013) An evidence-based consensus for the classification of arbuscular mycorrhizal fungi (Glomeromycota). Mycorrhiza 23:515–531

Reid CPP (1990) Mycorrhizas. In: Lynch JM (ed) The rhizosphere. Wiley, Chichester, pp 281–315

Remy W, Taylor TN, Hass H et al (1994) Four hundred-million-year-old vesicular arbuscular mycorrhizae. Proc Natl Acad Sci U S A 91(25):11841–11843

Repetto O, Bestel-Corre G, Dumas-Gaudot E et al (2003) Targeted proteomics to identify cadmium-induced protein modifications in *Glomus mosseae*-inoculated pea roots. New Phytol 157:555–567

Rillig MC (2004) Arbuscular mycorrhizae, glomalin, and soil aggregation. Can J Soil Sci 84:355–363

Rillig MC, Steinberg PD (2002) Glomalin production by an arbuscular mycorrhizal fungus: a mechanism of habitat modification. Soil Biol Biochem 34:1371–1374

Rivera-Becerril F, van Tuinen D, Martin-Laurent F et al (2005) Molecular changes in *Pisum sativum* L. roots during arbuscular mycorrhiza buffering of cadmium stress. Mycorrhiza 16:51–60

Rodriguez RJ, Redman RS, Henson JM (2004) The role of fungal symbioses in the adaptation of plants to high stress environments. Mitig Adapt Strateg Glob Chang 9:261–272

Sampangi RK (1989) Some recent advances in the study of fungal root diseases. Ind Phytopth 22:1–17

Schüßler A, Schwarzott D, Walker C (2001) A new fungal phylum the *Glomeromycota*: phylogeny and evolution. Mycol Res 105:1413–1142

Schwab AP, Banks MK (1994) Biologically mediated dissipation of polyaromatic hydrocarbons in the root zone. In: Anderson TA, Coats JR (eds) Bioremediation through rhizosphere technology. American Chemical Society, Washington, DC, pp 132–141

Shahid M, Khalid S, Abbas G et al (2015) Heavy metal stress and crop productivity. In: Hakeem K (ed) Crop production and global environmental issues. Springer, Cham, pp 1–25

Sharma P, Jha AB, Dubey RS et al (2012) Reactive oxygen species, oxidative damage, and antioxidative defense mechanism in plants under stressful conditions. J Bot. Article ID 217037

Shivakumar CK, Hemavani C, Thippeswamy B et al (2011) Effect of inoculation with arbuscular mycorrhizal fungi on green gram grown in soil containing heavy metal zinc. Journal of Experimental Sciences 2:17–21

Sieverding E, Oehl F (2006) Revision of *Entrophospora* and description of *Kuklospora* and *Intraspora*, two new genera in the arbuscular mycorrhizal Glomeromycetes. J Appl Bot Food Qual 80:69–81

Singh R, Tripathi RD, Dwivedi S et al (2010) Lead bioaccumulation potential of an aquatic macrophyte *Najas indica* are related to antioxidant system. Bioresour Technol 101:3025–3032

Spain JL, Sieverding E, Oehl F (2006) Appendicispora, a new genus in the arbuscular mycorrhizal-forming *Glomeromycetes*, with a discussion of the genus Archaeospora. Mycotaxon 97:163–182

Strobel NE, Sinclair WA (1991) Role of flavonolic wall infusions in the resistance induced by *Laccaria bicolor* to *Fusarium oxysporum* in primary roots of Douglas fir. Phytopathology 81:420–425

Sytar O, Kumar A, Latowski D et al (2013) Heavy metal-induced oxidative damage, defense reactions, and detoxification mechanisms in plants. Acta Physiol Plant 35:985–999

Takács T, Biro B, Vörös I (2001) Arbuscular mycorrhizal effect on heavy metal uptake of ryegrass (*Lolium perenne* L.) in pot culture with polluted soil. In: Horst WWJ, Scheck MK, Bürkert A et al (eds) Plant nutrition: food security and sustainability of agro-ecosystems through basic and applied research, developments in plant and soil sciences. Kluwer Academic, Dordrecht, pp 480–481

Tamayo E, Gómez-Gallego T, Azcón-Aguilar C et al (2014) Genome-wide analysis of copper, iron and zinc transporters in the arbuscular mycorrhizal fungus *Rhizophagus irregularis*. Plant Traffic Transp 5:1–13

Tao HQ (1997) Effect of arbuscular mycorrhiza on resistance of red clover to heavy metal Zn and Cd pollution. MS thesis, China Agricultural University, Beijing

Taylor J, Harrier LA (2001) A comparison of development and mineral nutrition of micropropagated *Fragaria×ananassa* cv. Elvira (strawberry) when colonized by nine species of arbuscular mycorrhizal fungi. Appl Soil Ecol 18(3):205–215

Thangavel P, Subhuram CV (2004) Phytoextraction-role of hyper accumulators in metal contaminated soils. Proc Indian Natl Sci Acad 70(1):109–130

Tisserant E, Malbreil M, Kuo A et al (2013) Genome of an arbuscular mycorrhizal fungus provides insight into the oldest plant symbiosis. Proc Natl Acad Sci 110:20117–20122

Tullio M, Pierandrei F, Salerno A et al (2003) Tolerance to cadmium of vesicular arbuscular mycorrhizae spores isolated from a cadmium-polluted and unpolluted soil. Biol Fertil Soils 37:211–214

Turnau K (1998) Heavy metal content and localization in mycorrhizal *Euphorbia cyparissias* from zinc wastes in southern Poland. Acta Soc Bot Poloniae 67:105–113

Turnau K, Kottke I, Oberwinkler F (1993) Element localization in mycorrhizal roots of *Pteridium aquilinum* (L.) Kuhn collected from experimental plots treated with cadmium dust. New Phytol 123:313–324

Walker C, Schüßler A (2004) Nomenclature clarifications and new taxa in the *Glomeromycota*. Mycol Res 108:979–982

Weissenhorn I, Leyval C, Belgy G et al (1995) Arbuscular mycorrhizal contribution to heavy metal uptake by maize (*Zea mays* L.) in pot culture with contaminated soil. Mycorrhiza 5:245–251

Weller DM, Raaijmakers JM, Gardener BB et al (2002) Microbial populations responsible for specific soil suppressiveness to plant pathogens. Annu Rev Phytopathol 40:309–348. https://doi.org/10.1146/annurev.phyto.40.030402.110010

Whipps JM (2004) Prospects and limitations for mycorrhizas in biocontrol of root pathogens. Can J Bot 1227:1198–1227

Wright SF, Upadhyaya A (1998) A survey of soils for aggregate stability and glomalin, a glycoprotein produced by hyphae of arbuscular mycorrhizal fungi. Plant Soil 198(1):97–107. https://doi.org/10.1023/A:1004347701584

Xiong T, Leveque T, Shahid M et al (2014) Lead and cadmium phytoavailability and human bioaccessibility for vegetables exposed to soil or atmospheric pollution by process ultrafine particles. J Environ Qual 43:1593–1600

Yang J, Kloepper JW, Ryu CM (2009) Rhizosphere bacteria help plants tolerate abiotic stress. Trends Plant Sci 14:1–4

Yang Y, Han X, Liang Y et al (2015) The combined effects of arbuscular mycorrhizal fungi (AMF) and lead (Pb) stress on Pb accumulation, plant growth parameters, photosynthesis, and antioxidant enzymes in Robinia pseudoacacia L. PLoS One 10(12):e0145726

Zhou JL (1999) Zn biosorption by *Rhizopus arrhizus* and other fungi. Appl Microbiol Biotechnol 51:686–693

Zhu YG, Christie P, Laidlaw AS (2001) Uptake of Zn by arbuscular mycorrhizal white clover from Zn-contaminated soil. Chemosphere 42:193–199

Environmental and Human Exposure to Antimicrobial Agent Triclosan: A Review

13

Rekha Kumari, Shashwati Ghosh Sachan, and Ashish Sachan

Abstract

Triclosan, commonly found in the ingredient list of several household and personal care products, exploded onto the market because of its broad-spectrum antimicrobial activity. Triclosan was introduced in the market owing to its antibacterial and antifungal properties but has now become a ubiquitous pollutant. Excessive use of triclosan has resulted in its presence not only in the environment but in the human body as well. The different mechanisms by which triclosan acts include efflux mechanisms, inhibition of fatty acid synthesis, and membrane destabilization. The major source of triclosan pollution in the environment is through wastewater. There are growing concerns regarding the possible harmful effects of triclosan on animal and human health. Triclosan, a known endocrine disruptor, has been reported to be a potent carcinogen, mutagen, and a teratogen. It is feared that its overuse may contribute to the growth of super-resistant bacteria. By the virtue of being similar in its mode of action to various antibiotics, it might confer antibiotic resistance among different bacterial species. This review article aims to present an overview of the advanced research on triclosan, primarily focusing on its toxicity and environmental persistence. It also highlights the potential hazards of triclosan including its link to developing cross-resistance. Keeping in view the limitation of the present knowledge base, future research directions have also been highlighted.

R. Kumari · S. G. Sachan
Department of Bio-Engineering, Birla Institute of Technology, Ranchi, Jharkhand, India
e-mail: ssachan@bitmesra.ac.in

A. Sachan (✉)
Department of Life Sciences, Central University of Jharkhand, Ranchi, Jharkhand, India
e-mail: ashish.sachan@cuj.ac.in

M. Kumar et al. (eds.), *In vitro Plant Breeding towards Novel Agronomic Traits*,
https://doi.org/10.1007/978-981-32-9824-8_13

Keywords
Competitive inhibitor · Enoyl-acyl carrier protein reductase · Homeostasis · Teratogenicity · Tumorigenesis

13.1 Introduction

Scientists have exploited triclosan since decades due to its wide range of antibacterial, antifungal, and anti-inflammatory properties. Triclosan can act as both bacteriostatic and bactericidal depending upon its concentration. The physicochemical properties of this compound have been listed in Table 13.1. It has been reported that triclosan acts as a bacteriostatic agent at its lower concentration (0.05–3.1 μg/mL), while the higher concentrations (25–100 μg/mL) make it bactericidal (Russell and McDonnell 2000). There are different trade names of triclosan registered: Irgasan

Table 13.1 Physicochemical properties of triclosan

Appearance	White crystalline powder
Chemical name	5-Chloro-2-(2,4-dichlorophenoxy) phenol
Molecular formula	$C_{12}H_7Cl_3O_2$
Average mass (Da)	289.54
Monoisotopic mass (Da)	287.95
CAS no.	3380–34–5
Structure	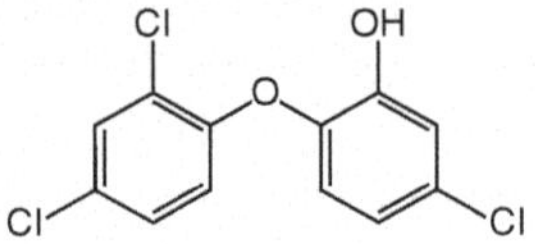
Nature	Hydrophobic, lipophilic
Other names	2,4,4′-Trichloro-2′-hydroxydiphenyl ether
	Ether, 2′-hydroxy-2,4,4′-trichlorodiphenyl
	Trichloro-2′-hydroxydiphenyl ether
	5-Chloro-(2,4-dichlorophenoxy)phenol
Bioactivity	Antibacterial, antifungal, anti-inflammatory
Solubility	6.05 mg/L in H_2O; readily soluble in ethanol, acetone, ethyl acetate, isopropanol, DMSO
Critical temperature (T_c)	1022.29 K
Boiling point temperature (T_{boil})	757.59 K
Critical pressure (P_c)	3333.53 KPa
Normal melting point (T_{fus})	539.11 K
Critical volume (V_c)	0.62 m^3/kg-Mol
Partition coefficient (octanol/water) (log $P_{oct/wat}$)	5.14
Ideal gas heat capacity C_p gas	411.53 J/Mol K @ 757.59 K
Vapor pressure	5.2×10^{-6} pa (mm hg @ 20 ° C)
Bioconcentration factor (BCF)	2500–8700

DP 300, Lexol 300, Microban, Biofresh, Ster-Zac, Chloxifenolum and, and others (Schmid and Kalpan 2004).

Triclosan exposure nowadays has increased to such a level that it has been found in the blood (Wolff et al. 2007), urine (Calafat et al. 2008), and breast milk (Allmyr et al. 2006) of people inhabiting the globe. While those who use triclosan products regularly have elevated levels of triclosan in their bodies, even consumers who do not use these products are exposed to it through household dust, food, and water (Canosa et al. 2007). Several reports have established the acute and chronic toxicity of triclosan (Orvos et al. 2002; Ishibashi et al. 2004; Tatarazako et al. 2004; Farre et al. 2008; Franz et al. 2008), its carcinogenic nature (Johnson et al. 2002; Lee et al. 2012, 2014; Winitthana et al. 2014; Wu et al. 2014; Yueh et al. 2014), and reproductive toxicology both in terms of human and animal studies (Kumar et al. 2009; Jung et al. 2012; Henry and Fair 2013; Wallet et al. 2013; Schiffer et al. 2014; Kumar et al. 2015; Ibtisham et al. 2016).

Triclosan is being used excessively over a past few decades that has raised alarm of its persistence in the environmental samples such as sediments, surface waters, and wastewaters (Miller et al. 2008) and in several biological samples including those from fish (Balmer et al. 2004), algae (Wilson et al. 2003), and human plasma (Hovander et al. 2002). This biocide is affecting the aquatic ecosystem adversely. The structure and functions of algal communities get affected by triclosan in contaminated aquatic ecosystems, which in turn affects the natural nutrient removal and cycling, potentially affecting the entire ecological balance of natural waters (Wilson and Salyers 2003). There are other health concerns associated with triclosan. Triclosan levels can build up in the human body over time and may cause long-term health risks (Calafat et al. 2008; Allmyr et al. 2006). The degradation products of triclosan are more toxic and fatal. The most concerning degradation products are dioxins, which are known carcinogens and have also been found to mutate DNA and cause birth defects in offspring (Geyer et al. 2002). The results of triclosan evaluation studies related to chronic oncogenicity in rats, mice, and hamsters revealed the presence of tumors in the livers of female and male mice (Rodricks et al. 2010).

Regular use of triclosan domestically causes more harm than benefits. Different methods used for antimicrobial treatment are toxic to eradicate all the microbes to which they are applied. This could result in increasing the toxicity risk to the host organism that is exposed to treatment against microbial infections. Triclosan could also prove to be toxic to humans when food items are kept in plastic wares containing this chemical (Cutter 1999; Vermeiren et al. 2002; Junker and Hay 2004; Efstratiou et al. 2007). Frequent exposure to triclosan might result in genetic mutations which cause drug resistance in microbes that are very harmful (Nester et al. 2001). The major disadvantage of using an antimicrobial agent is that its excessive use results in microbial resistance, thus making the product useless to those who are in actual need of it, like the people with compromised immune systems. There is a high possibility that due to the similarity of triclosan's target site in bacteria and its mode of action to that of other antibiotics, bacteria which will become resistant toward triclosan will also acquire resistance against different antibiotics. In an environment exposed to triclosan for a longer duration, triclosan kills the normal

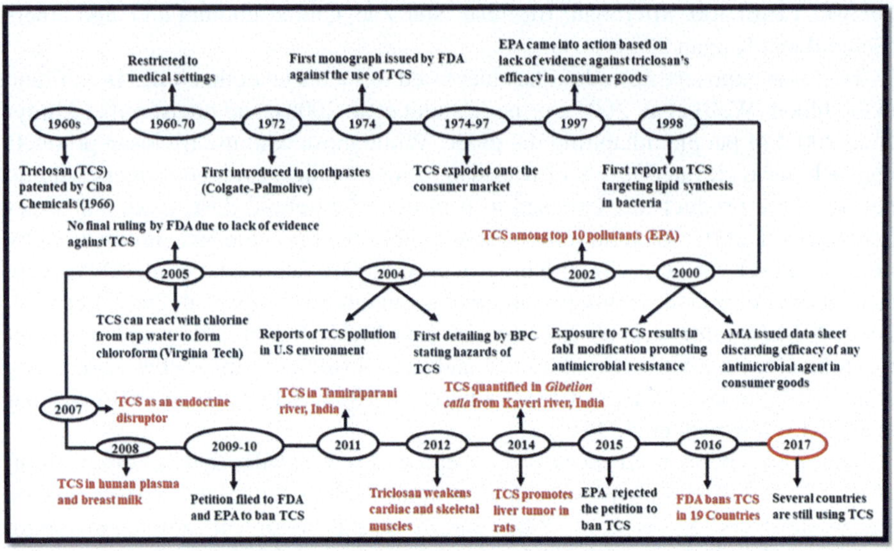

Fig. 13.1 Timeline concerning the use and occurrence of triclosan all over the world with special mention to FDA and EPA

bacteria, thereby favoring the survival and reproduction of bacteria which have acquired resistance against triclosan (Singer et al. 2002). The excessive use of triclosan is being continuously monitored by regulatory bodies such as Food and Drug Administration (FDA), USA, and Environmental Protection Agency (EPA), USA. Due to increasing environmental concerns, FDA issued a monograph on its 40th anniversary legalizing a ban on the use of triclosan (Fig. 13.1).

13.2 Synthesis and Use

Triclosan, a polychlorinated aromatic antimicrobial agent, was patented in the year 1964 for its immediate use in surgical scrubs to kill microorganisms non-specifically. But today, on a global scale, its production has exceeded the mark of 1500 tons per year with Europe being the sole producer covering 23% of its market (Singer et al. 2002). Nowadays, it has become an active ingredient in almost every household and personal care product (Fig. 13.2). Triclosan synthesis is a three-step process. The first step involves dehydration of 1-(2-Hydroxyethyl)-2-pyrrolidone in the presence of zinc or calcium oxide. The first intermediate formed is 1-Vinyl-2-pyrrolidinone, which is then treated with 5-Chloro-2-(2, 4-dichlorophenoxy) phenyl acrylate in the presence of n-heptane to form triclosan (Fig. 13.3).

One of the major concerns related to triclosan manufacturing is the toxic impurities that exist with the original compound produced. These unwanted by-products were reported to be dibenzodioxins and furans (Beck et al. 1989). These two impurities are not the only deciding factors of its enhanced toxicity, it also depends upon

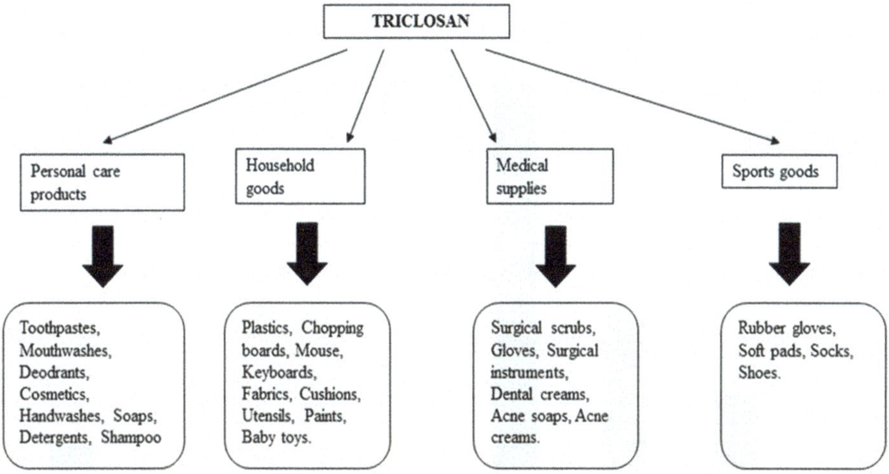

Fig. 13.2 Different products containing triclosan

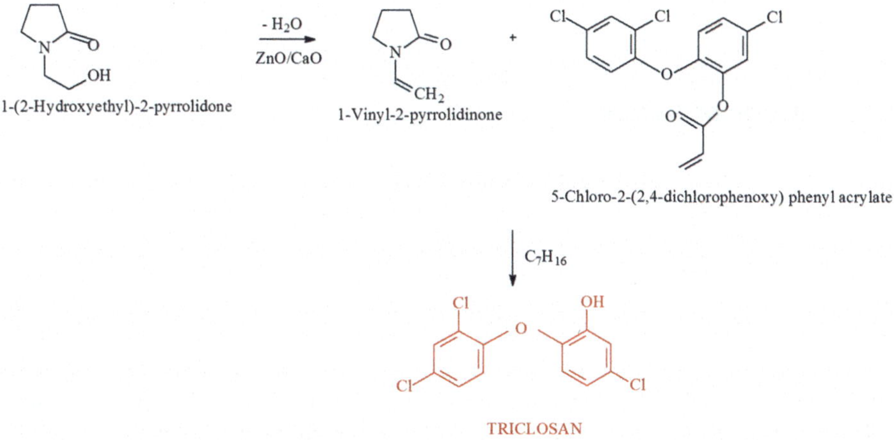

Fig. 13.3 Industrial synthesis of triclosan

the quality of chemicals being used to synthesize triclosan and moreover the conditions prevailing at the time of chemical reaction such as temperature and pressure (Ni et al. 2005). As regulated by the US Food and Drug Administration, the maximum concentration of triclosan admissible in dental products is 0.3% w/w (Rodricks et al. 2010). As per the regulations of the National Library of Medicine's Household Product Database, the recommended levels of triclosan in case of liquid soaps should fall in the range of 0.1–0.3% w/w (Fig. 13.4). The specified amount of triclosan that can be used in antiseptics and surgical scrubs should be limited to 1% w/w as regulated by FDA monograph for healthcare products (NIH, Household Products Database 2007).

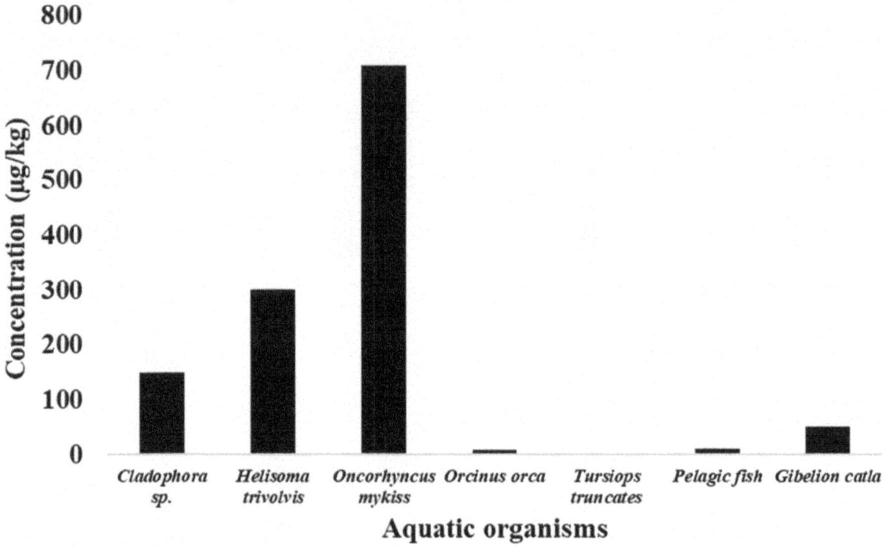

Fig. 13.4 Permissible concentration (% w/w) of triclosan in some commonly used products

13.3 Mode of Action

Since triclosan is a broad range antimicrobial agent, it targets many, but not all Gram-positive and non-sporulating Gram-negative bacterial strains. It is also effective against some protozoans like *Plasmodium falciparum* and *Toxoplasma gondii* (McLeod et al. 2001) and some fungi as well (Schweizer 2001). *Escherichia coli*, *Proteus* sp., staphylococci, some streptococci, and some mycobacteria are found to be the most sensitive organisms toward triclosan (Singer et al. 2002). Methicillin-resistant *Staphylococcus aureus* (MRSA) strains also show sensitivity toward triclosan (Al-Doori et al. 2003). One of the most effective remedies for the decolonization of MRSA strain from the skin of patients is to bath them vigorously with 2% triclosan (Brady et al. 1990). Over the last few decades, the susceptibility of MRSA strains has changed as a result of mutation in *fabI* gene through heterologous duplication and increased tolerance toward triclosan (Brenwald and Fraise 2003; Ciusa et al. 2012; Carey and MacNamara 2014). *Pseudomonas aeruginosa* is highly resistant, while enterococci were found to be less susceptible compared to staphylococci (Russell 2003). Table 13.2 presents the list of microbes showing a varying response to triclosan.

Triclosan acts similar to that of other antimicrobials which are considered to have multiple target sites against microorganisms and it has been evident that these interactions depend upon concentration (Russell et al. 1997). The different mechanisms by which triclosan acts include membrane destabilization, inhibition of fatty acid synthesis, and efflux mechanisms. These mechanisms depend on organism type and level of triclosan. The most important parameter which is indicative of

Table 13.2 Response of different microorganisms toward triclosan

Microorganisms	Highly resistant	Highly susceptible	Less susceptible
Enterococcus			√
Pseudomonas aeruginosa	√		
Proteus mirabilis		√	
MRSA strains			√
Klebsiella pneumonia		√	
Acinetobacter		√	
Plasmodium falciparum		√	
Clostridium difficile	√		
Toxoplasma gondii		√	
Staphylococcus		√	
Streptococcus		√	
Escherichia coli		√	

membrane damage is the immediate action of triclosan at a higher concentration (Villalain et al. 2001), and it is very much evident that the lethal effect of triclosan cannot solely be explained by targeting fatty acid synthesis (Gomez-Escalda et al. 2001). The functional integrity of cell membranes is compromised due to membranotropic effects of triclosan without inducing cell lysis (Villalain et al. 2001). Triclosan is likely to intercalate into bacterial cell membranes, thereby changing the functional and structural integrity of those membranes, and thus accounts for some of triclosan antibacterial effects (Guillén et al. 2004).

Triclosan has been found to specifically target fatty acid synthesis by inhibiting the enzyme enoyl reductase (enoyl-acyl carrier protein reductase, *FabI*) (McMurry et al. 1999a). Triclosan behaves structurally similar to the natural substrate (enoyl-acyl reductase-NAD^+ complex) of *fabI* gene, thereby acting as a potent irreversible inhibitor of the same gene (Heath et al. 1999; Bergler et al. 1994), and this sort of inhibition is said to be slow and competitive (Heath and Rock 1996). The natural tendency of triclosan to inhibit fatty acid synthesis in *Toxoplasma gondii* and *Plasmodium falciparum* (McLeod et al. 2001) has resulted in the development of several antimalarial and antibacterial prodrugs based on this synthetic antimicrobial agent (Mishra et al. 2008).

13.4 Triclosan: Bioaccumulation and Its Fate

13.4.1 Environmental Exposure

Majority of products having triclosan, after use, are released into residential drains (Reiss et al. 2002). Wastewater treatment plants are not capable of effectively removing triclosan, and it persists in the environment for a longer time (Allmyr et al. 2006). Therefore, it can be said that a major fraction of triclosan is pumped via waterways in the environment. In a research conducted on wastewater samples from US streams, it was shown that triclosan concentration (> 1.4 ng/L) in these streams

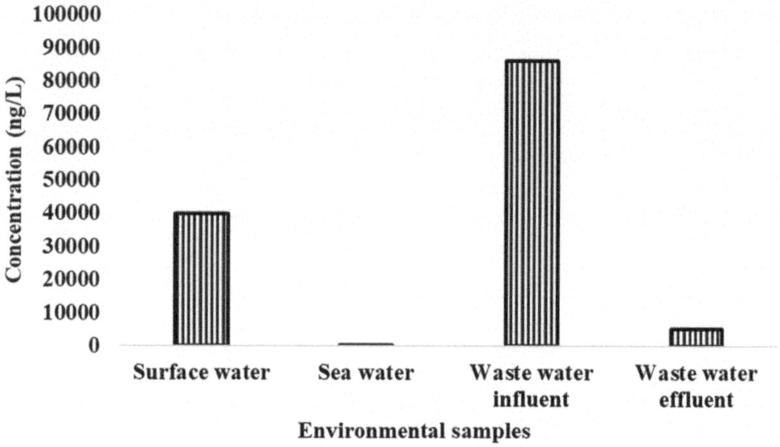

Fig. 13.5 Triclosan concentration (ng/L) in different environmental samples worldwide

was higher than any other chemical found in them (Kolpin et al. 2002; Coogan and La Point 2008). Even in India, surface waters of Tamraparni River showed a high concentration of triclosan (944 ng/L) in it. The studies suggested that the industries surrounding this river are the major contributors of triclosan in it (Ramaswamy et al. 2011). There have been reports of triclosan being released in huge amount (Fig. 13.5) worldwide which is thus adversely affecting the environment (SCCP 2010).

Triclosan adversely affects the aquatic ecosystem. This chemical has been found to affect the structure and function of algal communities (Yang et al. 2008). There are reports illuminating the toxic effect of triclosan on aquatic bacteria at different levels found in the environment, and it also affects photosynthesis in diatomaceous algae (Ricart et al. 2010). Since algae are the "first-step producers," any damage to algal life will disturb the ecological balance adversely. Additive and synergistic effect of triclosan is seen when it conglomerates with other contaminants (phthalates, parbenes) in waterways making triclosan way more hazardous to other aquatic organisms (DeLorenzo et al. 2008; Mathew et al. 2017). Even in snails, the bioaccumulation of triclosan was determined using isotope dilution GC-MS which showed a bioaccumulation factor of 500 (Coogan and La Point 2008). DeLorenzo and his co-workers reported that triclosan accumulation was not found in adult grass shrimp (*Palaemonetes pugio*) even though it was exposed to 100 μg/L of triclosan because of the fact that triclosan was converted to methyl triclosan on 14 days of exposure. As these aquatic organisms depend on algae for their source of nutrition, it is pretty much evident that triclosan concentration would increase as we move higher the food chain keeping in view the ubiquitous nature of triclosan (DeLorenzo et al. 2008).

Triclosan was quantified in *Gibelion catla* fish from the Kaveri River showing that triclosan is accumulative in fish. This bioaccumulation indicated triclosan presence throughout the river stretch (Shanguman et al. 2014). Several other fishes and aquatic organisms have been identified which are capable of accumulating triclosan

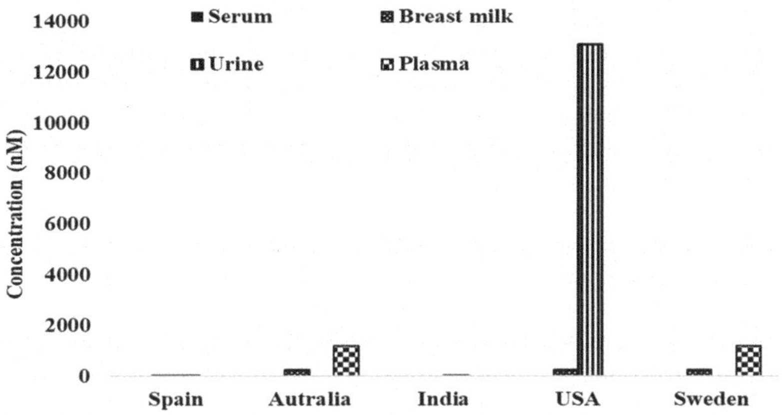

Fig. 13.6 The concentration (μg/kg) of triclosan in different aquatic organisms

and its transformation products (Fig. 13.6). In a study conducted by Houtman and his teammates, triclosan along with other xenobiotic compounds was quantified using GC-MS from the bile of male bream (*Abramis brama*) (Houtman et al. 2004). In freshwater fish samples collected from the Tokyo Bay and Tama River, methyl triclosan was determined using GC-MS which showed the levels in the range of 1–38 μg/kg body weight (Miyazaki et al. 1984). A certain amount of triclosan was also found in rainbow trout (*Oncorhynchus mykiss*) captured from a wastewater treatment plant in Sweden where these trout were exposed to treated water in a tank (Adolfsson-Erici et al. 2002). An effective concentration of methyl triclosan in the range of 130–2100 ng/g of lipid weight was detected in the body of a juvenile brown trout (*Salmo trutta fario*) collected from the effluent from wastewater treatment plant in Northern Switzerland (Buser et al. 2006). Thirteen different species of fish were collected from the Detroit River and were analyzed thoroughly for the presence of triclosan in them. The plasma samples from these fishes were found to contain triclosan in the range of 750–10,000 pg/g of wet weight (Valters et al. 2005). A similar study was conducted by Leiker et al. (2009) in which he collected male common carp from Las Vegas Bay and detected in them methyl triclosan in the range of 520–596 μg/kg of wet weight (Leiker et al. 2009).

The first study to document the presence of triclosan in a marine mammal was conducted by Fair and his fellow researchers, who detected triclosan in the plasma of *Tursiops truncatus* in the range of 0.12–0.27 μg/kg (Fair et al. 2009). Similarly, triclosan has also been detected in the plasma of killer whale at a concentration of 9.0 ng/g (Bennett et al. 2009). Even in estuarine water samples collected from two southeast states of the USA, triclosan has been detected in the range of 4.9–13.7 ng/l (Bennett et al. 2009).

Triclosan is capable of entering the terrestrial environment in the form of sewage sludge when it is spread on agricultural and industrial land (Fuchsman et al. 2010). In a study conducted by Mendez and his team, triclosan was detected in all edible parts of onion in the range of 115–435 ng/g, mostly in bulbs without affecting the biomass. They also reported accumulation of triclosan in tomato species that translocated to shoots and fruits. The main reason behind this bioaccumulation was the recycled water that was used to irrigate these crops and soil that had a certain amount of triclosan in it which was not removed during the treatment process in wastewater treatment plants (Mendez et al. 2016). Similarly, wheat plants showed bioaccumulation of triclosan higher in roots as compared to shoots. The soil which was used to grow these crops was spread with biosolids spiked with triclosan (Jachero et al. 2016). Triclosan uptake was also reported in bean plants, but the accumulation was limited to roots itself (Karnjanapiboonwong et al. 2011). Phytoaccumulation of triclosan has been observed in various plants, even in several food crops such as radish and lettuce when grown in soils applied with biosolids containing triclosan (Pannu et al. 2012). The documented concentration of triclosan in biosolids is in the range of 90–32,900 μg/kg, and in the activated sludge, the effective concentration was reported to be 580–14,700 μg/kg (Chalew and Halden 2009; Lozano et al. 2010). When the organic biosolids containing triclosan was applied to agricultural land, the triclosan concentration in the tissues of earthworm enhanced to 2610 μg/kg, resulting in a bioaccumulation factor of 27 (Kinney et al. 2008). Even the microbial communities inhabiting the agricultural land were found to be slightly affected by the triclosan concentration in biosolids. Due to the presence of triclosan in the soil, the relative abundance of several Gram-positive and Gram-negative bacteria reduced including some fungal species (Park et al. 2013). These studies further suggest the alarming rate of triclosan concentration in environment and also highlight the immediate monitoring and formulation of certain stringent laws governing the usage of triclosan in consumer goods.

13.4.2 Environmental Degradation of Triclosan

It has been established that triclosan is transported to wastewater treatment plants through our domestic and industrial waste streams (Fig. 13.7).

Depending upon the technological buildup of our wastewater treatment plants, triclosan removal efficiency varies from 50% to 90% (Singer et al. 2002; Thompson et al. 2005; Lishman et al. 2006). Since triclosan is capable of surviving the rigorous treatment in the wastewater treatment plants, it enters our drinking water systems whereupon bleaching it reacts with free chlorine to form chloroform (Rule et al. 2005). While undergoing the wastewater treatment process, it has been reported that some part of triclosan is transformed biologically into methyl triclosan which is much more lipophilic (Balmer et al. 2004) and resistant to degradation (Lindstorm et al. 2002). Triclosan reaches the aquatic environment through residential drains whereupon coming in contact of UV rays from sunlight, triclosan degrades into 2,8-dichlorodibenzodioxin and other dioxin derivatives which are considered as potent carcinogens (Lores et al. 2005; Sanchez-Prado et al. 2006; Aranami and Readman 2007). Buth and his fellow workers reported that the photolytic conversion of triclosan is dependent on speciation, i.e., the phenolate form of triclosan can easily be degraded compared to its phenol form (Buth et al. 2009). Several physical

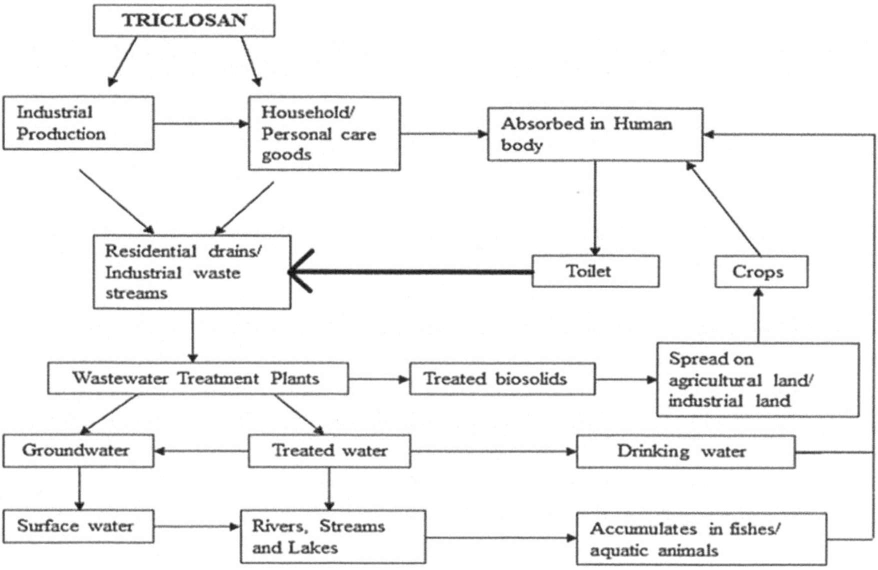

Fig. 13.7 A major route of triclosan pollution in environment and the human body

and chemical processes have been employed to remove triclosan from the wastewater systems. The pore size acts as a limiting factor in case of activated carbon used as a physical method for triclosan removal (Letterman et al. 2009). The chemical oxidation of triclosan results in its partial decomposition, thereby indicating the formation of toxic intermediates (Lores et al. 2005; Sanchez-Prado et al. 2006; Aranami and Readman 2007). Thus, the researchers are shifting their focus toward the biological ways for triclosan removal.

Aerobic hydrolysis by bacteria plays a pivotal role in triclosan degradation. A bacterial consortium was found capable of partially degrading triclosan, and one of its members was found to be a *Sphingomonas*-like microorganism (Hay et al. 2001). In another study, two different soil bacteria *Pseudomonas putida* TriRY and *Alcaligenes xylosoxidans* subsp. *denitrificans* TRI were capable of using triclosan as a sole carbon source (Meade et al. 2001). *Sphingomonas* or *Sphingopyxis*, a triclosan-degrading bacterium, was isolated from activated sludge. Zhao found that *Nitrosomonas europaea*, an important bacterium that carries out nitrification in wastewater treatment plants, is able to degrade triclosan via cometabolism (Zhao 2006). Under aerobic conditions, triclosan and its chlorinated derivatives were found to be degraded by *Burkholderia* sp. (SCCP 2010).

Triclosan is quite stable under anaerobic conditions and in dark. Triclosan readily adsorbs to particles and tends to aggregate, in sediments by virtue of its low water solubility. Digested sludge concentrations of triclosan varied from 0.5 to 15.6 μg/g (dry weight), where the aerobic digestion process showed less concentration of triclosan as compared to anaerobic digestion process of triclosan. These results depicted that triclosan degrades easily under aerobic conditions unlike anaerobic conditions (McAvoy et al. 2002). Biodegradation of triclosan has also been reported in fungal species like *Trametes versicolor* and *Pycnoporus cinnabarinus* (Hundt et al. 2000). These are white rot fungi which are capable of naturally degrading triclosan. Another fungus *Aspergillus versicolor* has recently been reported as being capable of triclosan conversion up to 2, 4-dichlorophenol and release of chloride ions at pH 5 and 7 (Tastana and Donmeza 2015). Even a green alga identified as *Nannochloris* sp. was found capable of 100% mineralization of triclosan within 7 days of incubation. It was reported that algae-mediated uptake of triclosan was the primary pathway of triclosan removal (Bai and Acharya 2016). *Dyella* sp. WW1 isolated from activated sludge was found capable of degrading triclosan in the range of 0.2–10 mg/L. The optimal conditions at which *Dyella* sp. degraded triclosan were at pH 7 and 15 °C. In the presence of an additional carbon source glucose, *Dyella* sp. WW1 firstly synthesized the biomass using glucose, and then triclosan was degraded (Wang et al. 2018). Another wastewater isolate *Pseudomonas aeruginosa* KS2002 was found capable of tolerating up to 2 g/L of triclosan (Kumari and Ghosh Sachan 2018). However, the knowledge regarding approaches independent of culture for the identification and ecophysiological studies of triclosan degrading bacterial strains in complex microbial systems is limited. Keeping in view the limited knowledge base on microbial degradation of triclosan, we tried to explore certain wastewater microorganisms capable of effective triclosan degradation.

Wastewater sample was collected from a housing society of Ranchi district. Master dilution of the sample was prepared in 0.9% saline (w/v) and was serially diluted up to 10^{-2} dilution. 100 μL of each dilution along with the master dilution and undiluted sample was spread onto tryptone soya agar plates supplemented with 0.2% triclosan. The plates were then incubated at 37 °C for 72 h. The colonies showing zone of clearance were purified by streaking them onto nutrient agar slants. The slants were kept at 4 °C for further studies.

A total of four morphologically distinct colonies were isolated which were found capable of surviving on 0.2% triclosan (Fig. 13.8). The isolation was based on the appearance of halo zone around the colonies. There could be two possible reasons for the appearance of zone of hydrolysis around the colonies. The first probable reason could be the conversion of triclosan to some other compound which is less turbid than triclosan itself. The other reason could be the decrease in the pH of the media resulting in the formation of zone of clearance around the colonies. The morphological characteristics of the isolated bacterial colonies are listed in Table 13.3.

For each isolate, 50 mL of M9 minimal salt media was prepared in 250 mL Erlenmeyer flask. The flasks were then inoculated with 4% of the starter cultures of respective isolates. The flasks were then incubated at 37 °C under static condition. At an interval of every 24 h, 2 mL of the sample was withdrawn and centrifuged at 10000 rpm for 20 min. The supernatant was collected, and it was analyzed spectrophotometrically at 280 nm using UV-visible spectrophotometer. The percentage degradation of triclosan was calculated using Eq. (13.1) (Tastana and Donmeza 2015):

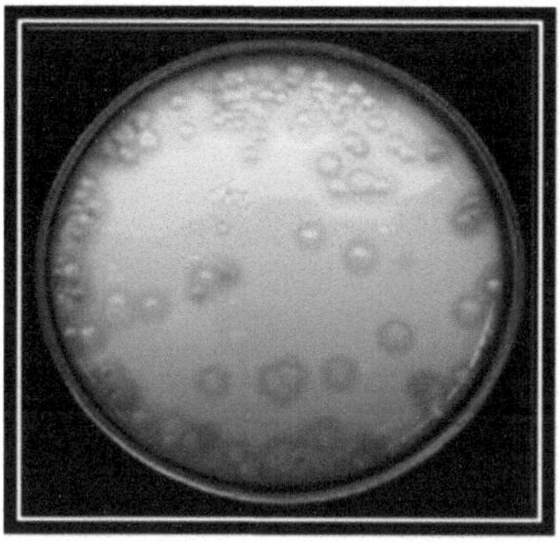

Fig. 13.8 TSA plates supplemented with 0.2% triclosan showing halazone around the colonies

Table 13.3 Morphological characteristics of different isolates

	Morphological characteristics				
Isolate	Size	Form	Elevation	Margin	Color
KS1001	Large	Circular	Umbonate	Entire	Cream
KS1002	Large	Irregular	Umbonate	Undulate	Off-white
KS1003	Large	Circular	Flat	Entire	Faded green
KS1004	Small	Circular	Convex	Entire	Light brown

$$\text{Degradation}(\%) = \left\{\frac{(C_0 - C_1)}{C_0}\right\} \times 100 \tag{13.1}$$

where

C_0 is the initial triclosan concentration and C_1 is the final triclosan concentration.

Out of the four presumptive isolates, isolate KS1003 was capable of maximally degrading triclosan within 4 days of incubation (Fig. 13.9). Absolute degradation was observed in case of isolate KS1003. All the isolates were capable of utilizing triclosan as a sole carbon and energy source. After 4 days of incubation, the supernatant obtained from isolate KS1003 showed no trace of triclosan left in it. The degradation data is supported with the scan obtained from UV-visible spectrophotometer showing decrease in the intensity of triclosan within 4 days of incubation (Fig. 13.10).

13.4.3 Human Exposure of Triclosan

According to the European Commission (2010), the approved level of triclosan for topical administration is 0.3% in humans (Larsson et al. 2014; MacIsaac et al. 2014). Humans do not depend on a single product in our day-to-day lives. A large number of products containing triclosan are being used by human beings on a regular basis which is resulting in the deposition of a large amount of triclosan in our

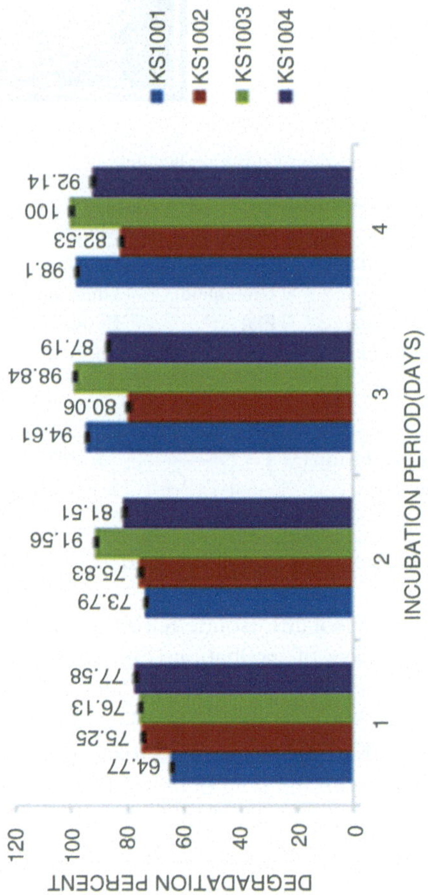

Fig. 13.9 Bar graph depicting percentage degradation of triclosan by different isolates within 4 days of incubation. Error bars depict standard deviation

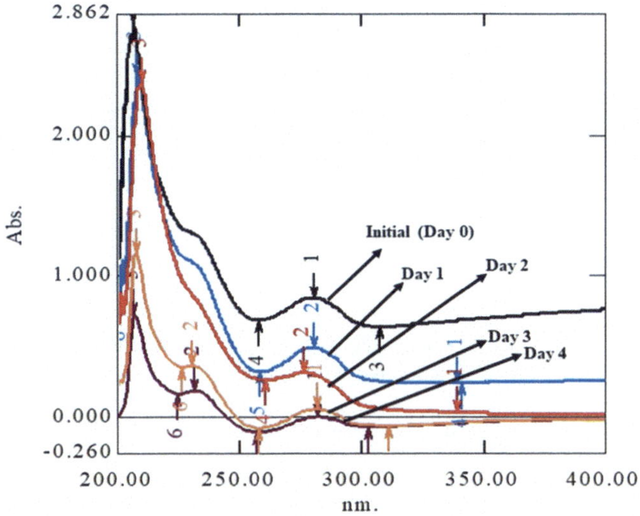

Fig. 13.10 UV-visible scan of isolate KS1003 showing decrease in intensity of triclosan over 4 days of incubation

bodies gradually. There are several routes through which triclosan is absorbed in humans and animals. It can be absorbed dermally (Moss et al. 2000; Fang et al. 2016) or through gastrointestinal tract (Bagley and Lin 2000). It can also be absorbed by our mucous membranes of oral cavity (Lin 2000) and thereby reaches the circulatory systems of the human body (Hovander et al. 2002).

Dermal absorption of triclosan generally varies between 3 and 7% as reported by Queckenberg et al. (2010). Another report by Moss and his co-workers in 2000 showed absorption via human skin up to 6.7% (Moss et al. 2000). When human beings use the mouthwash containing triclosan, it was found that 7.33% of it was being absorbed by the mucous membrane (Lin 2000). The absorption rates were found higher in the gastrointestinal tract when the mouthwash was being swallowed by the user (Bagley and Lin 2000). Various reports have shown significant levels of triclosan in body fluids of human beings (Calafat et al. 2008; Allmyr et al. 2006; Adolfsson-Erici et al. 2002; Xue et al. 2015; Azzouz et al. 2016) (Fig. 13.11). When Sandborgh-Englund and his co-workers conducted a controlled exposure study, it was found that 70% of the triclosan was in the conjugated form and primarily voided through urine, thus resulting in the balance of unconjugated form which could trigger physiological responses (Sandborgh-Englund et al. 2006). Triclosan has been found in breast milk (Dayan 2007). The maximum triclosan concentration to be found in human blood, breast milk, and urine were 41.4, 7.3, and 7.3 nM, respectively (Azzouz et al. 2016). The presence of triclosan was investigated in the urine and nail samples of Chinese population. About 80% of the population was detected positive for the presence of triclosan in their bodies. Females showed higher geometric mean (GM) values of triclosan in urinary samples (0.53 μg/g) as compared to that of males (0.31 μg/g). The gender differences as observed in case of fingernail

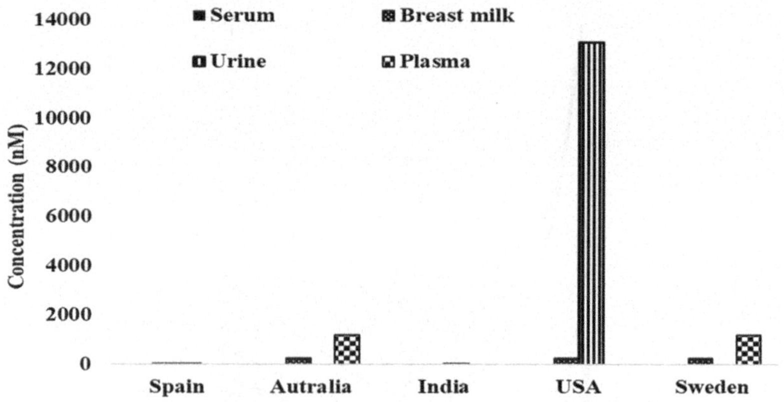

Fig. 13.11 The concentration (nM) of triclosan in body fluids as reported in different countries

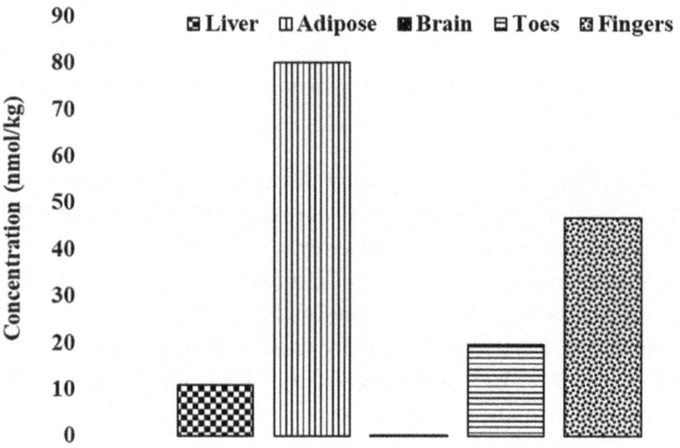

Fig. 13.12 The concentration (nmol/kg) of triclosan in different body tissues

samples were similar to those of urine samples. Females demonstrated higher triclosan concentrations as that of males (Yin et al. 2016).

High levels of triclosan were found in three out of five human milk samples as high as 300 μg/kg lipid weight indicating that triclosan gets easily absorbed in the body (Adolfsson-Erici et al. 2002). Triclosan being lipophilic is capable of bioaccumulating in fatty tissues. Higher triclosan concentration has been found in different body tissues (Yin et al. 2016; Geens et al. 2012; Wang et al. 2015) (Fig. 13.12) in comparison with different environmental samples. From this report it can be inferred that triclosan more readily bioaccumulates in the human body. Triclosan disturbs the thyroid homeostasis in the human body (Veldhoen et al. 2006; Kodavanti and Curras-Collazo 2010). Triclosan has been reported to intervene the thyroid hormone metabolism in the human body, resulting in lowering the body temperature and thus causing a "non-specific depressant effect on the central nervous system" of mice (Miller et al. 1983). Exposure to lower levels (0.03 mcg/l) of triclosan with altered thyroid hormone-associated gene expression levels in tadpoles encourages them to prematurely metamorphose into frogs (Veldhoen et al. 2006), while another study showed a reduction in sperm production in male rats when exposed to triclosan (Kumar et al. 2009). All the studies led to hypothesizing that triclosan blocks the thyroid metabolism by chemically mimicking thyroid hormone and binding to the hormone receptor site and thus blocking it. Triclosan acts as a competitive inhibitor of the enzyme placental estrogen sulfotransferase, which blocked the delivery of estrogen from the placenta to the fetus in sheep during pregnancy (James et al. 2010). Triclosan negatively affects the muscle health by weakening cardiac and skeletal muscles contractility (Cherednichenko et al. 2012).

Triclosan toxicity has also been reported in different cells including human liver and cancerous cells (Wallet et al. 2013; Arancibia et al. 2009). According to a study conducted in 2014, it was found that triclosan acts as a liver tumor promoter and the mechanism by which triclosan induces cancer in the mouse may be relevant to that in humans (Yueh et al. 2014). As reported by Wu et al. in 2014, triclosan was found to cause hepatic tumorigenesis in mice which were mediated by peroxisome proliferator-activated receptor alpha signaling pathway (Wu et al. 2014). A more recent study showed that TCS is capable of inducing cancer progression in MCF-7 human breast cancer cell through estrogen receptor-mediated signaling pathway indicating that triclosan contributes in breast cancer advancement through multiple mechanisms (Lee et al. 2014). Triclosan has been reported to be a prooxidant (Yueh et al. 2014; Tamura et al. 2012; Ma et al. 2013). An inverse relation exists between triclosan-mediated oxidation of deoxyguanosine and that of DNA methylation (Ma et al. 2013), a potent biomarker for tumor development and progression. The significant levels of reactive oxygen species are generated in β cells, which are the site for low expression of antioxidant enzymes (Pi et al. 2007) which definitely affects insulin synthesis and ultimately diabetes pathogenesis.

There are also pieces of evidence showing that triclosan might be a causative agent of photoallergic contact dermatitis (PACD), which occurs on being exposed to sunlight after a brief exposure toward triclosan (Durbize et al. 2003). It has also been found to cause allergic reactions in children who are exposed to it (Clayton et al. 2011). Similarly, blisters were observed in the mouth and on the lips of human subjects who were continuously using triclosan-containing toothpaste (Robertshaw and Leppard 2007). It was also observed that increased triclosan levels in urine resulted in immune dysfunction (Clayton et al. 2011), allergic reactions, and asthma in children (Bertelsen et al. 2013; Spanier et al. 2014). Triclosan-containing personal care products leave their effects for hours after use (Glaser 2004).

13.4.4 Effect of Triclosan in the Development of Cross-Resistance in Bacteria

Several laboratory studies have been carried out to establish the fact that triclosan resistance might lead to resistance toward other antimicrobial agents. Different mechanisms responsible for inducing cross-resistance among bacterial species have been reported through such laboratory studies. It was observed in case of *Salmonella enterica* that the strain which became resistant to triclosan overexpressed an efflux pump (*acr*AB), because of which it became less susceptible to other antibiotics than its wild strain (Karatzas et al. 2007). Similar results were obtained when *S. enterica* serovar *typhimurium* was exposed to 0.5 mg/L of triclosan, where it resulted in growth of ciprofloxacin-resistant strains (Birosova and Mikulasova 2009). Sanchez along with his fellow researchers examined the effect of triclosan on the emergence of mutants in *Stenotrophomonas* overexpressing the efflux pump SmeDEF which is involved as a mechanism in both innate and extrinsic resistance to antimicrobials. The authors showed that out of 12 mutants, triclosan was capable of selecting 5 mutants that overexpressed this pump. This mechanism of overexpression was responsible for conferring resistance to different antibiotics (Sanchez et al. 2005). Triclosan adapted strains of *E. coli* O157:H7, and *E. coli* O55 were exposed to eight different antibiotics. *E. coli* O157:H7 was found to be resistant toward these eight antibiotics, whereas *E. coli* O55 exhibited resistance toward only one antibiotic, i.e., trimethoprim (Braoudaki and Hilton 2005).

The conditions responsible for conferring resistance toward triclosan drive cross-resistance to different antibiotics. Bailey and his co-workers showed that triclosan acted as a stimulus that triggered the enunciation of several genes directly involved in antibiotic resistance and other regulatory genes responsible for the control of antimicrobial resistance gene cascade (Bailey et al. 2009). Any transformation in *InhA* gene in *M. smegmatis* followed by susceptivity to triclosan showed impedance to isoniazid (McMurry et al. 1999b). Cottell and his co-workers found no change in the susceptibility of triclosan-tolerant *S. aureus*, *E. coli*, and *Acinetobacter johnsonii* toward other antibiotics in clinical settings (Cottell et al. 2009). In another study, 3% of 732 clinical isolates of *Acinetobacter baumannii* collected from hospitals showed reduced susceptibility toward triclosan having a MIC greater than 1 mg/L. An increased tolerance toward levofloxacin, tetracycline, imipenem, and amikacin was found in those isolates which could tolerate 4 mg/L of triclosan (Chen et al. 2009). More research monitoring the effect of triclosan on antimicrobial resistance needs to be carried out to further establish the link between triclosan exposure and bacterial resistance universally.

13.5 Conclusion and Future Directions

Due to the fact that triclosan is being used in every household across the globe, there are emerging concerns regarding the subsistence of triclosan remnants in the ecosystem and its conceivable privative impact on the biotic and abiotic environment. Scientists still doubt the impregnability and adequacy of its use in various consumer goods using it as an active component. Though triclosan is of prime importance in the medical field, its negativity is restraining its use only to those who actually need it as in the case of immune-compromised patients.

Based on the limited literature review on the existing usage of triclosan, its efficacy and safety are debatable. Since the reports on the emergence of cross-resistance among different bacterial strains in clinical settings are limited, research needs to be carried out in order to find out its effective concentration responsible for such resistance in the environment. This review indicates the overall effect of triclosan both in terms of environment and human exposure. Further work can be carried out in terms of biodegradation kinetics of triclosan and the pathway which is being followed by different microbes to degrade this xenobiotic compound.

Though FDA has put a ban on the usage of triclosan in most of the consumer goods resulting in its slow accumulation, heavy use of triclosan in the past few decades has left us with several contaminated environments exerting a negative impact for years to come. Moreover, there are still several products in the market that contain triclosan, likely to add onto the existing chemical pressure on the environmental bacteria leading to the survival of more antibiotic-resistant bacteria. Further, strict laws need to be executed while considering what sort of chemicals is being used in everyday products. We are of the opinion that everyday use of triclosan should be discouraged and must be limited to healthcare and clinical settings.

Acknowledgments The authors are highly thankful to the Council of Scientific and Industrial Research (Scheme no. 24(0340)/16/EMR-II) for providing us with the financial assistance for the work, and we also acknowledge Birla Institute of Technology, Mesra, for providing us with the infrastructure to carry out our work.

Conflict of Interests The authors declare no conflict of interest.

References

Adolfsson-Erici M, Patterson M, Parkkonen J, Sturve J (2002) Triclosan, a commonly used bactericide found in human milk and in the aquatic environment in Sweden. Chemosphere 46:1485–1489

Al-Doori Z, Morrison D, Edwards G, Gemmell C (2003) Susceptibility of MRSA to triclosan. J Antimicrob Chemother 51:185–186

Allmyr M, Adolfsson-Erici M, McLachlan MS, Sandborgh-Englund G (2006) Triclosan in plasma and milk from Swedish nursing mothers and their exposure via personal care products. Sci Total Environ 372:87–93

Aranami K, Readman JW (2007) Photolytic degradation of triclosan in freshwater and seawater. Chemosphere 66:1052–1056

Arancibia R, Caceres M, Martinez J, Smith PC (2009) Triclosan inhibits tumor necrosis factor-stimulated urokinase production in human gingival fibroblasts. J Periodontal Res 44:726–735

Azzouz A, Rascon AJ, Ballesteros E (2016) Simultaneous determination of parabens, alkylphenols, phenylphenols, bisphenol A and triclosan in human urine, blood and breast milk by continuous solid-phase extraction and gas-chromatography-mass spectrometry. J Pharm Biomed Anal 119:16–26

Bagley DM, Lin YJ (2000) Clinical evidence for the lack of triclosan accumulation from daily use in dentifrices. Am J Dent 13:148–152

Bai X, Acharya K (2016) Removal of trimethoprim, sulfamethoxazole and triclosan by the green alga *Nannochloris* sp. J Hazard Mater 315:70–75

Bailey AM, Constantinidou C, Ivens A, Garvey MI, Webber MA et al (2009) Exposure of *Escherichia coli* and *Salmonella enterica* serovar *typhimurium* to triclosan induces a species-specific response, including drug detoxification. J Antimicrob Chemother 64:973–985

Balmer ME, Poiger T, Droz C, Romanin K, Bergqvist PA et al (2004) Occurrence of methyl triclosan, a transformation product of the bactericide triclosan, in fish from various lakes in Switzerland. Environ Sci Technol 38:390–395

Beck H, Droβ A, Eckart K, Mathar W, Wittkowski R (1989) Determination of PCDDs and PCDFs in Irgasan DP 300. Chemosphere 19:167–170

Bennett ER, Ross PS, Huff D, Alaee M, Letcher RJ (2009) Chlorinated and brominated organic contaminants and metabolites in the plasma and diet of a captive killer whale (*Orcinus orca*). Mar Pollut Bull 58:1078–1083

Bergler H, Wallner P, Ebeling A, Leitinger B, Fuchsbichler S et al (1994) Protein EnvM is the NADH-dependent enoyl-ACP reductase (FabI) of *Escherichia coli*. J Biol Chem 269:5493–5496

Bertelsen RJ, Longnecker MP, Løvik M, Calafat AM, Carlsen KH et al (2013) Triclosan exposure and allergic sensitization in Norwegian children. Allergy 68:84–91

Birosova L, Mikulasova M (2009) Development of triclosan and antibiotic resistance in *Salmonella enterica* serovar *typhimurium*. J Med Microbiol 58:436–441

Brady LM, Thomson M, Palmer MA, Harkness JL (1990) Successful control of endemic MRSA in a cardiothoracic surgical unit. Med J Aust 152:240–245

Braoudaki M, Hilton AC (2005) Mechanisms of resistance in *Salmonella enterica* adapted to erythromycin, benzalkonium chloride and triclosan. Int J Antimicrob Agents 25:31–37

Brenwald N, Fraise A (2003) Triclosan resistance in methicillin-resistant Staphylococcus aureus (MRSA). J Hosp Infect 55:141–144

Buser HR, Balmer ME, Schmid P, Kohler M (2006) Occurrence of UV filters 4-methylbenzylidene camphor and octocrylene in fish from various Swiss rivers with inputs from wastewater treatment plants. Environ Sci Technol 40:1427–1431

Buth JM, Grandbois M, Vikesland PJ, McNeill K, Arnold WA (2009) Aquatic photochemistry of chlorinated triclosan derivatives: potential source of polychlorodibenzo-p-dioxins. Environ Toxicol Chem 28:2555–2563

Calafat AM, Ye X, Wong LY, Reidy JA, Needham LL (2008) Urinary concentrations of triclosan in the U.S. population: 2003–2004. Environ Health Perspect 116:303–307

Canosa P, Rodriguez I, Rubi E, Cela R (2007) Determination of parabens and triclosan in indoor dust using matrix solid-phase dispersion and gas chromatography with tandem mass spectrometry. Anal Chem 79:1675–1681

Carey DE, MacNamara PJ (2014) The impact of triclosan on the spread of antibiotic resistance in the environment. Front Microbiol 5:780–786

Chalew TE, Halden RU (2009) Environmental exposure of aquatic and terrestrial biota to triclosan and triclocarban. J Am Water Works Assoc 45:4–13

Chen Y, Pi B, Zhou H, Yu Y, Li L (2009) Triclosan resistance in clinical isolates of *Acinetobacter baumannii*. J Med Microbiol 58:1086–1091

Cherednichenko G, Zhang R, Bannister RA (2012) Triclosan impairs excitation – contraction coupling and Calcium ions dynamics in striated muscle. PNAS 109:14158–14163

Ciusa ML, Furi L, Knight D, Decorosi F, Fondi M et al (2012) A novel resistance mechanism to triclosan that suggests horizontal gene transfer and demonstrates a potential selective pressure for reduced biocide susceptibility in clinical strains of Staphylococcus aureus. Int J Antimicrob Agents 40:210–220

Clayton EM, Todd M, Dowd JB, Aiello AE (2011) The impact of bisphenol A and triclosan on immune parameters in the US population, NHANES 2003-2006. Environ Health Perspect 119:390–396

Coogan MA, La Point TW (2008) Snail bioaccumulation of triclocarban, triclosan and methyltriclosan in a North Texas, USA, stream affected by wastewater treatment plant runoff. Environ Toxicol Chem 27:1788–1793

Cottell A, Denyer SP, Hanlon GW, Maillard JY (2009) Triclosan-tolerant bacteria: changes in susceptibility to antibiotics. J Hosp Infect 72:71–76

Cutter CN (1999) The effectiveness of triclosan-incorporated plastic against bacteria on beef surfaces. J Food Prot 62:474–479

Dayan AD (2007) Risk assessment of triclosan (Irgasan R) in human breast milk. Food Chem Toxicol 45:125–129

DeLorenzo ME, Keller JM, Arthur CD, Finnegan MC, Harper HE et al (2008) Toxicity of the antimicrobial compound triclosan and formation of the metabolite methyl-triclosan in estuarine systems. Environ Toxicol 23:224–232

Durbize E, Vigan M, Puzenat E (2003) Spectrum of cross-photosensitization in 18 consecutive patients with contact photoallergy to ketoprofen: associated photoallergies to non-benzophenone-containing microbe. Contact Dermatitis 48:144–149

Efstratiou M, Papaioannou W, Nakou M, Ktenas E, Vrotsos IA et al (2007) Contamination of a toothbrush with antibacterial properties by oral microorganisms. J Dent 35:331–337

Fair PA, Lee HB, Adams J, Darling C, Pacepavicius G et al (2009) Occurrence of triclosan in plasma of wild Atlantic bottlenose dolphins (*Tursiops truncatus*) and in their environment. Environ Pollut 157:2248–2254

Fang JL, Vanlandin M, DaCosta GG, Beland FA (2016) Absorption and metabolism of triclosan after application to the skin of B6C3F1 mice. Environ Toxicol 31:609–623

Farre M, Asperger D, Kantiani L, Gonzalez S (2008) Assessment of the acute toxicity of triclosan and methyl triclosan in wastewater based on the bioluminescence inhibition of *Vibrio fischeri*. Anal Bioanal Chem 390:1999–2007

Franz S, Altenburger R, Heilmeier H, Schmitt-Jansen M (2008) What contributes to the sensitivity of microalgae to triclosan? Aquat Toxicol 90:102–108

Fuchsman P, Lyndall J, Bock M, Lauren D, Barber T et al (2010) Terrestrial ecological risk evaluation for triclosan in land-applied biosolids. Integr Environ Assess Manag 6:405–418

Geens T, Neels H, Covaci A (2012) Distribution of bisphenol A, triclosan and n-nonylphenol in human adipose tissue, liver and brain. Chemosphere 87:796–802

Geyer HJ, Schramm KW, Feicht EA, Behechti A, Steinberg C et al (2002) Half-lives of tetra-, penta-, hexa-, hepta-, and octachlorodibenzo-p-dioxin in rats, monkeys, and humans—a critical review. Chemosphere 48:631–644

Glaser A (2004) The ubiquitous triclosan. A common antibacterial agent exposed. Pestic You 24:12–17

Gomez-Escalda M, Maillard JY, Russell AD (2001) Effects of triclosan-sensitive and resistant strains of gram-negative bacteria. Soc Appl Microbiol 30:9–13

Guillén J, Bernabeu A, Shapiro S, Villalaín J (2004) Location and orientation of Triclosan in phospholipid model membranes. Eur Biophys J 33:448–453

Hay AG, Dees PM, Sayler GS (2001) Growth of bacterial consortium on triclosan. FEMS Microbiol Ecol 36:105–112

Heath RJ, Rock CO (1996) Inhibition of graphic-ketoacyl-acyl carrier protein synthase III (FabH) by acyl-acyl carrier protein in *Escherichia coli*. J Biol Chem 271:10996–11000

Heath RJ, Rubin JR, Holland DR, Zhang E, Snow ME et al (1999) Mechanism of triclosan inhibition of bacterial fatty acid synthesis. J Biol Chem 274:11110–11114

Henry ND, Fair PA (2013) Comparison of in vitro cytotoxicity, estrogenicity and anti-estrogenicity of triclosan, perfluorooctane sulfonate and perfluorooctanoic acid. J Appl Toxicol 33:265–272

Houtman CJ, Van Oostveen AM, Brouwer A, Lamoree MH, Legler J (2004) Identification of estrogenic compounds in fish bile using bioassay directed fractionation. Environ Sci Technol 38:6415–6423

Hovander L, Malmberg T, Athanasiadou M, Athanassiadis I, Rahm S et al (2002) Identification of hydroxylated PCB metabolites and other phenolic halogenated pollutants in human blood plasma. Arch Environ Contam Toxicol 42:105–117

Hundt K, Martin D, Hammer E, Jonas U, Kindermann MK et al (2000) Transformation of triclosan by *Trametes versicolor* and *Pycnoporus cinnabarinus*. Appl Environ Microbiol 66:4157–4160

Ibtisham F, Nawab A, Zhao Y, Li G, Xiao M et al (2016) Effect of antimicrobial triclosan on reproductive system of male rat. Pharm Anal Acta 7(1000516):1–5
Ishibashi H, Matsumura N, Hirano M, Matsuoka M, Shiratsuchi H et al (2004) Effects of triclosan on the early life stages and reproduction of *medaka Oryzias latipes* and induction of hepatic vitellogenin. Aquat Toxicol 67:167–179
Jachero L, Ahumada I, Fuentes E, Richter P (2016) Decrease in the bioconcentration of triclosan in wheat plants according to increasing amounts of biosolids added to soil. Geoderma 276:19–25
James MO, Li W, Summerlot DP, Rowland-Faux L, Wood CE (2010) Triclosan is a potent inhibitor of estradiol and estrone sulfonation in sheep placenta. Environ Int 36:942–949
Johnson EF, Hsu MH, Savas U, Griffin KJ (2002) Regulation of P450 4A expression by peroxisome proliferator activated receptors. Toxicology 181:203–206
Jung EM, An BS, Choi KC, Jeung EB (2012) Potential estrogenic activity of triclosan in the uterus of immature rats and rat pituitary GH3 cells. Toxicol Lett 208:142–148
Junker LM, Hay AG (2004) Effects of triclosan incorporation into ABS plastic on biofilm communities. J Antimicrob Chemother 53:989–996
Karatzas KA, Webber MA, Jorgensen F, Woodward MJ, Piddock LJ et al (2007) Prolonged treatment of *Salmonella enterica* serovar *typhimurium* with commercial disinfectants selects for multiple antibiotic resistance, increased efflux and reduced invasiveness. J Antimicrob Chemother 60:947–955
Karnjanapiboonwong A, Chase DA, Canas JE, Jackson WA, Maul JD et al (2011) Uptake of 17 α-ethynylestradiol and triclosan in pinto bean, Phaseolus vulgaris. Ecotoxicol Environ Saf 74:1336–1342
Kinney CA, Furlong ET, Kolpin DW, Burkhardt MR, Steven D et al (2008) Bioaccumulation of pharmaceuticals and other anthropogenic waste indicators in earthworms from agricultural soil amended with biosolid or swine manure. Environ Sci Technol 42:1863–1870
Kodavanti PRS, Curras-Collazo MC (2010) Neuroendocrine actions of organohalogens: thyroid hormones, arginine vasopressin, and neuroplasticity. Front Neuroendocrinol 31:479–496
Kolpin DW, Furlong ET, Meyer MT, Thurman EM (2002) Pharmaceuticals, hormones and other organic wastewater contaminants in U.S. Streams, 1999–2000: a national reconnaissance. Environ Sci Technol 36:1202–1211
Kumar V, Chakraborty A, Kural MR, Roy P (2009) Alteration of testicular steroidogenesis and histopathology of reproductive system in male rats treated with triclosan. Reprod Toxicol 27:177–185
Kumar KR, Rohini P, Md AR, Devi YP (2015) A review on occurrence, fate and toxicity of Triclosan. World J Pharm Sci 4:336–369
Kumari R, Ghosh Sachan S (2018) Bioconversion of toxic micropollutant triclosan to 2,4-dichlorophenol using a wastewater isolate *Pseudomonas aeruginosa* KS2002. Int J Environ Sci Technol. https://doi.org/10.1007/s13762-018-2129-5
Larsson K, Bjorklund KL, Palm B, Wennberg M, Kaj L et al (2014) Exposure determinants of phthalates, parabens, bisphenol A and triclosan in Swedish mothers and their children. Environ Int 73:323–333
Lee H, Park M, Yi B, Choi K (2012) Octylphenol and triclosan induced proliferation of human breast cancer cells via an estrogen receptor-mediated signaling in vitro. Endocr Abstr 29:749
Lee HR, Hwang KA, Nam KH, Kim HC, Choi KC (2014) Progression of breast cancer cells was enhanced by endocrine disrupting chemicals, triclosan and octylphenol, via an estrogen receptor-dependent signaling pathway in cellular and mouse xenograft models. Chem Res Toxicol 27:834–842
Leiker TJ, Abney SR, Goodbred SL, Rosen MR (2009) Identification of methyl triclosan and halogenated analogues in male common carp (*Cyprinus carpio*) from Las Vegas Bay and semipermeable membrane devices from Las Vegas Wash, Nevada. Sci Total Environ 407:2102–2114
Letterman RD, Amirtharajah A, O'Melia CR (2009) Coagulation and flocculation in water quality and treatment. In: Letterman RD (ed) A handbook of community water supplies, 2nd edn. McGraw-Hill, New York, pp 312–313

Lin YJ (2000) Buccal absorption of triclosan following topical mouth rinse application. Am J Dent 13:215–217

Lindstorm A, Buerge IJ, Poiger T, Bergovist PA, Muller MD et al (2002) Occurrence and environmental behavior of the bactericide triclosan and its methyl derivative in surface waters and in wastewater. Environ Sci Technol 36:2322–2329

Lishman L, Smyth SA, Sarafin K, Kleywegt S, Toito J et al (2006) Occurrence and reductions of pharmaceuticals and personal care products and estrogens by municipal wastewater treatment plants in Ontario, Canada. Sci Total Environ 367:544–558

Lores M, Llompart M, Sanchez-Prado L, Garcia-Jares C, Cela R (2005) Confirmation of the formation of dichlorodibenzo-p-dioxin in the photodegradation of triclosan by photo-SPME. Anal Bioanal Chem 381:1294–1298

Lozano N, Rice CP, Ramirez M, Torrents A (2010) Fate of triclosan in agricultural soils after biosolid applications. Chemosphere 78:760–766

Ma H, Zheng L, Li Y, Pan S, Hu J et al (2013) Triclosan reduces the levels of global DNA methylation in HepG2 cells. Chemosphere 90:1023–1029

MacIsaac JK, Gerona RR, Blanc PD, Apatira L, Friesen MW et al (2014) Health care worker exposures to the antibacterial agent triclosan. J Occup Environ Med 56:834–839

Mathew J, Joy NS, Kuppuswamy S (2017) A review on "Triclosan a controversial antibacterial". Int J Pharm Pharm Sci 8:200–216

McAvoy DC, Schatowitz B, Jacob M, Hauk A, Eckhoff WS (2002) Measurement of triclosan in wastewater treatment systems. Environ Toxicol Chem 21:1323–1329

McLeod R, Muench SP, Rafferty JB (2001) Triclosan inhibits the growth of Plasmodium falciparum and Toxoplasma gondii by inhibition of apicompexan FabI. Int J Parasitol 31:109–113

McMurry LM, McDermott PF, Levy SB (1999a) Genetic evidence that InhA of *Mycobacterium smegmatis* is a target for triclosan. Antimicrob Agents Chemother 43:711–713

McMurry LM, McDermott PF, Lev SB (1999b) Genetic evidence that InhA of *Mycobacterium smegmatis* is a target for triclosan. Antimicrob Agents Chemother 43:711–713

Meade MJ, Waddell RL, Callahan TM (2001) Soil bacteria *Pseudomonas putida* and *Alcaligenes xylosoxidans* subsp. *denitrificans* inactivate triclosan in liquid and solid substrates. FEMS Microbiol Lett 204:45–48

Mendez MO, Valdez EM, Martinez EM, Sancedo M, Wilson BA (2016) Fate of triclosan in irrigated soil: degradation in soil and translocation into onion and tomato. J Environ Qual 45:1029–1035

Miller TL, Lorusso DJ, Walsh ML, Deinzer ML (1983) The acute toxicity of penta-, hexa- and heptachlorohydroxydiphenyl ethers in mice. J Toxicol Environ Health 12:245–253

Miller TR, Heidler J, Chillrud SN, Delaquil A, Ritchie JC et al (2008) Fate of triclosan and evidence for reductive dechlorination of triclocarban in estuarine sediments. Environ Sci Technol 42:4570–4576

Mishra S, Karmodiya K, Parasuraman P, Surolia A, Surolia N (2008) Design, synthesis, and application of novel Triclosan prodrugs as potential antimalarial and antibacterial agents. Bioorg Med Chem 16:5536–5546

Miyazaki T, Yamagishi T, Matsumoto M (1984) Residues of 4-chloro- 1-(2,4-dichlorophenoxy)-2-methoxybenzene (triclosan methyl) in aquatic biota. Bull Environ Contam Toxicol 32:227–232

Moss T, Howes D, Williams F (2000) Percutaneous penetration and dermal metabolism of triclosan (2,4,4′-trichloro-2′-hydroxydiphenyl ether). Food Chem Toxicol 38:361–370

Nester EW, Anderson DG, Robert CE Jr, Pearsall NN, Nester MT (2001) Microbiology: a human perspective. McGraw-Hills, New York

Ni Y, Zhang Z, Zhang Q, Chen J, Wu Y et al (2005) Distribution patterns of PCDD/Fs in chlorinated chemicals. Chemosphere 60:779–784

NIH, Household Products Database (2007) Available online: https://householdproducts.nlm.nih.gov/cgi-bin/household/brands?tbl=chem&id=75. Accessed 20 Feb 2018

Orvos DR, Versteeg DJ, Inauen J, Capdevielle M, Rothenstein A et al (2002) Aquatic toxicity of triclosan. Environ Toxicol Chem 21:1338–1349

Pannu MW, Toor GS, O'Connor GA, Wilson PC (2012) Toxicity and bioaccumulation of biosolids-borne triclosan in food crops. Environ Toxicol Chem 31:2130–2137
Park I, Zhang N, Ogunyoku AT, Young TM, Scow KM (2013) Effects of triclosan and biosolids on microbial community composition in an agricultural soil. Water Environ Res 85:2237–2242
Pi J, Bai Y, Zhang Q, Wong V, Floering LM et al (2007) Reactive oxygen species as a signal in glucose-stimulated insulin secretion. Diabetes 56:1783–1791
Queckenberg C, Meins J, Wachall B, Doroshyenko O, Tomalik-Scharte D et al (2010) Absorption, pharmacokinetics and safety of triclosan after dermal administration. Antimicrob Agents Chemother 54:570–572
Ramaswamy BR, Shanmugam G, Velu G, Rengarajan B, Larsson DG (2011) GCMS analysis and ecotoxicological risk assessment of Triclosan, carbamazepine and parabens in Indian rivers. J Hazard Master 186:1586–1593
Reiss R, Mackay N, Habig C, Griffin J (2002) An ecological risk assessment for triclosan in lotic systems following discharge from wastewater treatment plants in the United States. Environ Toxicol Chem 21:2483–2492
Ricart M, Guasch H, Alberch M, Barcelo D, Bonnineau C et al (2010) Triclosan persistence through wastewater system plants and its potential toxic effects on river biofilms. Aquat Toxicol 100:346–353
Robertshaw H, Leppard B (2007) Contact dermatitis to triclosan in toothpaste. Contact Dermatitis 57:383–384
Rodricks JV, Swenberg JA, Borzelleca JF, Maronpot RR, Shipp AM (2010) Triclosan: a critical review of the experimental data and development of margins of safety for consumer products. Crit Rev Toxicol 40:422–484
Rule KL, Ebbett VR, Vikesland PJ (2005) Formation of chloroform and chlorinated organics by free-chlorine-mediated oxidation of Triclosan. Environ Sci Technol 39:3176–3185
Russell AD (2003) Similarities and differences in the responses of microorganisms to biocides. J Antimicrob Chemother 52:750–763
Russell AD, McDonnell G (2000) Concentration: a major factor in studying biocidal action. J Hosp Infect 44:1–3
Russell AD, Furr JR, Maillard JY (1997) Microbial susceptibility and resistance to biocides: an understanding. ASM News 63:481–487
Sanchez P, Moreno E, Martinez JL (2005) The biocide triclosan selects *Stenotrophomonas maltophilia* mutants that overproduce the SmeDEF multidrug efflux pump. Antimicrob Agents Chemother 49:781–782
Sanchez-Prado L, Llompart M, Lores M, Fernandez-Alvarez M, Garcia-Jares C et al (2006) Further research on the photo-SPME of triclosan. Anal Bioanal Chem 384:1548–1557
Sandborgh-Englund G, Adolfsson-Erici M, Odham G, Ekstrand J (2006) Pharmacokinetics of triclosan following oral ingestion in humans. J Toxicol Environ Health 69:1861–1873
SCCP. Scientific Committee on Consumer Products opinion on: Triclosan (SCCP/1251/09) Adopted by the SCCP during the 7th plenary meeting on 22 June 2010
Schiffer C, Muller A, Egeberg CL, Alvarez L, Brenker CA et al (2014) Direct action of endocrine disrupting chemicals on human sperm. EMBO Rep 15:758–765
Schmid MD, Kalpan N (2004) Reduced triclosan susceptibility in methicillin-resistant *Staphylococcus epidermidis*. Antimicrob Agents Chemother 48:1397–1399
Schweizer HP (2001) Triclosan: a widely used biocide and its link to antibiotics. FEMS Microbiol Lett 202:1–7
Shanguman G, Ramasamy K, Selvaraj KK, Sampath S, Ramaswamy BR (2014) Triclosan in fresh water fish *Gibelion catla* from the Kaveri river, India, and its consumption risk assessment. Environ Forensics 15:207–212
Singer H, Muller S, Tixier C, Pillonel L (2002) Triclosan: occurrence and fate of a widely used biocide in the aquatic environment: field measurements in wastewater treatment plants, surface waters, and lake sediments. Environ Sci Technol 36:4998–5004
Spanier AJ, Fausnight T, Camacho TF, Braun JM (2014) The associations of triclosan and paraben exposure with allergen sensitization and wheeze in children. Allergy Asthma Proc 35:475–481

Tamura I, Kanbara Y, Saito M, Horimoto K, Satoh M et al (2012) Triclosan, an antibacterial agent, increases intracellular Zn+2 concentration in rat thymocytes: its relation to oxidative stress. Chemosphere 86:70–75

Tastana BE, Donmeza G (2015) Biodegradation of pesticide triclosan by *A. versicolor* in simulated wastewater and semi-synthetic media. Pestic Biochem Physiol 118:33–37

Tatarazako N, Ishibashi H, Teshima K, Kishi K, Arizono K (2004) Effects of triclosan on various organisms. Environ Sci 11:133–140

Thompson A, Griffin P, Stuetz R, Cartmell E (2005) The fate and removal of triclosan during wastewater treatment. Water Environ Res 77:63–67

Valters K, Li HX, Alaee M, D'Sa I, Marsh G (2005) Polybrominated diphenyl ethers and hydroxylated and methoxylated brominated and chlorinated analogues in the plasma of fish from the Detroit River. Environ Sci Technol 39:5612–5619

Veldhoen N, Skirrow RC, Osachoff H, Wigmore H, Clapson DJ et al (2006) The bactericidal agent triclosan modulates thyroid hormone-associated gene expression and disrupts postembryonic anuran development. Aquat Toxicol 80:217–227

Vermeiren L, Devlieghere F, Debevere J (2002) Effectiveness of some recent antimicrobial packaging concepts. Food Addit Contam 19:163–171

Villalain J, Mateo CR, Aranda FJ, Shapiro S, Micol V (2001) Membranotropic effects of the antibacterial agent triclosan. Arch Biochem Biophys 390:128–136

Wallet MA, Calderon NL, Alonso TR, Choe CS, Catalfamo DL et al (2013) Triclosan alters antimicrobial and inflammatory responses of epithelial of epithelial cells. Oral Dis 19:296–302

Wang L, Asimakopoulos AG, Kannan K (2015) Accumulation of 19 environmental phenolic and xenobiotic heterocyclic aromatic compounds in human adipose tissue. Environ Int 78:45–50

Wang S, Yin Y, Wang J (2018) Microbial degradation of triclosan by a novel strain of *Dyella* sp. Appl Microbiol Biotechnol 102:1997–2006

Wilson BA, Salyers AA (2003) Is the evolution of bacterial pathogens an out-of-body experience? Trends Microbiol 11:347–350

Wilson BA, Smith VH, DeNoyelles F, Larive CK (2003) Effects of three pharmaceutical and personal care products on natural freshwater algal assemblages. Environ Sci Technol 37:1713–1719

Winitthana T, Lawanprasert S, Chanvorachote P (2014) Triclosan potentiates epithelial-to-mesenchymal transition in anoikis-resistant human lung cancer cells. PLoS One 9:110851

Wolff MS, Teitelbaum SL, Windham G, Pinney SM, Britton JA et al (2007) Pilot study of urinary biomarkers of phytoestrogens, phthalates, and phenols in girls. Environ Health Perspect 115:116–121

Wu Y, Wu Q, Belanda FA, Gea P, Manjanathab MG et al (2014) Differential effects of triclosan on the activation of mouse and human peroxisome proliferator-activated receptor alpha. Toxicol Lett 231:17–28

Xue J, Wu Q, Sakhtivel S, Pavithran PV, Vasukutty JR et al (2015) Urinary levels of endocrine-disrupting chemicals, including bisphenols, bisphenol A diglycidyl ethers, benzophenones, parabens and triclosan in obese and non-obese Indian children. Environ Res 137:120–128

Yang LH, Ying GG, Su HC (2008) Growth-inhibiting effects of 12 antibacterial agents and their mixtures on the freshwater microalga *Pseudokirchneriella subcapitata*. Environ Toxicol Chem 27:1201–1208

Yin J, Wei L, Shi Y, Zhang J, Wu Q et al (2016) Chinese population exposure to triclosan and triclocarban as measured via human urine and nails. Environ Geochem Health 38:1125–1135

Yueh MF, Taniguchi K, Chen S, Evans RM, Hammock BD et al (2014) The commonly used antimicrobial additive triclosan is a liver tumor promoter. PNAS 111:172005

Zhao F (2006) Biodegradation of triclosan by a triclosan degrading isolate and an ammonia oxidizing bacterium. PhD thesis, Texas A&M University

A Simple Procedure for Isolation, Culture of Protoplast and Plant Regeneration

14

Indu Kumari

Abstract

This chapter presents different steps involved in isolation and culture of protoplast; discusses basic steps involved in the developmental sequence from protoplast to plant; and overviews a simple procedure for protoplast isolation, culture and plant regeneration. Protoplasts are naked plant cells without the cell wall, but they possess plasma membrane and other components of cell. Protoplasts are potentially totipotent individuals at the single-cell level. They have the potential to regenerate a cell wall; dedifferentiate, divide mitotically and form unlimited growing cell clones; and differentiate shoot and root meristems or embryos which regenerate whole plant.

Keywords

Isolation · Culture · Embryos · Cell clones · Protoplast

14.1 Introduction

The protoplasts are isolated, single and naked plant cells. Protoplast is the content which is enclosed by plasma membrane. The term protoplast means spherical plasmolysed content of plant cell covered by plasma membranes or naked cell without cell wall. Protoplasts are naked plant cells without the cell wall, but they possess plasma membrane and all other cellular components. Protoplast represent the functional plant cell without cell wall. Protoplasts of different species can be fused to generate a hybrid, and this process is known as protoplast fusion or somatic hybridization.

I. Kumari (✉)
Department of Botany, Nirmala College, Ranchi, Jharkhand, India

M. Kumar et al. (eds.), *In vitro Plant Breeding towards Novel Agronomic Traits*,
https://doi.org/10.1007/978-981-32-9824-8_14

The term protoplast was introduced by Hanstein in 1880. The first isolation of protoplasts was achieved by Klercker (1892) employing a mechanical method. A real research in protoplast was started by Cocking in 1960. He used an enzymatic method for the removal of cell wall. Protoplasts can be isolated directly from the different parts of whole plant which contains parenchymatous tissue or indirectly from the in vitro grown plant tissue or callus tissue.

Takebe et al. (1971) regenerated complete plants from leaf protoplasts of tobacco increased the potential of protoplast culture methods. Several species of plants have been regenerated by using protoplasts of different plants (Bajaj et al. 1974; Davey and Short 1973; Frearson et al. 1973; Grun and Chu 1978; Hess et al. 1973; Sink and Power 1977; Wilson et al. 1980; Wallin and Eriksson 1972).

Protoplast cultures have been used in genetic engineering for the transfer of DNA and extrachromosomal bodies like chloroplast, mitochondria, plasmids and *nif* (nitrogen fixing) genes.

14.2 Materials and Methods

14.2.1 Sources of Explant for Protoplast Isolation

The protoplasts can be isolated from a variety of tissues including leaves, roots, in vitro shoot cultures, callus, cell suspension and pollen. Among these, the mesophyll tissue of fully expanded leaves of young plants or new shoots is used most frequently. In addition, callus and suspension cultures also serve as good sources for protoplast isolation. Young cell suspensions are ideal for isolation of protoplasts in large quantities. Cell suspension cultures may provide a very good source of protoplast.

14.2.2 Techniques of Isolation, Culture and Regeneration of Protoplast

The basic techniques of isolation, culture and regeneration of protoplast are the following:

(i) *Sterilization of Samples*: Samples like mature leaves are collected from healthy plants which are washed in tap water to remove adhering soil particles and sterilized with sodium hypochlorite solution.

(ii) *Rinsing in Suitable Osmoticum*: After 10 min, sample is properly washed with sterile distilled water or MS medium adjusted to a suitable pH and buffer to maintain osmotic pressure. Washing should be done for about six times to remove the traces of sodium hypochlorite.

(iii) *Plasmolysis of Cells*: The lower epidermis covered by thin wax cuticle is removed with a forceps. Stripping should be done from midrib to margin of lamina. The stripped surface of leaf is kept in mannitol solution (13% W/V) for 3 h to allow plasmolysis of cells.

(iv) *Peeling of Lower Epidermis*: Thereafter, leaves of plant are peeled off and used for isolation of protoplasts.

(v) *Isolation of Protoplast*: Each plant cell is enclosed by cellulose wall which is attached with each other by middle lamella, made up of pectin. Protoplast is the content which is enclosed by plasma membrane. It is essential to get isolated protoplast, to remove the pectin material to obtain the single cells and then to remove cell wall (Figs. 14.1 and 14.2). Protoplasts may be isolated by any one of the two following ways: (a) mechanical method and (b) enzymatic method.

14.2.3 Mechanical Method

In mechanical method, cells are kept in a suitable plasmolyticum (in plasmolysed cells, protoplasts shrink away from cell wall) and cut with a fine knife, so that

Fig. 14.1 Isolated protoplasts

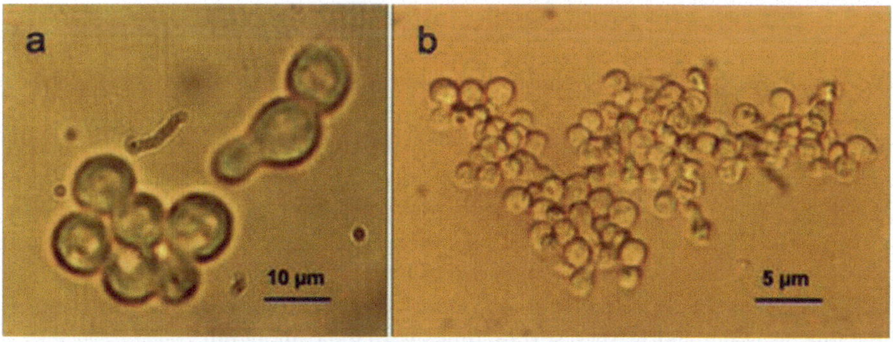

Fig. 14.2 Protoplasts obtained from *Trichoderma harzianum* by means of enzymatic action

protoplasts are released from cells cut through the cell wall, when the tissue is again deplasmolysed. Mechanical method is suitable for isolation of protoplasts from vacuolated cells such as onion bulbs, scales and radish roots. The mechanical method was used as early as 1892. Klercker (1892) was the first to isolate protoplasts from plasmolysed cell of *Stratiotes aloides*. But now it is rarely used for isolation of protoplasts. This technique involves the following stages:

Step 1. A small piece of epidermis from a plant is selected.

Step 2. The cells are subjected to plasmolysis. This causes protoplasts to shrink away from the cell walls.

Step 3. The tissue is dissected to release the protoplasts.

Limitation of mechanical method: Mechanical method for isolation of protoplasts is restricted to certain tissues with vacuolated cells. This method is not suitable for isolating protoplast from meristematic and less vacuolated cells. It gives poor yield of protoplasts, and also the viability of protoplasts is low. It is laborious and tedious.

14.2.4 Enzymatic Method

This method involves the use of enzymes to dissolve the cell wall for releasing protoplasts. The enzymatic method is almost invariably used now for the isolation of protoplasts. Protoplasts can be isolated from plant tissues or cultured cells by enzymatic digestion to remove the cell walls. It gives large quantities of protoplasts, where cells are not broken and osmotic shrinkage is minimum. Sometimes mechanical and enzymatic methods are combined, where cells are first separated mechanically and later protoplast isolated through enzymatic method. The credit of developing high-yield protoplast isolation through enzymatic method goes to Cocking (1960). He used crude cellulase from the fungus *Myrothecium verrucaria* to dissolve cell wall and release the protoplasts from tomato roots. Later this method with suitable modifications and using purified enzymes has been extensively used by other scientists.

14.2.5 Enzymes for Protoplast Isolation

The enzymes that can digest the cell walls are required for protoplast isolation. Chemically, the plant cell wall is mainly composed of cellulose, hemicellulose and pectin which can be, respectively, degraded by the enzymes cellulase, hemicellulase and pectinase. The enzymes are usually used at a pH 4.5 to 6.0 and temperature 25–30 °C with a wide variation in incubation period that may range from half an hour to 20 h. The success of protoplast isolation depends especially on the condition of the tissue and the combination of enzymes being used. After release of protoplast into the suspension, for removal of enzymes, the protoplasts are collected in centrifuge tube as pellet and washed several times with the osmoticum.

The enzymatic method could be used as a one-step method (direct method) or as a two-step method (sequential method):

1. *The one-step method*: Protoplasts are isolated directly from the tissue by using two enzymes, cellulase and pectinase, simultaneously.
2. *The two-step method*: Cells are first isolated from callus or tissue by using pectinase, and to this cell suspension, cellulase is added to digest the cell wall and release protoplasts.

An enzymatic solution used for protoplast isolation contains the enzymes and a sugar as an osmoticum to prevent the plasma membrane from rupturing. Some salts and nutrients are also used as osmoticum. Generally 50 mM CaCl2 is added to increase the stability of released protoplasts (Rose 1980).

Advantages of enzymatic method—Enzymatic method is a very widely used technique for the isolation of protoplasts. It includes good yield of viable cells and minimal or no damage to the protoplasts.

14.2.5.1 Purification of Protoplasts

Leaf debris are removed with forceps, and enzyme solution containing protoplasts is filtered with a steel or a nylon mesh (45 mm). The cell clumps and undigested tissues can be removed by filtration. This is followed by centrifugation and washings of the protoplasts (Gregory and Cocking 1965; Power and Cocking 1969; Evans et al. 1980a, b; Schenk and Hilderbrandt 1969). Filtrate is centrifuged at 75 × g for 5 min and supernatant is decanted. Again a fresh MS medium plus 13% mannitol is added to centrifuge. The protoplasts settle to the bottom of the centrifuge tube, while the supernatant is removed with the help of a pipette. Repeated washings with nutrient medium, centrifugation and decantation are done for about three times. Traces of enzyme are removed by washing the protoplasts twice or thrice with the medium (Fig. 14.3). Finally specific concentration of protoplast suspension is prepared.

14.2.5.2 Viability of Protoplasts

It is essential to ensure that the isolated protoplasts are healthy and viable so that they are capable of undergoing sustained cell divisions and regeneration.

14.2.5.3 Assessment of the Protoplast Viability

1. Fluorescein diacetate (FDA) staining method—Fluorescence microscopy detects the dye accumulates inside viable protoplasts.
2. Phenosafranine stain is used which is specific for dead protoplasts that shows red in colour, while the viable cells remain unstained.
3. Exclusion of Evans blue dye by intact membranes.
4. Cell wall formation can be detected by staining with 0.1% calcofluor white (CPW) fluorescent stain. Calcofluor white (CFW) stain binds to the newly formed cell walls which emit fluorescence.

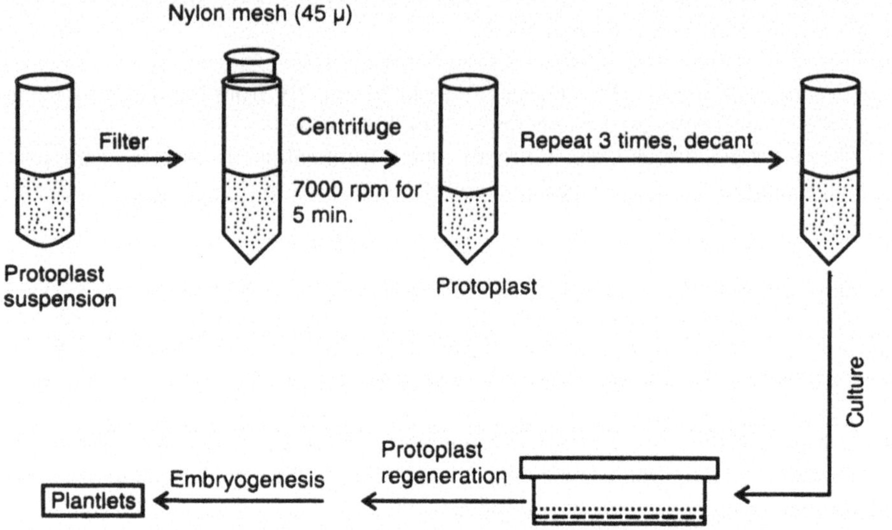

Fig. 14.3 Protoplast purification and regeneration. (Source: biocyclopedia.com)

5. Oxygen uptake by protoplasts can be measured by oxygen electrode.
6. The ability of protoplasts to undergo continuous mitotic divisions (direct measure).
7. Photosynthetic activity of protoplasts.

14.2.5.4 Protoplast Culture and Regeneration

Protoplast regeneration, which may also be regarded as protoplast development, occurs in two stages: first, formation of cell wall and, second, development of callus/whole plant.

14.2.5.4.1 Formation of Cell Wall

The process of cell wall formation in cultured protoplasts starts within a few hours after isolation. From the protoplast solution of known density (about 10^5 protoplast/ml), about 1 ml suspension is poured on sterile and cooled down nutrient medium in Petri dishes. The plates are incubated at 25 °C in a dim white light. Protoplast develops cell wall within 2 to several days under suitable conditions. The process of cell wall development requires continuous supply of nutrients, particularly carbon source such as sucrose.

As the cell wall development occurs, the protoplasts lose their characteristic spherical shape. The newly developed cell wall by protoplasts can be detected by staining with 0.1% calcofluor white (CPW) fluorescent stain. The presence of a proper wall is essential for a regular division. On the other hand, protoplasts with poorly regenerated cell wall show budding and fail to undergo normal mitosis.

14.3 Development of Callus/Whole Plant

Protoplasts are cultured either in semisolid agar or liquid medium. Sometimes, protoplasts are first allowed to develop cell wall in liquid medium and then transferred to agar medium. Agarose is the most frequently used agar to solidify the culture media. In agar cultures, the protoplasts remain in a fixed position, divide and form cell clones. The advantage with agar culture is that clumping of protoplasts is avoided. But liquid culture is easy to dilute and transfer. Density of the cells can be manipulated as desired. For some plant species, the cells cannot divide in agar medium; therefore liquid medium is the only choice. Osmotic pressure of liquid medium can be altered as desired.

14.3.1 Culture Media

The culture media for protoplast culture contain nutritional components and osmoticum.

14.3.1.1 Nutritional Components

The nutritional requirements of protoplasts are similar to callus and suspension cultures. Mostly, MS and B5 media with suitable modifications are used. The medium for protoplast culture should be devoid of ammonium, and the quantities of iron and zinc should be less. The concentration of calcium should be 2–4 times higher used for cell cultures. This is needed for membrane stability. The vitamins used for protoplast cultures are the same as used in standard tissue culture media. Glucose is the preferred carbon source by protoplasts although a combination of sugars, such as glucose and sucrose, can be used. High auxin/kinetin ratio is suitable to induce cell divisions, while high kinetin/auxin ratio is required for regeneration.

14.3.1.2 Osmoticum

Osmoticum broadly refers to the reagents or chemicals that are added to increase the osmotic pressure of a liquid. The isolation and culture of protoplasts require osmotic protection until they develop a strong cell wall. In fact, if the freshly isolated protoplasts are directly added to the normal culture medium, they will burst. Thus, addition of an osmoticum is essential for both isolation and culture media of protoplast to prevent their rupture. The osmotica are of two types: non-ionic and ionic.

14.3.1.2.1 Non-ionic Osmotica

The non-ionic substances most commonly used are soluble carbohydrates such as mannitol, sorbitol, glucose, fructose, galactose and sucrose. Mannitol, being metabolically inert, is most frequently used.

14.3.1.2.2 Ionic Osmotica

Potassium chloride, calcium chloride and magnesium phosphate are the ionic substances in use to maintain osmotic pressure. When the protoplasts are transferred to a culture medium, the use of metabolically active osmotic stabilizers (e.g. glucose, sucrose) along with metabolically inert osmotic stabilizers (mannitol) is advantageous. As the growth of protoplasts and cell wall regeneration occurs, the metabolically active compounds are utilized, and this results in the reduced osmotic pressure so that proper osmolarity is maintained.

14.4 Culture and Regeneration of Protoplast

As the cell wall formation around protoplasts is complete, the cells increase in size. The protoplasts, which are capable of dividing, undergo first division within 2–7 days and form small cell colonies after 2–3 weeks. With suitable manipulations of nutritional and physiological conditions, the cell colonies may be grown continuously and form visible colonies (macroscopic colonies). These colonies are then transferred to an osmotic-free (mannitol or sorbitol-free) medium for further development to form callus. The callus can also be subcultured. With induction and appropriate manipulations, the callus can undergo organogenic or embryogenic differentiation to finally form the whole plant. A general view of the protoplast isolation, culture and regeneration is represented in Fig. 14.4.

Plant regeneration can be done from the callus obtained either from protoplasts or from the culture of plant organs. There are, however, certain differences in these two calluses. The callus derived from plant organs carries preformed buds or organized structures, while the callus from protoplast culture does not have such structures.

The first success of regeneration of plants from protoplast cultures of Nicotiana tabacum was achieved by Takebe et al. (1971). Since then, several species of plants have been regenerated by using protoplasts (Binding et al. 1978; Cocking and Pojnar 1969; Cocking 1972; Dos et al. 1980; Harada 1973; Vasil and Vasil 1974, 1980; Zieg and Outka 1980). Some examples of plant species regenerated from protoplasts are listed in Table 14.1.

14.5 Conclusions

The isolation, culture and fusion of protoplasts are a fascinating field in plant research. Protoplasts have a wide range of applications. A whole plant can be regenerated by protoplast culture. It is easy to perform single-cell cloning with protoplasts. Protoplast fusion or somatic hybridization can produce hybrids. Protoplasts are excellent materials for studies of ultrastructure of cell. Protoplasts are useful for studies of membrane function like transport and uptake processes. Isolation of cell organelles and chromosomes is easy from protoplasts. Genetic transformations can be achieved through genetic engineering of protoplast DNA. It is easy to isolate mutants from protoplast cultures.

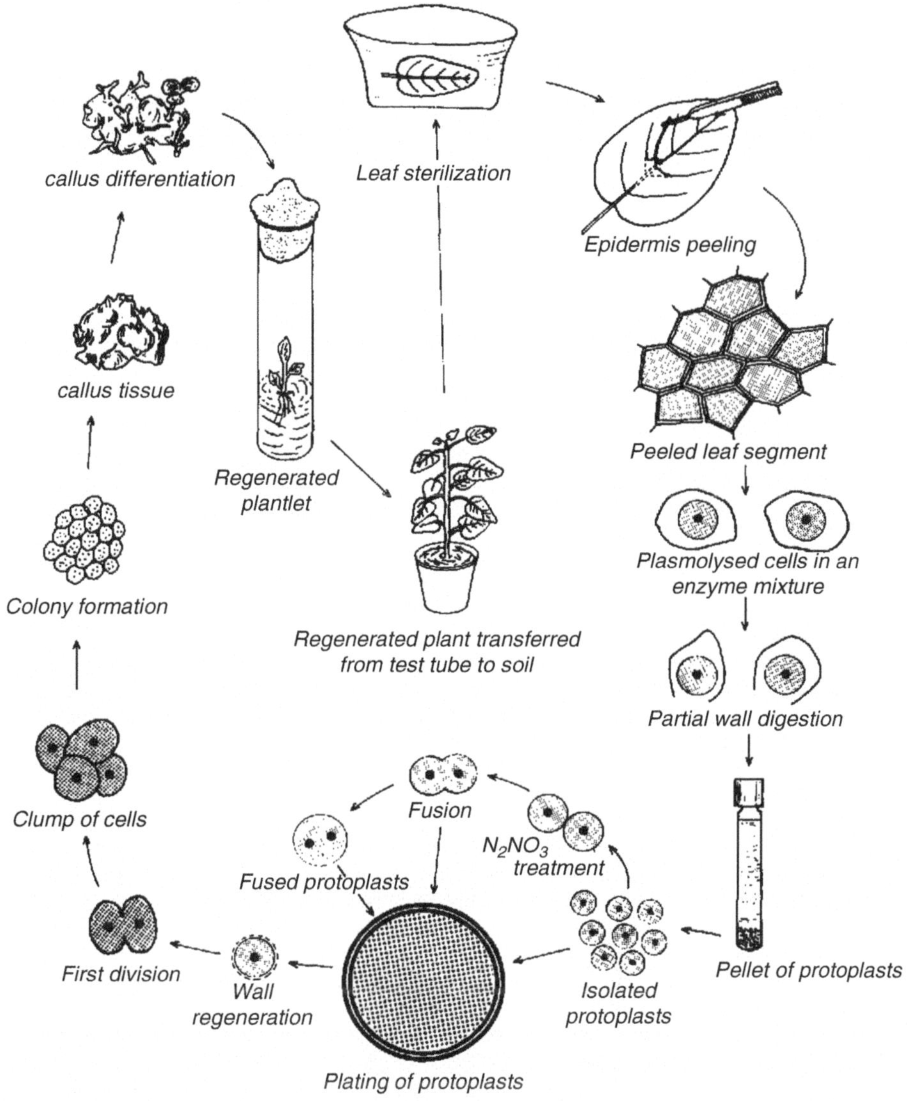

Fig. 14.4 Major steps of protoplast isolation, culture and regeneration of plant. (Source: Bhojawani and Rajdan 1996)

Table 14.1 Selected examples of plant species regenerated from protoplasts

Category	Plant species
Cereals	*Oryza sativa*, *Zea mays*
Vegetables	*Brassica oleracea*, *Capsicum annuum*
Ornamentals	*Chrysanthemum* sp., *Rosa* sp.
Woody trees	*Prunus avium*, *Larix eurolepsis*
Oil crops	*Brassica napus*, *Helianthus annuus*
Legumes	*Glycine max*

References

Bajaj YPS (1974) Potentials of protoplast culture work in agriculture. Euphytica 23:633–649

Binding H, Behls R, Schieder O, Sopory SK, Wenzel G (1978) Regeneration of mesophyll protoplasts isolated from dihaploid clones of *Solanum tuberosum*. Physiol Plant 43:52–54

Bhojawani SS, Rajdan MK (1996) Plant tissue culture theory and practice, vol 5, p 766

Cocking EC (1960) A method for the isolation of plant protoplasts and vacuoles. Nature (London) 187:927–929

Cocking EC (1972) Plant cell protoplasts—isolation and development. Annu Rev Plant Physiol 23:29–50

Cocking EC, Pojnar E (1969) An electron microscopic study of the infection of isolated tomato fruit protoplasts by tobacco mosaic virus. J Gen Virol 4:305–312

Davey MR, Short KC (1973) The isolation, culture, and fusion of legume and corn flower leaf protoplasts. In: Protoplastes et fusion de cellules somatiques végétales. Colloq Int CNRS, Paris, pp 437–444

Dos Santos AVP, Outka DE, Cocking EC, Davey MR (1980) Organogenesis and somatic embryogenesis in tissues derived from leaf protoplasts and leaf expiants of *Medicago sativa*. Z Pflanzenphysiol 99:261–270

Evans DA, Crocomo OJ, De Carvalho MTV (1980a) Protoplast isolation and subsequent callus regeneration in sugarcane. Z Pflanzenphysiol 98:355–358

Evans DA, Wetter LR, Gamborg OL (1980b) Somatic hybrid plants of *Nicotiana glauca* and *Nicotiana tabacum* obtained by protoplast fusion. Physiol Plant 48:225–230

Frearson EM, Power JB, Cocking EC (1973) The isolation, culture, and regeneration of *Petunia* leaf protoplasts. Dev Biol 33:130–137

Gregory DW, Cocking EC (1965) The large scale isolation of protoplasts from immature tomato fruit. J Cell Biol 24:143–146

Grun P, Chu LJ (1978) Development of plants from protoplasts of *Solanum* (Solanaceae). Am J Bot 65:538–543

Harada H (1973) A new method for obtaining protoplasts from mesophyll cells. Z Pflanzenphysiol 69:77–80

Hess D, Potrykus I, Doon G, Durand T, Hoffmann F (1973) Transformation experiments in higher plants: prerequisites for the use of isolated protoplasts (isolation from mesophyll and callus cultures, uptake of proteins and DNA, and regeneration of whole plants). In: Protoplastes et fusion de cellules somatiques végétales. Colloq Int CNRS, Paris, pp 343–351

Power JB, Cocking EC (1969) A simple method for the isolation of very large numbers of leaf protoplasts. Biochem J 111:33

Rose RJ (1980) Factors that influence the yield, stability in culture and cell wall regeneration of. Spinach mesophyll protoplasts. Aust J Plant Physiol 7:71–725

Schenk RU, Hilderbrandt AC (1969) Production of protoplasts from plant cells in liquid culture using purified commercial cellulases. Crop Sci 9:629–631

Sink KC, Power JB (1977) The isolation, culture, and regeneration of leaf protoplasts of *Petunia parviflora* Juss. Plant Sci Lett 10:335–340

Takebe I, Labib G, Melchers G (1971) Regeneration of whole plants from isolated mesophyll protoplasts of tobacco. Naturwissenschaften 58:318–320

Vasil V, Vasil IK (1974) Regeneration of tobacco and *Petunia* plants from protoplasts and culture of corn protoplasts. In Vitro 10:83–96

Vasil V, Vasil IK (1980) Isolation and culture of cereal protoplasts II. Embryogenesis and plantlet formation from protoplasts of *Pennisetum americanum*. Theor Appl Genet 56:97–99

von Klercker J (1892) Eine Methode zur isolier lebender protoplasten. Ofvers Vetensk Akad Forhandl 49:463–474

Wallin A, Eriksson T (1972) Plating of protoplasts from cell suspension cultures of *Daucus carota*. In: Protoplastes et fusion de cellules somatiques végétales. Colloq Int CNRS, Paris, pp 301–307

Wilson HM, Styer DJ, Conard PL, Durbin RD, Helgeson JP (1980) Isolation of sterile protoplasts from unsterilized leaves. Plant Sci Lett 18:151–154
Zieg RG, Outka DE (1980) The isolation, culture and callus formation of soybean pod protoplasts. Plant Sci Lett 18:105–114